John J. Qu
Wei Gao
Menas Kafatos
Robert E. Murphy
Vincent V. Salomonson

Earth Science Satellite Remote Sensing
Vol. 2: Data, Computational Processing, and Tools

John J. Qu
Wei Gao
Menas Kafatos
Robert E. Murphy
Vincent V. Salomonson

Earth Science Satellite Remote Sensing

Vol. 2: Data, Computational Processing, and Tools

With 119 figures (62 in color)

TSINGHUA UNIVERSITY PRESS

Springer

EDITORS:

Prof. John J. Qu
Technical Director of EastFIRE Lab
School of Computational Sciences
George Mason University
USA
E-mail: jqu@gmu.edu

Dr. Wei Gao
Group Leader/Research Scientist
Natural Resource Ecology Laboratory
Colorado State University
USA
E-mail: wgao@uvb.nrel.colostate.edu

Prof. Menas Kafatos
Director, Center for Earth Observing and Space Research
Dean, School of Computational Sciences
George Mason University
USA
E-mail: mkafatos@gmu.edu

Dr. Robert E. Murphy
Project Scientist
NPOESS Preparatory Project (NPP), NOAA
Code 920, NASA GSFC
Greenbelt, MD 20771
USA
E-mail: robert.murphy@noaa.gov

Dr. Vincent V. Salomonson
MODIS Science Team Leader
NASA/Goddard Space Flight Center
Greenbelt, MD 20771
USA
E-mail: vincent.v.salomonson@nasa.gov

ISBN 10 7-302-12855-3 **Tsinghua University Press, Beijing**
ISBN 10 3-540-35630-4 **Springer Berlin Heidelberg New York**
ISBN 13 978-3-540-35630-1 **Springer Berlin Heidelberg New York**

Library of Congress Control Number: 2006930102

Co-published by Tsinghua University Press, Beijing and Springer-Verlag GmbH Berlin Heidelberg

Springer is a part of Springer Science+Business Media
springer.com

Cover design: Joseph Piliero, New York
Printed on acid-free paper

This book is dedicated to Dr. Vincent V. Salomonson

Few individuals have had such profound impact on the development of Earth remote sensing as Dr. Vincent V. Salomonson. We, his co-editors of this volume, have been chosen to dedicate it to him in recognition of his many contributions to our field. There is not a topic discussed in the book that has not been strongly influenced either by his personal research or his leadership.

After completing his undergraduate studies, he began his career as a weather officer in the US Air Force. He then returned to graduate school, earning a PhD in Atmospheric Science from Colorado State University in 1968. The bulk of his career was spent at the NASA Goddard Space Flight Center (1968 – 2005) where he conducted research and served as a branch head, laboratory chief, and, for 11 years, as the Director of Earth Sciences. He was deeply engaged in mission development, serving as the Project Scientist for Landsat-4 and -5 (1977 – 1989), and as the team leader for the Moderate Resolution Imaging Spectroradiometer (MODIS) from 1989 to the present. Under his leadership as a laboratory chief and as the Director of Earth Sciences, the men and women of the NASA Goddard Space Flight Center built the foundation for the study of global climate and environmental change using space-based systems and theoretical modeling.

He has served as the president of the American Society for Photogrammetry and Remote Sensing (ASPRS), and is a fellow of the ASPRS, as well as of the Institute for Electrical and Electronics Engineers (IEEE). He has served as an associate editor of several journals and has twice received the NASA Exceptional Scientific Achievement Medal. He has been recognized for career achievements twice, first with the William T Pecora award for his work on Landsat and the NASA Outstanding Leadership Medal for his role in establishing the Earth Sciences Directorate as an internationally recognized entity performing interdisciplinary Earth System Science.

Dr. Salomonson is now a Research Professor at the University of Utah, and the Director of Earth Sciences (Emeritus) at NASA Goddard Space Flight Center.

Foreword

Volume 2 of this book complements Volume 1 which was devoted primarily to missions, instruments, and selected applications by describing several related activities and developments focused primarily on data processing and attendant tools. Chapters 1 – 7 provide background and insights to the EOS Data and Information System (EOSDIS) providing EOS data to millions of users in science and applications throughout the world, EOS Direct Broadcast capabilities delivering MODIS data at over 100 sites spread worldwide and supporting hundreds of users and agencies, the MODIS Rapid Response System that has been so successfully used for applications and outreach to the public, as well as the extensive efforts devoted to calibration and performance characterization of the MODIS instrument. Chapters 8 – 13 describe various formats and tools that are available to everyone and facilitate the use of data from the EOS series. Chapters 14 – 17 go on to describe the use of spaceborne products in a GIS context along with selected applications. In total the content of this volume along with the materials in Volume 1 provides the reader with a view not only of the extensive context and development activity in the NASA EOS effort, but also a glimpse into not only the present applications or utility of the data, but also the potential for future use and advancement in science and applications using spaceborne observations.

Vincent V. Salomonson

Senior Scientist and MODIS Science Team Leader
NASA, Goddard Space Flight Center, Greenbelt, MD

Preface

Earth science satellite remote sensing has seen rapid expansion during the last decade. NASA's Earth Observing System (EOS) program is providing data for in-depth scientific understanding of the functioning of the Earth as a system through a constellation of satellites. The National Polar-orbiting Operational Environmental Satellite System (NPOESS) Preparatory Project (NPP) mission, the latest in the series, will provide NASA with a continuation of global change observations following EOS Terra and Aqua missions. NPP will provide NPOESS with risk reduction demonstration and validation for the four critical NPOESS sensors, algorithms, and processing. NPOESS will provide a national, operational, polar-orbiting remote-sensing capability by converging Department of Defense (DoD) and National Oceanic and Atmospheric Administration (NOAA) satellite programs while incorporating new technologies from NASA. Scientists and students have expressed great interest in these missions. However, there is currently no textbook for graduate students to learn about the EOS, NPP and NPOESS missions, or the current and potential applications of the resulting data.

The core of this book arose from the Workshop for Earth Science Satellite Remote Sensing held at George Mason University (GMU) from Oct. 15 to 22, 2002. Updated information is included in this book. This book is designed to give readers having limited remote sensing background a thorough introduction to current and future NASA, NOAA and other Earth science remote sensing. The book is supported by the Center of Earth Observing and Space Research (CEOSR) at GMU and NASA/GSFC MODIS and NPP projects.

The goals of this volume are to (1) provide information on Earth science remote sensing data information, tools and data formats such as HDF-EOS; (2) discuss MODIS calibration and characterization; (3) evaluate current data processing approaches; (4) introduce data searching and ordering from different public domains; and (5) explore remote sensing and GIS products and WebGIS applications. Detailed information about MODIS data products, NPP and NPOESS missions can be found in volume 1 of Eearth Science Satellite Sensing.

There are many people who assisted with this book. First, the editorial team would like to thank all authors involved in contributing chapters for the Earth Science Satellite Remote Sensing. Each author has spent extra hours in addition to existing workloads and ongoing commitments. Second, we would like to thank over eighty anonymous reviewers for their constructive comments and suggestions. The most chapters in this book were originally presented at the Second Workshop of the Earth Science Satellite Remote Sensing at George Mason University (GMU). We would also like to thank many of the Moderate Resolution Imaging Spectroradiometer (MODIS) and the NPOESS Preparatory Project (NPP) science team members who contributed their MODIS and NPOESS/NPP chapters. Much appreciation also goes the Center for Earth Observation and Space Research (CEOSR) at GMU for supporting the workshop and this book. Special thanks and appreciation go to Mr. Manny Smith for providing editing assistance and tracking chapter status with leading authors. We would like to acknowledge Ms. Lingli Wang, Ms. Bockhwa Kim and Ms. Wanting Wang of the School of Computational Sciences (SCS) at GMU spending tremendous effort working on templates, tables and figures for this book. The efforts of many individuals including Prof. George Taylor, Dr. William Sommers, Prof. Ruixin Yang, Dr. Xianjun Hao and Mr. Hank Wolf at GMU and Dr. Xiaoxiong Xiong at NASA/GSFC, who supported this book, are highly appreciated.

John J. Qu
George Mason University
Fairfax, VA
October 25, 2005

Contents

List of Contributors

Prof. John J. Qu	EastFIRE Lab George Mason University 4400 University Drive Fairfax, VA 22030, USA Phone: 703-993-3958 Fax: 703-993-1993 E-mail: jqu@scs.gmu.edu
Dr. Wei Gao	USDA UV-B Monitoring and Research Program Natural Resource Ecology Laboratory Colorado State University Fort Collins, CO 80523, USA Phone: 970-491-3609 Fax: 970-491-3601 E-mail: wgao@uvb.nrel.colostate.edu
Prof. Menas Kafatos	School of Computational Sciences George Mason University 4400 University Drive, Fairfax, VA 22030, USA Phone: (703)993-3616 Fax: (703)993-1993 E-mail: mkafatos@gmu.edu
Dr. Robert E. Murphy	Earth Sciences Directorate NASA/Goddard Space Flight Center Greenbelt, MD 20771, USA Phone: (301)713-4875 E-mail: robert.murphy@noaa.gov
Dr. Vincent V. Salomonson	University of Utah, Salt Lake City UT 84112, Eearth-Sun Division, Code 600 NASA/Goddard Space Flight Center Greenbelt, MD 20771, USA Phone: (301)614-5631 Fax: (301)614-5808 E-mail: vincent.v.salomonson@nasa.gov

Dr. Carl A. Reber	NASA Goddard Space Flight Center, Code 423.0, Greenbelt, MD 20771 E-mail: skip.reber@gmail.com
Dr. H.K. Ramapriyan	NASA Goddard Space Flight Center, Code 423.0, Greenbelt, MD 20771 E-mail: ramapriyan@gsfc.nasa.gov
Mr. John L. Dwyer	Project Scientist, Land Processes DAAC, Science Applications International Corporation (SAIC), USGS Center for EROS Sioux Falls, SD 57198 E-mail: dwyer@usgs.gov
Mr. Patrick L. Coronado	NASA Goddard Space Flight Center Greenbelt MD 20771 E-mail: patrick.l.coronado@nasa.gov
Mr. Kelvin W. Brentzel	NASA Goddard Space Flight Center Greenbelt MD 20771 E-mail: kelvin.w.brentzel@gsfc.nasa.gov
Dr. Xiaoxiong (Jack) Xiong	Earth Sciences Directorate, NASA/Goddard Space Flight Center, Greenbelt, MD 20771 E-mail: jxiong@ltpmail.gsfc.nasa.gov
Dr. William Barnes	University of Maryland, Baltimore County, Baltimore, MD 21250 E-mail: wbarnes@neptune.gsfc.nasa.gov
Mr. Robert Eplee	Science Applications International Corperation, Beltsville, MD 20705 E-mail: eplee@seawifs@gsfc.nasa.gov
Dr. Junqiang Sun	NASA GSFC E-mail: junqiang_sun@ssaihq.com
Mr. Cheng-Hsuan Lyu	Science Systems and Applications, Inc., 10210 Greenbelt Rd., Lanham, MD
Dr. Wenli Yang	Center for Spatial Information Science and Systems College of Sciences George Mason University, 4400 University Drive, Fairfax, VA, 22030, USA (LAITS), SCS/GMU E-mail: yang@rattler.gsfc.nasa.gov

Dr. Christopher Lynnes	NASA/GSFC, Code 902, Greenbelt, MD 20771 E-mail: clynnes@pop600.gsfc.nasa.gov
Gail Schmidt	Science Applications International Corporation USGS EROS Data Center Sioux Falls, SD 57198
Dr. Larry Klein	RS Information system, Inc.7515 Mission Dr. Suite A1C1 Lanham, MD 20706 E-mail: lkaj@verizon.net
Mr. Ray Milburn	L-3 Communications, Government Services, Inc. 1801 McCormick Dr. #170 Largo, MD 20774 E-mail: ray.milburn@L-3com.com
Mr. Cid Praderas	L-3 Communications, Government Services, Inc. 1801 McCormick Dr. #170 Largo, MD 20774 E-mail: cid.praderas@l-3com.com
Dr. Abe Taaheri	L-3 Communications, Government Services, Inc. 1801 McCormick Dr. #170 Largo, MD 20774 E-mail: abe.taaheri@l-3com.com
Dr. Louis Gonzalez	Laboratoire d'Optique Atmospherique U.S.T.L., Lille, 59665 Villeneuve d'Ascq, Cedex France E-mail: louis@loa.univ-lillel.fr
Dr. Dimitar Ouzounov	NASA GES DAAC/SSAI, MS 902, Goddard Space Flight Center, Greenbelt, MD, 20771, USA E-mail: ouzounov@core2.gsfc.nasa.gov
Mr. Darryl Nickless	NASA GES DAAC/SSAI, MS 902, Goddard Space Flight Center, Greenbelt, MD, 20771, USA E-mail: darry1.nickless@gsfc.nasa.gov
Dr. Gregory Leptoukh	NASA GES DAAC, MS 902, Goddard Space Flight Center, Greenbelt, MD, 20771, USA E-mail: leptoukh@daac.gsfc.nasa.gov
Dr. Long S. Chiu	NASA/GES Data and Information Services, Code 610.2, Greenbelt, MD 20771 and Assoc. Professor, Center for Earth Observing and Space Research, George Mason University Fairfax Virginia 22030- 4444 E-mail: lchiu@gmu.edu

Dr. Zhong Liu	Center for Earth Observing and Space Research, George Mason University Fairfax Virginia 22030- 4444 E-mail: zliu@eosdata.gsfc.nasa.gov
Ms. Hualan Rui	NASA NASA/GES Data and Information Services Center, Code 610.2, Greenbelt, Maryland, 20771 E-mail: rui@daac.gsfc.nasa.gov
Dr. William L. Teng	NASA NASA/GES Data and Information Services Center, Code 610.2, Greenbelt, Maryland, 20771 and Science Systems and Applications., Lanhman, MD 20706 E-mail: teng@daac.gsfc.nasa.gov
Dr. Liping Di	Laboratory for Advanced Information Technology and Standards (LAITS), George Mason University, 6301 Ivy Lane, Suite 620, Greenbelt, MD 20770, USA E-mail: lpd@rattler.gsfc.nasa.gov
Dr. Xianjun Hao	Center for Earth Observing and Space research (CEOSR), School of Computational Sciences, Fairfax, VA 20224, USA
Dr. Swarvanu Dasgupta	Center for Earth Observing and Space research (CEOSR), School of Computational Sciences, Fairfax, VA 20224, USA
Dr. Kwang-Sụ Yang	Center for Earth Observing and Space research (CEOSR), School of Computational Sciences, Fairfax, VA 20224, USA E-mail: kyang@scs.gmu.edu
Dr. Kenneth R. McDonald	NASA Goddard Space Flight Center, Code 423.0, Greenbelt, MD 20771
Dr. Chaowei (Phil) Yang	Earth Systems and GeoInformation Sciences, School of Computational Sciences, George Mason University, 4400 University Drive, Fairfax, VA, 22030, USA
Dr. David Wong	Earth Systems and GeoInformation Sciences, School of Computational Sciences, George Mason University, 4400 University Drive, Fairfax, VA, 22030, USA

Dr. Ruixin Yang	Earth Systems and GeoInformation Sciences, School of Computational Sciences, George Mason University, 4400 University Drive, Fairfax, VA, 22030, USA E-mail: yang@yang.gmu.edu
Ms. Jiang Tang	Center for Earth Observing and Space Research (CEOSR), School of Computational Scicnces, George Mason University E-mail: jtang@scs.gmu.edu
Dr. Zhiqiang Gao	Institute of Geographical Sciences and Natural Resources Research, Chinese Academy of Sciences, Beijing, China
Dr. Xiaoling Pan	International Center for Desert Affairs-Research for Sustainable Development in Arid and Semi-arid Land, Urumqi, China
Dr. James Slusser	USDA UV-B Monitoring and Research Program, Natural Resource Ecology Laboratory, Colorado State University, Fort Collins, CO
Dr. Hamid Yimit	College of Resources and Environment Xinjiang University, Urumqi, China
Dr. Yingjun Ma	International Center for Desert Affairs-Research for Sustainable Development in Arid and Semi-arid Land, Urumqi, China
Dr. Steven J. Kempler	NASA/Goddard Space Flight Center, Code 610.2 Greenbelt, MD 20771

1 Introduction to Data, Computational Processing, and Tools of Satellite Remote Sensing

John J. Qu and Menas Kafatos

Since late 1980s, NASA has been developing the Earth Observing System (EOS), which is an integrated, multi-satellite, long-term program designed for monitoring Earth's land, ocean, and atmosphere as an integrated system (Asrar and Dozier, 1994; King, 1999). Since the Terra and Aqua satellites launched in December 1999 and May 2002, we have entered a new era of Earth science satellite remote sensing. Particularly, the Moderate Resolution Imaging Spectroradiometer (MODIS) applications are of interest not only to land, ocean, and atmosphere discipline researchers but also to application, interdisciplinary and environmental scientists because MODIS senses all of the Earth's surface in 36 spectral bands spanning the visible (VIS, 0.415 μm) to infrared (14.235 μm) spectrum at nadir spatial resolutions of 1 km, 500 m and 250 m. It is also a new challenge to produce and distribute the EOS data to end-users. Barriers to wider use of National Aeronautics and Space Administration (NASA) data include: (1) MODIS datasets are too large for most users; (2) Native MODIS data files are stored in Hierarchical Data Format Earth Observing System (HDF-EOS), a file format that does not currently have wide support; (3) MODIS land and ocean product imagery is in a new map projection called the Integerized Sinusoidal (ISIN) projection; (4) most application users utilize Geographic Information System (GIS) system for input, management, analysis and visualization of data, but most GIS software can not handle HDF-EOS and ISIN projection currently (Qu et al., 2005). Many prospective users of the data with diverse needs, the large file sizes and the associated complex data format have posed difficulties, including insufficient bandwidth for downloading the data, insufficient storage on the users' platforms, and incompatibility of the format with visualization packages, including most GIS packages.

In response to these issues, NASA has engaged in facilitating new applications and distributions for Earth science remote sensing data. The second volume of this book concentrate on three major sections: (1) remote sensing data and calibration (Chapters 2 – 6); (2) data format and tools (Chapters 7 – 13); and local applications and GIS immigrated products (Chapters 14 – 18). We also include two appendices "Earth Science Remote Sensing Data and Services and Information Technology at the NASA/GES DISC DAAC" and the other is the source code for Chapter 7.

In Chapter 2, Dr. Carl A. Reber and Dr. H. K. Ramapriyan provide an introduction of the Earth Observing System Data and Information System (EOSDIS), which is NASA's largest and most complex data system. This system was developed and is being operated by the Earth Sciences Data and Information System project (ESDIS) at the Goddard Space Flight Center (GSFC). The development of EOSDIS started in the early 1990's along with the beginning of the EOS program. The current EOSDIS supports EOS missions and Landsat-7 in operational mode. The EOSDIS Core System (ECS), the EOS Clearinghouse (ECHO) and data pool are all discussed in Chapter 2. There are three major ECS elements, which are the Flight Operations Segment (FOS); the Science Data Processing Segment (SDPS); and the Communications and System Management Segment (CSMS). The relationships between each element and the system infrastructure are discussed in Chapter 2. The ECS data pool can provide large on-line data access capabilities to a user community. A web site that summarizes EOS missions, product definitions, and data availability can be found at http://eosdatainfo.gsfc.nasa.gov. Many of the Distributed Active Archive Centers (DAACs) have developed data access interfaces to support their own user communities.

Following Chapter 2, Mr. John L. Dwyer provides a summary of "Remotely Sensed Data Available from the US Geological Survey EROS Data Center". The Earth Resources Observation Systems (EROS) Data Center (EDC) is a national center (http://edc.usgs.gov/) for archive aerial and satellite remote sensing data products. The EDC has two World Data Center-A designations: Land Remote Sensing and Land Cover. It was established in the Geography Discipline within the US Geological Survey (USGS) of the Department of the Interior (DOI). EDC also established the Land Processes Distributed Active Archive Center (LP DAAC) in support of NASA's EOS. Because the customer base using EDC products is international and interdisciplinary in nature, the scope of services provided to support the user community is comprehensive. The EDC products and services are discussed in Chapter 3.

Many remote sensing users need real-time satellite measurements, so satellite Direct Broadcast (DB) system become very important part in the Earth science satellite remote sensing. NASA Direct Readout (DR) for its Polar Orbiting Satellites is discussed by Dr. Patrick Coronado and Dr. Kelvin Brentzel in Chapter 4. The two overarching concepts that encompass the content of the Chapter 4 are Direct Broadcast and DR (http://directreadout.gsfc. nasa.gov/). Direct Broadcast is the real-time transmission of satellite data to the ground. As the Earth is observed by satellite instruments, data is formatted and broadcast omni-directionally in a hemispheric pattern to Earth in real-time. Users who have compatible ground receiving equipment and are in direct line of sight to the satellite may receive these transmissions. Direct readout is the process of acquiring freely transmitted live satellite data. This was not done by design, but by the human need to simplify terminology. In Chapter 4, DB will be described briefly to differentiate it from the primary method of acquiring instrument data from Earth remote sensing satellites, such as POES, EOS series, NPP, and future NPOESS satellites. Instead, DR will be the focus where history, methods of acquisition, supporting technologies, and its roadmap will be described. Further, this Chapter 4 address only NASA's satellites including its involvement in the POES DR system's development and how it has, and still is providing a bridge between NASA DB Earth science missions and the broader user community. Most MODIS software/algorithms are released from the NASA/GSFC Direct Readout Webpage, http://directreadout.gsfc.nasa.gov/index.cfm?section = downloads.

Instrument calibration and characterization are critical for Earth science satellite remote sensing. Dr. Xiaoxiong Xiong and Dr. William Barnes provide the MODIS calibration and characterization information. MODIS has 36 spectral bands with center wavelengths from 0.41 μm to 14.5 μm and located on four focal plane assemblies (FPAs): visible (VIS), near infrared (NIR), short- and mid-wave infrared (SMIR), and long-wave infrared (LWIR). For each 1 km cross-track integration period (frame), there are 4 sub-frames for the 250 m resolution bands and 2 sub-frames for the 500 m resolution bands. The MODIS on-board calibrators (OBCs) include a solar diffuser (SD), a blackbody (BB), and a spectro-radiometric calibration assembly (SRCA). In addition, there is a space view (SV) port that is used to provide a zero reference. The on-orbit degradation of the SD is monitored by the solar diffuser stability monitor (SDSM). The on-orbit instrument calibration and characterization are also discussed in Chapter 5.

Dr. William Barnes discusses the "Use of the Moon for Calibration and Characterization of MODIS, SeaWiFS, and VIRS" in Chapter 6. Establishing the long-term high quality of the Earth System Data Records (ESDRs) with satellite remote sensing measurements will rely on the best efforts on instrument calibrations, so cross-sensor absolute radiometric calibrations are very critical and fundamental to ensure satellite measurements in the Climate Data Record (CDR) quality. Approaches for establishing the calibration and characterization have been developed using the moon light. For the visible through shortwave-infrared portion of the electromagnetic spectrum the moon meets this requirement. Chapter 6 examines how lunar data is being used by three different imaging radiometers, the results from each sensor and how these techniques will be used by the next generation of operational environmental satellites being developed by the joint DOD/NOAA/NASA Integrated Program Office (IPO).

In Chapter 7, Dr. Wenli Yang provides a detailed review of remote sensing data formats for Earth science remote sensing measurements. Earth observing remote sensing involves a variety of diverse instruments and platforms, ranging from hand-held close-range spectrometers to imagers and sounders on board satellites tens of thousands kilometres above the earth. There is not a unique data format for all remote sensing missions. It is usually based on a number of considerations, including requirements of the sensing system, mission objectives, the design and technology of data processing, archiving, and distribution systems, as well as community data standards. Currently, Earth observing data, or geospatial data, are stored in a wide variety of formats. Several common formats, such as Hierarchical Data Format-Earth Observing System (HDF-EOS) and HDF are introduced. Some software is also provided (see Appendix B).

Dr. Christopher Lynnes introduces "A Simple, Scalable, Script-based Science Processor (S4P)" system in Chapter 8. In 1999, the impending launch of the Terra satellite, combined with concerns about the ability of commercial software to process its high data volumes, led the Goddard Earth Sciences Distributed Active Archive Center (GES DAAC) to develop a contingency science processing system. Severe time and money constraints forced the GES DAAC towards a minimalist architecture that eventually became the Simple, Scalable, Script-based Science Processor (S4P). It was named Simple for its architecture and small size (a few thousand lines of code); Scalable for its ability to scale up to heavy processing loads or down to modest automation tasks; Script-based for its reliance on the Perl scripting language for its infrastructure; and Science Processor for its genesis in running scientific algorithms. This S4P took on processing for the forward MODIS data stream and the processing for the Atmospheric Infrared Sounder (AIRS), Humidity Sounder for Brazil (HSB), and Advanced Microwave Sounding Unit (AMSU). The updated information about S4P can be obtained from GES DAAC web page, http://daac.gsfc.nasa.gov/ techlab/s4pm/index.shtml.

As we discussed before, the MODIS ISIN project is one of the obstacles of the wider use of NASA data. The MODIS Reprojection Tool (MRT) has been developed to help individuals work with MODIS Level-2G, Level-3, and Level-4 land data products. Dr. John Dwyer and Dr. Gail Schmidt introduce the MRT in Chapter 9. The software may also be used to convert MODIS products to file formats (generic binary and GeoTIFF) that are more readily compatible with existing software packages. The MRT can be used for re-projecting, sub-setting, re-formatting. The MODIS land products distributed by the Land Processes Distributed Active Archive Center are in the HDF-EOS and developed Metadata are embedded in the HDF file as well as contained in a .met file that is associated with each HDF-EOS file. The MRT software is highly portable and currently runs under four different operating platforms. The look and feel of the software is consistent across different platforms. The MRT may be invoked from a user-friendly graphical user interface (GUI) or from a powerful command-line interface (http://edcdaac.usgs.gov/landdaac/tools /modis/index.asp).

In Chapter 10, Dr. Larry Klein et al. introduce a new tool entitled "A Tool for Conversion of Earth Observing System Data Products to GIS Compatible Formats and for the Provision of Post-Processing Functionality". The standard format for EOS instrument data is HDF-EOS, which is based on the National Center for Supercomputing Applications (NCSA) HDF (HDF 4.1R3 Users Guide, 2003). HDF is a self-describing, portable format for scientific data. HDF-EOS applies standards for attaching geolocation and temporal information to scientific data. The format was developed to provide a convention for geo-locating data from disparate instruments. Most EOS data products are stored and distributed in this format. HDF-EOS files also contain core and product-specific metadata. Core metadata are also stored in databases for user search and access. In order that the data are useful to GIS users, the most commonly used data must be accessible to application tools commonly used by that community. The HDF-EOS to GIS Format Conversion Tool (HEG) has been developed in response to these needs. It allows a user working on his/her workstation to reformat, re-project and perform operations such as subsetting and stitching on selectable HDF-EOS objects stored in EOS products. The tool described in this section is "stand-alone" and down-loadable to a variety of workstations. A version of HEG is also available at EOS archive centers. This version can access a subset of EOS archived data directly (http://eosweb.larc.nasa.gov/PRODOCS/misr/tools/geotiff_tool.html). Detailed information about HEG can be obtained in Chapter 10.

Chapter 11 is another tool chapter: HDFLook-multifunctional HDF-EOS tool. Drs. Louis Gonzales, Dimitar Ouzounov and Darryl Nickless introduce the HDFLook for MODIS and AIRS data processing at the Goddard Earth Sciences (GES) Data and Information Services Center (DISC) Distributed Active Archive Center (DAAC). The GES DISC DAAC Data Support Teams provide expert assistance to users in accessing data, including information on visualization tools and documentation for data products. To provide easy access, processing and visualization of the science data, the data support teams have additionally developed many desktop and online tools (http://daac.gsfc.nasa.gov/). The MODIS Data Support has developed a simple and quick MODIS data processing tool on the base of the existing HDFLook tool originally developed by the Laboratoire d'Optique Atmospherique (LOA) U. S. T. L., Lille, France. Several other data tools are also provided by GES DISC DAAC (http://daac.gsfc.nasa. gov/tools/index.shtml).

In Chapter 12, Dr. Long Chiu et al. introduce the "Tropical Rainfall Measuring Mission (TRMM) Data and Access Tools". TRMM provides visible, infrared, and microwave observations of tropical and subtropical rain systems, as well as lightning and cloud and radiation measurements. The TRMM Online Visualization and Analysis System (TOVAS), developed by the GES DAAC, provides user communities with a friendly web-based interface for visualization and analysis of TRMM gridded rainfall products and other precipitation data. Chapter 12 provides a summary of the TRMM data and access tool. The GES DAAC has developed a new tool The GES-DISC Interactive Online Visualization and Analysis Infrastructure (Giovanni). It is a powerful online analysis tool that is helping scientists better understanding Earth science data without downloading data. The Giovanni is the underlying infrastructure for a growing family of Web interfaces that allows users to analyze gridded data interactively online without having to download any data. Through Giovanni, users are invited to discover and explore our data using sophisticated analyses and visualizations (http://daac.gsfc.nasa.gov/techlab/giovanni/index.shtml). It also can be used for TRMM on-line data access.

In Chapter 13, Dr. Liping Di discusses "The Open GIS Web Service Specifications for Interoperable Access and Services of NASA EOS Data". Data is a representation of objects or phenomena in a formalized manner subject to interpretation or to which meaning may be assigned (ISO/IEC 2382-1, 1993). Geospatial data is the data that can be associated with location on Earth. Geospatial data is the dominant form of data in terms of data volume. Geospatial data are widely used in many socio-economic activities, ranging from environmental management to military operations. Because of the importance of geospatial data, both public and private sectors have invested huge amounts of resources and money in collecting, managing, archiving, and distributing geospatial data. The data collected by EOS are processed, distributed, archived, and managed by EOS Data and Information System (EOSDIS), and are distributed at nine Distributed Active Archival Centers (DAACs). Although EOS data are intended for use in studying global climate and environmental change, they have also been increasingly used for local and regional applications. In the past several years, individual data producers and many national and international organizations have worked on the issue of interoperability for remote sensing data and information and have set various levels of standards. Chapter 13 discusses the NASA EOS data environment, summarizes Open Geospatial Consortium (OGC) specifications applicable for accessing NASA EOS data in the web environment, and presents an approach for applying OGC specifications to NASA EOS data.

Dr. John Qu et al. discuss the "Global MODIS Remote Sensing Data for Local Usage: VAccess/MAGIC". Although intended as an instrument focusing on global change research, MODIS, along with global Earth observing and remote sensing (RS) sensors, has been shown to be at least very useful for environmental applications. One such consortium that heavily utilizes Earth observing data has been assembled for the Mid-Atlantic region: Termed Virginia Access—Middle Atlantic Geospatial Information Consortium, hereinafter called, MAGIC, which is a consortium of Middle-Atlantic universities to develop a remote sensing, applications-oriented, geospatial/ temporal data and information system (http://cpgis.gmu.edu/vaccess/mission.htm). It focuses on two primary aspects: the usage of such data for national priorities, following the emphasis of NASA's Earth Science Applications; and the dissemination of such data through interoperable information systems, that couple to NASA's systems and promote open source solutions and standards. The MODIS data compressing and pre-processing (sub-setting and sub-sampling), data format conversion (easy accessing data format such as GIS compatible formats), GIS and Open GIS applications, and simple real time data processing for future Earth observing satellite systems are discussed in Chapter 14. In this chapter, they emphasize some usage of MODIS products for the Mid-Atlantic region, which is the main focus of MAGIC (http://cpgis.gmu.edu/vaccess/mission.htm).

Following Chapter 14, the NASA HDF-EOS Web GIS Software Suite (NWGISS) is discussed by Dr. Liping Di and Dr. Kenneth R. McDonald. The

NASA Earth Science Enterprise (ESE) is generating a huge volume of remote sensing data to support Earth system science and applications research through its Earth Observing System (EOS) program. Most of the data are stored in HDF-EOS, the standard format for NASA's EOSDIS. EOS data are widely used not only for global change research but also for local and regional applications. There are two typical groups of EOS data users. For this group of users, EOSDIS provides special access interfaces with a dedicated fast network connection. The second group of users is a large number of small data users with limited resources. They typically use a GIS tool to analyze data from EOS for a variety of applications. For most of them, obtaining the needed data from EOSDIS and pre-processing them into a form ready to be analyzed by their in-house GIS represents a major technical challenge that takes a significant amount of their time and resources. Therefore, the development of capabilities to deliver GIS-ready EOS data directly to a variety of users' GIS systems through the Internet based on the users' requirements and to provide services that make user-specific products on demand will greatly enhance the interoperability and increase the public use of EOS data. Chapter 15 describes a NASA-funded project to develop such capabilities in a system called the NASA HDF-EOS Web GIS Software Suite (NWGISS), which provides interoperable access and services for HDF-EOS data using (OGC) protocols. Chapter 15 first presents the current functionalities and components of the system and then discusses the current efforts to extend NWGISS, including the integration of NWGISS with Grid technology and the development of geospatial Web service middleware.

In Chapter 16, Dr. Chaowei Yang et al. describe a "Network Geographic Information System". Since its interception in the 1960s (GIS World, 1996; Tomlinson, 1998), Geographic Information System has been well established as an information system for acquiring, storing, processing, analyzing and visualizing spatial data. The success of GIS is reflected from the following developments: (1) numerous GIS software have been developed and are widely used; (2) a large amount of spatial datasets have been collected and used; (3) many books and papers have been written to introduce and discuss different aspects of GIS. To address these issues and to provide a systematic introduction of Network GIS, the rest of this chapter is organized in the following manner: Section 16.2 describes the communication foundation of Network GIS, i.e., the underlying network infrastructure. Section 16.3 reviews basic GIS functions, discusses another critical issue of Network GIS, that is, how to distribute these functions over a network. Consequently, a flexible Network GIS framework is proposed. Section 16.4 introduces Distributed GIS as a new envision of Network GIS. Section 16.5 proposes a taxonomy of Network GIS and discusses the usage of different types of Network GIS. Two examples of Network GIS are given in Section 16.6 to illustrate how Network GIS can be used in disseminating data/ information and providing services. Finally, in Section 16.7, we present several research areas of Network GIS.

In Chapter 17, "A Content-Based Search Method and Its Application for EOS" is discussed by Dr. Ruixin Yang et al. Advances in Earth observing from space and associated Earth systems numerical simulations have resulted in rapidly growing data volumes for Earth science communities. The NASA's EOS generate huge data per day. To find interesting data, data users need an effective and efficient way to search through the data. Traditionally, metadata are provided in a database which support data searching by commonly used criteria such as spatial coverage, temporal coverage, spatial resolution, temporal resolution, etc. However, metadata search itself may still result in large amounts of data, which data users need to know about before determining the usefulness of the underlying data. Based on the algorithms for both query types, a prototype system has been developed, and sample EOS data are used in this system. This prototype is a web-based system. The rest of this chapter is organized as follows. In the next section, Section 17.2, the details of the query algorithms are described. The prototype system is described in Section 17.3. Experiment results are given in Section 17.4. Finally, discussion and future potential work are presented in Section 17.5.

In the last Chapter (18), "The Impact of Precipitation and Temperatue on Net Primary Productivity (NPP) in Xinjiang, China from 1981 to 2000" is discussed by Dr. Wei Gao et al. It provides another example for global remote sensing for local usage. It focuses on global change impacts on ecosystem over Xinjiang regions. While biogeochemical models address the problems related to primary productivity response in light of detailed eco-physiological mechanisms, they tend to simplify the structural aspect of responses. It has also become possible to investigate the global spatial and temporal dynamics of NPP by a combination of ecosystem process modelling and monitoring by remote sensing. Satellite observations provide a useful and powerful database for evaluating the dynamics of terrestrial biosphere because of its global coverage and short revisit interval. NPP is a quantifiable characteristic of the biosphere and is a fundamental ecosystem variable modelled at the global scale. Practical considerations for estimating NPP exist in its utility to measure crop yield, forest production and other economically and socially significant products of vegetation growth. In the Chapter 18, they use GLObal Production Efficiency Model (GLO-PEM), comparing with Carbon Exchange in the Vegetation-Soil- Atmosphere system (CEVSA) estimating interannual variations and spatial patterns of NPP caused by climate (precipitation and temperature) in Xinjiang during 1981 – 2000 periods.

This volume covers various topics about Earth science data processing, calibration, data accessing and tools, GIS applications. It is almost impossible for one book to cover the entire all topics of remote sensing data, tools and processing. Here, we focus on providing a synopsis of this volume. We use materials provided in these chapters themselves. The material is surely not ours; they are the authors' works, which we have relied on. The reader is guided to references and more detailed discussion in the chapters themselves, written by the experts in the relevant areas.

References

Asrar G, Dozier J (1994) EOS: Science Strategy for the Earth Observing System. New York: American Institute of Physics Press, Woodbury, 11,797

Barnes W, Xiong X, Eplee R, Sun J, Lyu C-H (2006) Use of the Moon for calibration and characterization of MODIS, SeaWiFS, and VIRS. In: Qu, Gao, Kafatos, Murphy and Salomonson (eds) Earth Science Satellite Remote Sensing: Vol II. Tsinghua University Press, Springer-Verlag (this publication)

Chiu LS, Liu Z, Rui H, Teng WL (2006) Tropical Rainfall Measurement Mission (TRMM) Data and Access Tools. In: Qu, Gao, Kafatos, Murphy and Salomonson (eds) Earth Science Satellite Remote Sensing: Vol II. Tsinghua University Press, Springer-Verlag (this publication)

Coronado PL, Brentzel KW (2006) NASA Direct Readout for its Polar Orbiting Satellites. In: Qu, Gao, Kafatos, Murphy and Salomonson (eds) Earth Science Satellite Remote Sensing: Vol II. Tsinghua University Press, Springer-Verlag (this publication)

Di L (2006) The Open GIS Web Service specifications for Interoperable Access and Services to NASA EOS Data. In: Qu, Gao, Kafatos, Murphy and Salomonson (eds) Earth Science Satellite Remote Sensing: Vol II. Tsinghua University Press, Springer-Verlag (this publication)

Di L, McDonald KR (2006) The NASA HDF-EOS Web GIS Software Suite (NWGISS). In: Qu, Gao, Kafatos, Murphy and Salomonson (eds) Earth Science Satellite Remote Sensing: Vol II. Tsinghua University Press, Springer-Verlag (this publication)

Dwyer J (2006) Remotely sensed data available from the US Geological Survey EROS Data Center. In: Qu, Gao, Kafatos, Murphy and Salomonson (eds) Earth Science Satellite Remote Sensing: Vol II. Tsinghua University Press, Springer-Verlag (this publication)

Dwyer J, Schmidt G (2006) The MODIS Reprojection Tool (MRT). In: Qu, Gao, Kafatos, Murphy and Salomonson (eds) Earth Science Satellite Remote Sensing: Vol II. Tsinghua University Press, Springer-Verlag (this publication)

Gao W, Gao Z, Pan X, Slusser J, Ma Y (2006) The impact of precipitation and temperature on Net Primary Productivity (NPP) in Xinjiang, China from 1981 – 2000. In: Qu, Gao, Kafatos, Murphy and Salomonson (eds) Earth Science Satellite Remote Sensing: Vol II. Tsinghua University Press, Springer-Verlag (this publication)

Gonzales L, Ouzounov D, Nickless D, Leptoukh G (2006) HDFLook-multifunctional HDF-EOS tool for MODIS and AIRS data processing at GES DISC DAAC. In: Qu, Gao, Kafatos, Murphy and Salomonson (eds) Earth Science Satellite Remote Sensing: Vol II. Tsinghua University Press, Springer-Verlag (this publication)

King MD (ed) (1999) EOS Science Plan. National Aeronautics and Space Administration, Washington DC NP-1998-12-069-GSFC

Klein L, Milburn R, Praderas C, Taaheri A (2006) A tool for conversion of Earth Observing System Data Products to GIS Compatible Formats and for the provision of post-processing functionality. In: Qu, Gao, Kafatos, Murphy and Salomonson (eds) Earth Science Satellite Remote Sensing: Vol II. Tsinghua University Press, Springer-Verlag (this publication)

Lynnes C (2006) A Simple, Scalable, Script-based Science Processor. In: Qu, Gao, Kafatos, Murphy and Salomonson (eds) Earth Science Satellite Remote Sensing: Vol II. Tsinghua University Press, Springer-Verlag (this publication)

Qu JJ, Kafatos M, Hao X, Dasgupta S, Yang K-S (2006) Global MODIS Remote Sensing Data for Local Usage: VAccess/MAGIC. In: Qu, Gao, Kafatos, Murphy and Salomonson (eds) Earth Science Satellite Remote Sensing: Vol II. Tsinghua University Press, Springer-Verlag (this publication)

Reber CA, Ramapriyan HK, (2006) Earth Science Satellite Remote Sensing Data from the EOS Data and Information System (EOSDIS). In: Qu, Gao, Kafatos, Murphy and Salomonson (eds) Earth Science Satellite Remote Sensing: Vol II. Tsinghua University Press, Springer-Verlag (this publication)

Xiong X, Barnes W (2006) MODIS Calibration and Characterization. In: Qu, Gao, Kafatos, Murphy and Salomonson (eds) Earth Science Satellite Remote Sensing: Vol II. Tsinghua University Press, Springer-Verlag (this publication)

Yang CP, Kafatos M, Wong DW, Yang R (2006) Network Geographic Information System (GIS). In: Qu, Gao, Kafatos, Murphy and Salomonson (eds) Earth Science Satellite Remote Sensing: Vol II. Tsinghua University Press, Springer-Verlag (this publication)

Yang R, Yang K-S, Tang J, Qu J, Kafatos M (2006) A content-based Search method and its Application for EOS. In: Qu, Gao, Kafatos, Murphy and Salomonson (eds) Earth Science Satellite Remote Sensing: Vol II. Tsinghua University Press, Springer-Verlag (this publication)

Yang W (2006) A review of Remote Sensing Data Formats for Earth System Observations. In: Qu, Gao, Kafatos, Murphy and Salomonson (eds) Earth Science Satellite Remote Sensing: Vol II. Tsinghua University Press, Springer-Verlag (this publication)

2 Earth Science Satellite Remote Sensing Data from the EOS Data and Information System

Carl A. Reber and H. K. Ramapriyan

2.1 Introduction

The Earth Observing System Data and Information System (EOSDIS), NASA's largest and most complex data system, was developed and is being operated by the Earth Sciences Data and Information System project (ESDIS) at the Goddard Space Flight Center in Greenbelt, Maryland. It was conceived in the late 1980s for ground-based spacecraft operations, and for collection, processing, archiving, and distribution of Earth science data, with a focus on the EOS missions (Fig. 2.1). The development of EOSDIS started in the early 1990's along with the beginning of the EOS program. An initial version of EOSDIS (called Version 0 and used to manage non-EOS, legacy datasets) grew to operational status in 1994. The version of EOSDIS supporting EOS missions became operational in incremental steps—first in 1997 supporting EOS instruments on TRMM, next in 1999 supporting Landsat-7 and Terra, and subsequently, the other EOS missions. For this discussion, focusing on the data themselves, the pertinent components of the EOSDIS include the EOSDIS Core System (ECS), the Distributed Active Archive Centers (DAACs), the Science Computing Facilities (SCFs, funded separately from the ESDIS Project), the Science Investigator-led Processing Systems (SIPS), and the various data access interfaces (Fig. 2.2). There are eight DAACs, each with expertise in one or more specific disciplines in Earth science, which archive and distribute EOS data (Table 2.1). The Land Processes Distributed Archive Center (LP DAAC) in Sioux Falls, South Dakota, for example, specializes in land imagery. The hardware and software needed to support the functions at four of these DAACs (ASDC at NASA Langley, GES DAAC at NASA Goddard, LP DAAC at EDC, and NSIDC) were developed as part of the ECS and later installed and integrated at the DAACs. The primary format employed is the Hierarchical Data Format (HDF) from the National Center for Supercomputing Applications (NCSA). HDF-EOS, developed through ESDIS-NCSA collaboration, defines standard "profiles" for the use of HDF in the domain of Earth observing data (i.e. attributes and services for point, grid and swath based data) as recommended by the Committee on Earth Observing Satellites (CEOS).

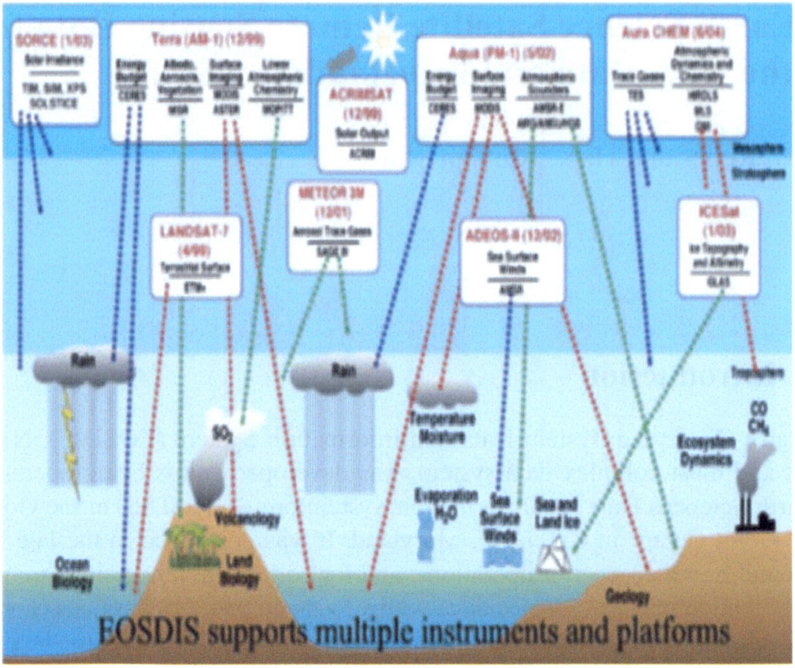

Figure 2.1 Some of the geophysically-oriented satellite missions supported by the EOSDIS, and the components of the Earth system on which they focus

Table 2.1 EOSDIS DAACs

DAAC	Discipline	Website	E-mail Address
Alaska Satellite Facility DAAC (ASF DAAC), University of Alaska	Synthetic Aperture Radar (SAR), Sea Ice, Polar Processes, Geophysics	http://www.asf .alaska.edu	asf@eos.nasa.gov
*Goddard Space Flight Center Earth Sciences DAAC (GES DAAC)	Upper Atmosphere, Atmospheric Dynamics, Global Land Biosphere, Global Precipitation, Ocean Color, Sea Surface Temperature	http://daac.gsfc . nasa.gov	gsfc@eos.nasa.gov
*Langley Research Center DAAC (Atmospheric Sciences Data Center (ASDC))	Radiation Budget, Clouds, Aerosols, Tropospheric Chemistry	http://eosweb. larc.nasa.gov	larc@eos.nasa.gov
*Land Processes DAAC (LP DAAC)-USGS EROS Data Center	Land Processes	http:// LPDAAC. usgs.gov	edc@eos.nasa.gov

Continued

DAAC	Discipline	Website	E-mail Address
*National Snow and Ice Data Center DAAC (NSIDC DAAC), University of Colorado	Snow and Ice, Cryosphere and Climate	http://nsidc.org	nsidc@eos.nasa.gov
Oak Ridge National Laboratory DAAC (ORNL DAAC)	Biogeochemical Dynamics, Ecological Data for Studying Environmental Processes	http://www.daac.ornl.gov	ornl@eos.nasa.gov
Physical Oceanography DAAC (PO DAAC)-Jet Propulsion Laboratory	Oceanic Processes, Air-Sea Interactions	http://podaac.jpl.nasa.gov	jpl@eos.nasa.gov
Soclo-Economic Data & Applications Center (SEDAC)	Population, Sustainability, Geospatial Data, Multilateral Environmental Agreements	http://sedac.clesin.columbia.edu	sedac@eos.nasa.gov

* Indicates a DAAC that incorporates the EOSDIS Core System.

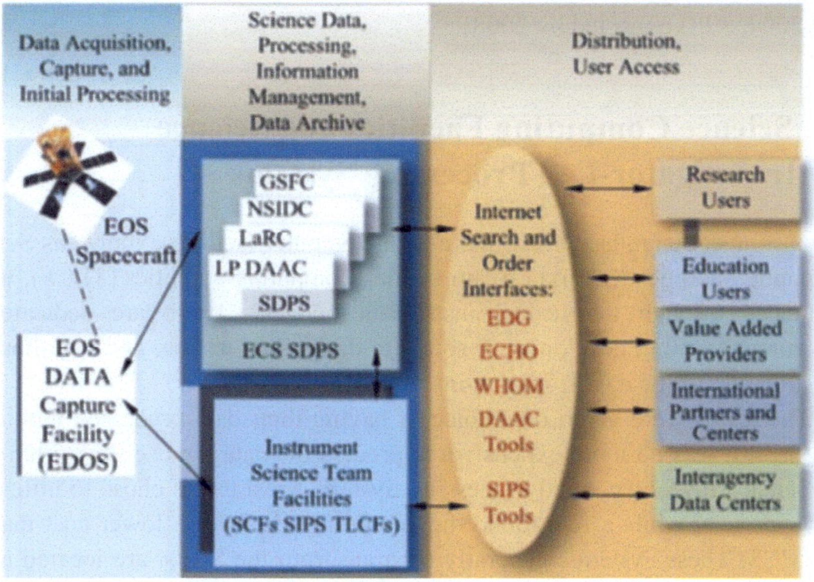

Figure 2.2 Overview of the EOSDIS, showing the main functions, components (not including the non-ECS DAACs), and interfaces

The algorithms and software for producing geophysical data products from the satellite measurements are the responsibility of the individual instrument teams. Initially these teams were to provide the software to one of the four ECS DAACs, which would then perform the operational data processing, archive the data, and be responsible for their distribution. Later, some of the higher level processing functions were taken over by the appropriate instrument teams themselves, at ESDIS-funded SIPS; the SIPS forward the higher level data to the DAACs, which then archive and distribute them. Since most of the EOS data are archived and distributed by the ECS DAACs the discussion here is focused on them.

2.2 EOSDIS Core System

The EOSDIS core system has three elements: the Flight Operations Segment (FOS); the Science Data Processing Segment (SDPS); and the Communications and System Management Segment (CSMS). The FOS, as the name implies, controls the spacecraft, and has responsibility for mission planning and health and safety of the spacecraft, while the CSMS provides internal networks connecting the DAACs to each other and to other parts of the system. The SDPS provides the infrastructure and operating system that support data product generation and data management, archival, and distribution for the EOSDIS. The ECS currently handles data from 9 spacecraft and 22 instruments, more than 1,700 types of products, and extremely large volumes of data (more than 4 TB per day; 3 PB archived in April 2004, 8 PB expected by 2008).

2.3 Science Computing Facilities and Science Investigator-Led Processing Systems

Each instrument team has one or more local computer facilities, known as Science Computing Facilities (SCF), or Team Leader Computing Facilities (TLCF), which are funded through the teams' instrument budgets. These are dedicated to supporting algorithm development, software design and testing, and to calibration and validation of the data from that instrument.

The teams were given the choice of having their data processed within the ECS, or, if it was no more expensive, to process the data themselves within their own Science Investigator-led Processing System. Most teams chose to utilize the SIPS approach, while a few found they could not do it for a lower cost than by using ECS. These systems, generally separate from the SCFs, are located at the scientists' home institutions, and are funded by the ESDIS Project. They, like the SCFs, are designed, developed, maintained, and operated by the individual science teams. Their input data are received from the DAACs and processed to higher-level products, which are then returned to the appropriate DAAC for archiving and distribution. Several of the SIPSs also make their data directly available to members of their scientific community.

2.4 Data Access

Independently from the ECS, the ESDIS Project collaborated with the DAACs during 1990 through 1994 to develop a system known as Version 0 (V0) to include capabilities such as interoperability among distributed systems. The V0 Information Management System (IMS) was incorporated on top of the existing data and services at each DAAC to provide this interoperability. Initially designed as a working prototype with operating elements, V0 has been adapted for use with the later, operational, versions of EOSDIS that include ECS. This capability, now called the EOS Data Gateway (EDG), provides single-point access through a common interface to data located at any of the various EOSDIS DAACs. In addition to archives in the US, EOSDIS has interagency and international partners who supply data through interoperable links and collaborate on other EOSDIS activities. The URL for EDG is http://eos.nasa.gov/imswelcome.

The EOS Clearinghouse (ECHO) has been incorporated into the EOSDIS to make the architecture sufficiently flexible so that alternative clients may be built to provide tailored interfaces for discipline-specific communities. ECHO is an Applications Program Interface (API)—based framework that allows members of the scientific community to share their data, services, and tools with the rest of the community. As a metadata clearinghouse, ECHO supports multiple data access paradigms that make data access quite efficient for end users. ECHO allows data providers to control visibility and access to their own data and, because of the clearinghouse approach, can serve to improve availability and reliability. For more details, see http://eos.nasa.gov/echo.

Many of the DAACs have developed data access interfaces designed with their own particular user communities in mind. The Goddard DAAC's Web Hierarchical Ordering Mechanism (WHOM), for example, supports on-the-fly processing (subsetting, analysis, etc.), spatial, temporal and parameter searches, and graphical representation of the geographical coverage in the inventory. The LP DAAC has the US Geological Survey (USGS) Global Visualization Viewer, a search and order tool that provides map-based searches, instrument selection, selection of maximum cloud cover, and limited "zoom" in and out capability. This tool supports Landsat, ASTER, and MODIS land data. In general, these DAAC-specific interfaces can be accessed through links from the appropriate DAAC home page shown in Table 2.1.

The Data Pools, a relatively recent addition to the EOSDIS, use a large (i.e. many tens of TB) disk cache at each DAAC to hold EOS data for an extended period (e.g. a month or more). The contents of the cache and the time retained are tuned by the respective DAACs to the needs of their user community. Utilization of disk storage for the data pools rather than the "near-line" archival tapes greatly decreases the time needed for access.

A Web site that summarizes EOS missions, product definitions, and data availability can be found at http://eosdatainfo.gsfc.nasa.gov. This site consists of a time chart of EOS missions from 1997 through 2004 and has links to each mission, instrument, and data center (Fig. 2.3).

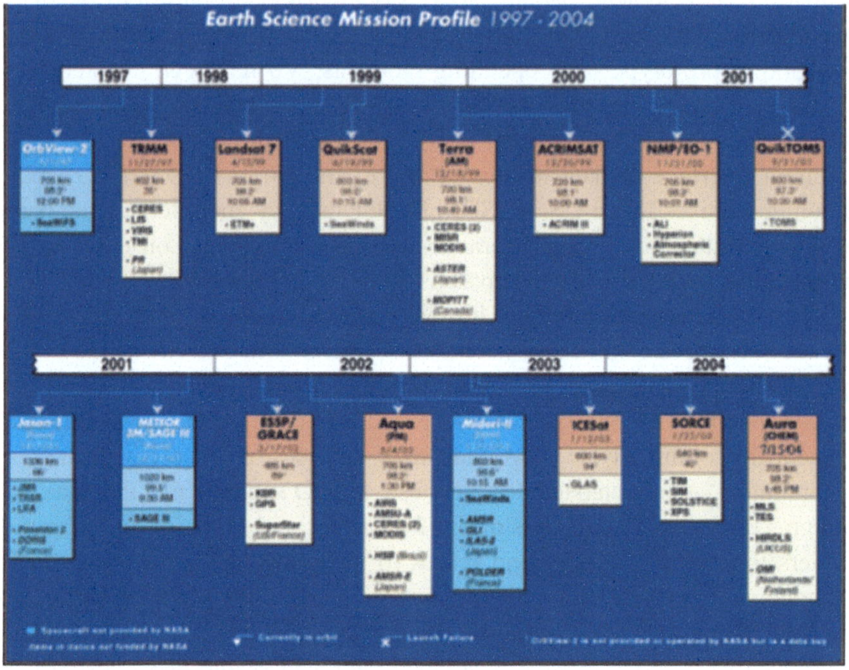

Figure 2.3 EOS missions from 1997 through 2004 from the web site http://eosdatainfo.gsfc.nasa.gov. From here, links are available to each mission, instrument, and data center

2.5 Perspectives

The Earth sciences data systems are still evolving. The current trend is toward more distributed systems, for processing as well as for archiving and distribution, than the relatively centralized architecture used for the EOSDIS. One approach being explored is grouping data systems by measurement type (e.g. ocean color, precipitation, or stratospheric chemistry), rather than by missions. The thought is that this will provide greater continuity in the algorithms and validation processes for Earth science data and make these data even more useful for long -term climate studies.

In addition, data archives for long term storage are in development under a NASA-NOAA (National Oceanic and Atmospheric Administration) Agreement.

Such an approach assures data and data product access by the scientific communities for many years into the future.

Whatever direction the new Earth sciences data systems may take, the current EOSDIS remains the largest scientific data system in use today, and is providing many lessons for designing and developing those scientific data systems of the future.

3 Remotely Sensed Data Available from the US Geological Survey EROS Data Center

John L. Dwyer

3.1 Introduction

The Center for Earth Resources Observation Systems (EROS) is a field center of the geography discipline within the US geological survey (USGS) of the Department of the Interior. The EROS Data Center (EDC) was established in the early 1970s as the nation's principal archive of remotely sensed data. Initially the EDC was responsible for the archive, reproduction, and distribution of black-and-white and color-infrared aerial photography acquired under numerous mapping programs conducted by various Federal agencies including the USGS, Department of Agriculture, Environmental Protection Agency, and NASA. The EDC was also designated the central archive for data acquired by the first satellite sensor designed for broad-scale earth observations in support of civilian agency needs for earth resource information. A four-band multispectral scanner (MSS) and a return-beam vidicon (RBV) camera were initially flown on the Earth Resources Technology Satellite-1, subsequently designated Landsat-1. The synoptic coverage, moderate spatial resolution, and multi-spectral view provided by these data stimulated scientists with an unprecedented perspective from which to study the Earth's surface and to understand the relationships between human activity and natural systems.

The EDC has two World Data Center-A designations: Land Remote Sensing and Land Cover. The EDC manages the National Satellite Land Remote Sensing Data Archive (NSLRSDA), as established by the Land Remote Sensing Policy Act (PL 102-555) on Oct. 28, 1992. Co-located at EDC are the North American node for the United Nations Environment Programme Global Resource Information Database (UNEP-GRID) and the Land Processes Distributed Active Archive Center (LP DAAC) in support of NASA's Earth Observing System (EOS). Although archive management, product generation, and data distribution are core to the mission of EDC, significant work is also performed in systems engineering, architecture and technology, software engineering, scientific research and applications, and educational outreach. Because the customer base using EDC products is international and interdisciplinary in nature, the scope of services provided to support the user community is comprehensive.

3.2 Data Products

The EDC is a leading source of land information for exploring our changing planet (http://edcwww.cr.usgs.gov/). Aerial photography, map-based (cartographic) information, elevation data sets and derivatives, and satellite data products are the major categories of data and information available from the USGS/EDC (see Tables 3.1 – 3.5). Many of these products are available through The National Map.

The National Map is a consistent framework for geographic knowledge needed by the Nation. It provides public access to high-quality, geospatial data and information from multiple partners to help support decision making by resource managers and the public. The National Map is the product of a consortium of Federal, State, and local partners who provide geospatial data to enhance America's ability to access, integrate, and apply geospatial data at global, national, and local scales.

Table 3.1 Aerial photography products available from the USGS/EDC

Aerial Products	Description	Search/Order
National Aerial Photography Program (NAPP)	Recent, high-quality aerial photos covering the conterminous United States on five-to seven-year cycles (1987 – present).	PhotoFinder Earth Explorer
National High Altitude Photography (NHAP)	High-altitude aerial photos for the conterminous United States (1980 – 1989).	Earth Explorer
Survey Photography	Aerial photos from a variety of sources (1939 – present).	Earth Explorer
Index/Mapping Photography	Aerial photos referenced as photo indexes from a variety of sources, such as EPA, USDA, NASA (1940s – 1980s).	Contact us
Digital Orthophoto Quadrangles (DOQs)	Digital images of aerial photos, which combine the image characteristics of the photo with the georeferenced qualities of a map (1987 – present).	Earth Explorer The National Map
Space Acquired Photography	Photos taken from the Space Shuttle, Skylab, Gemini, and Apollo missions (1965 – present).	Earth Explorer NASA Earth Sciences & Image Analysis (Shuttle and ISS)
Scientific Committee on Antarctic Research (SCAR)	A large collection of aerial photography over Antarctica (1946 – present).	Contact SCAR

Table 3.2 Map products available from the USGS/EDC

Map Products	Description	Search/Order
USGS Printed Maps	A wide variety of USGS printed maps, including 7.5-min topographic quadrangles.	USGS Store
Digital Raster Graphics (DRGs)	Scanned digital images of USGS topographic quadrangles.	Earth Explorer
Digital Line Graphs (DLGs)	Digital vector data derived from USGS maps and related sources.	Earth Explorer USGS Geographic Data
National Hydrography Dataset (NHD)	A national hydrologic database derived from USGS DLGs and other sources.	National Hydrography Dataset

Table 3.3 Elevation data sets available from the USGS/EDC

Elevation Products	Description	Search/Order
National Elevation Dataset (NED)	Seamless 10- and 30-meter digital raster elevation data covering the conterminous United States, Alaska, Hawaii, Puerto Rico, and Virgin Islands. Features periodic updates to incorporate the best available source data (primarily USGS 10-and 30-meter DEMs).	The National Map Earth Explorer NED Index Map
Shuttle Radar Topography Mission (SRTM)	1-arc second of continental United States. Periodic updates will include 3-arc second of international continents received in order of North America, South America, Australia, Eurasia, Africa, North & South Pacific, and North & South Atlantic.	The National Map
Digital Elevation Models (DEMs)	Digital raster elevation data based on USGS topographic quads.	Earth Explorer USGS Geographic Data
Global 30-Arc Second Elevation Dataset (GTOPO30)	Global 1-km digital raster data derived from a variety of sources.	GTOPO30 EOS Data Gateway
HYDRO1K	A global hydrologic database derived from 1996 GTOPO30 data.	HYDRO1K

Table 3.4 Satellite data products available from the USGS/EDC

Satellite Products	Description	Search/Order
Advanced Spaceborne Thermal Emission and Reflection Radiometer (ASTER)	High-resolution (15-to 90-meter) multispectral data from the Terra satellite (2000 to present).	EOS Data Gateway ASTER On-Demand Data Gateway Data Pool GloVis
Moderate Resolution Imaging Spectroradiometer (MODIS)	Moderate-resolution (250- to 1000-meter) multispectral data from the Terra satellite (2000 to present).	MODIS Direct Broadcast EOS Data Gateway Data Pool GloVis
Hyperion and Advanced Land Imager (ALI)	10-to 30-meter multipectral and hyperspectral data from the Earth Observing-1 (EO-1) Extended Mission (2000 to present).	Data Acquisition Request (DAR) Earth Explorer
Enhanced Thematic Mapper Plus (ETM +)	High-resolution (15- to 60-meter) multispectral data from Landsat-7 (1999 to present).	Earth Explorer GloVis
Thematic Mapper (TM)	30-to 120-meter multispectral data from Landsat-4 and -5 (1982 to present).	Earth Explorer GloVis
MSS	80-meter multispectral data from Landsats-1 to -5 (1972 to 1992).	Earth Explorer
Multi-Resolution Land Characteristics 2001 (MRLC 2001)	Selected ETM + and TM scenes (Landsats-7 and -5) from the National Land Cover Characterization 2000 project.	Land Cover Characterization Program (MRLC2001) Earth Explorer
Landsat Orthorectified Dataset	Orthorectified Landsat MSS (1980s), TM (1990s) and ETM + (2000), Landsat Orthorectified TM mosaic.	EOS Data Gateway
NASA Landsat Data Collection (NLDC)	Selected MSS and TM scenes (Landsats-1 to -5) from the NASA Landsat Data Collection (1975 to present).	EOS Data Gateway
North American Landscape Characterization (NALC)	Time-series triplicates of selected MSS scenes (Landsats-1 – -5) for the United States and Mexico. Average acquisition dates are 1973, 1986, and 1991.	EOS Data Gateway
Advanced Very High Resolution Radiometer (AVHRR)	1-km multispectral data from the NOAA satellite series (1979 to present).	NOAA Archive LP DAAC Web Site Earth Explorer

Continued

Satellite Products	Description	Search/Order
Declassified Satellite Imagery-1	Photographic imagery from the CORONA, ARGON and LANYARD satellites (1959 to 1972).	Earth Explorer
Declassified Satellite Imagery - 2	Photographic imagery from KH-7 surveillance and KH-9 mapping system (1963 to 1980).	Earth Explorer
Spaceborne Imaging Radar C-band (SIR-C)	Imaging radar data (C-band and L-band) from two Space Shuttle missions (1994).	LP DAAC (SIR-C)

Table 3.5 Land cover data sets available from the USGS/EDC

Land Cover Products	Description	Search/Order
National Land Cover Dataset 1992 (NLCD 92)	A US land cover classification product based primarily on 1992 Landsat Thematic Mapper data.	National Land Cover Characterization Program (NLCD 92) Seamless Data Distribution System
(Multi-Resolution Land Characteristics 2001 (MRLC2001)	Selected ETM + and TM scenes (Landsats-7 and -5) from the National Land Cover Characterization 2000 project.	Land Cover Characterization Program (MRLC2001)
Land Use and Land Cover Data (LULC)	Historical US land use and land cover data derived from 1970s and 1980s aerial photography.	USGS Geographic Data Download
Global Land Cover Characterization (GLCC)	A global land cover database primarily derived from 1992 to 1993 1-km AVHRR data.	LP DAAC (GLCC)
AVHRR Normalized Difference Vegetation Index (NDVI) Composites	Weekly and biweekly NDVI composites based on 1-km AVHRR data (1980 to present).	Contact EDC User Services

The USGS/EDC has archived thousands of rolls of film containing several million frames of historical aerial and satellite photographs dating back to the 1930's gathered from various Federal agencies. This archive includes the rather well known, often utilized NHAP and NAPP film, but the largest portion of this archive consists of original film acquired in support of the USGS 1:24,000 topographic quadrangle map series, taken during the 1940s through the 1970s.

Most of this photography is reasonably large scale to support the production of the quadrangle maps.

In September 2004, the USGS/EDC discontinued the generation and distribution of photographic products. Instead, a process has been initiated by which the original film products will be scanned or digitized for distribution as digital products. The high-resolution files are created at approximately 1,200 dpi with an output image size of around nine inches square. The file sizes are approximately 120 MB for black and white images and 360 MB for color images and are stored in TIFF format. At this time, only 9 × 9 inch and smaller original film is available for scanning. The medium-resolution files are created at approximately 600 dpi with an output image size of around five inches square. The file sizes are approximately 15 MB for black and white images and 45 MB for color images. The files are stored in a TIFF format. If customers are interested in obtaining film products from the digital files, they will be provided with information on USGS Business Partners that may provide such services.

3.2.1 Aircraft Scanners

The Aircraft Scanners data sets contain digital imagery acquired from several multispectral scanners including NS-001 multispectral scanner, Daedalus thematic mapper simulator (TMS), and the thermal infrared multispectral scanner (TIMS). These data are collected over selected areas of the conterminous United States, Alaska, and Hawaii by NASA ER-2 and NASA C-130B aircraft operating from the NASA Ames Research Center (ARC) in Moffett Field, California, and by NASA Learjet aircraft operating from Stennis Space Center (SSC) in Bay St. Louis, Mississippi. Limited international acquisitions also are available.

3.2.1.1 NS-001 Multispectral Scanner

The NS-001 multispectral scanner (see Table 3.6) was flown aboard a NASA C-130B aircraft and is a line-scanning device designed to simulate Landsat TM sensor performance. The scanner collects multispectral data in the seven Landsat-4 and -5 thematic mapper bands plus a band from 1.13 μm to 1.35 μm.

Table 3.6 NS-001 aircraft scanner sensor characteristics

Band	Wavelength (μm)		
1	0.458 – 0.519	Instantaneous Field of View (IFOV)	2.5 mrad
2	0.529 – 0.603	Field of View	100°
3	0.633 – 0.697	Pixels/Scan Line	699

Continued

Band	Wavelength (μm)		
4	0.767 – 0.910	Ground Resolution	7.6 m at 300 m (10,000 ft)
5	1.13 – 1.35	Swath Width	7.26 km (3.9 nmi)* at 3,000 m
6	1.57 – 1.71	Aperture	72.4 cm^2
7	2.10 – 2.38		
8	10.9 – 12.3		

* 1 nmi = 1.8532 km

3.2.1.2 Thermal Infrared Multispectral Scanner

NASA SSC, the Jet Propulsion Laboratory (JPL) and the Daedalus Corporation have developed the TIMS for exploiting mineral signature information. TIMS is a multispectral scanning system using a dispersive grating and a six-element mercury-cadmium-telluride (HgCdTe) detector array to produce six discrete channels in the 8.2- to 12.2-μm region (see Table 3.7). Used as an airborne geologic remote sensing tool, the TIMS acquires emitted and reflected thermal data that are useful in the discrimination of silicate, carbonate and hydrothermally altered rocks. TIMS data have been used extensively in geology and volcanology research in the western United States, Hawaiian Islands, Europe and Russia.

Table 3.7 Thermal infrared multispectral scanner (TIMS) sensor characteristics

Band	Wavelength (μm)		
1	8.2 – 8.6	IFOV	2.5 mrad
2	8.6 – 9.0	Field of View	76.564 degrees
3	9.0 – 9.4	Pixels/Scan Line	638
4	9.4 – 10.2	Scan Rate	7.3, 8.7, 12, or 25 scans/s
5	10.2 – 11.2	Ground Resolution	50 m at 20 km (65,000 ft)
6	11.2 – 12.2	Swath Width	31.3 km (16.9 nmi)

3.2.1.3 Thematic Mapper Simulator

The TMS is a Daedalus AADS-1268 scanner that is flown on the ER-2 aircraft and simulates the Landsat TM instrument, with slightly higher spatial resolution (see Table 3.8). The TMS is used for collecting data similar to data collected by the Landsat TM for applications such as earth resources mapping, vegetation/landcover mapping and geologic studies. The sensor acquires data in 12 spectral bands.

Table 3.8 Thematic mapper simulator (TMS) sensor characteristics and performance specifications

Channel	TM Band	Wavelength (μm)		
1	A	0.42 – 0.45	IFOV	1.25 mrad
2	1	0.45 – 0.52	Field of View	42.5 degrees
3	2	0.52 – 0.60	Pixels/Scan Line	716
4	B	0.60 – 0.62	Scan Rate	12.5 scans/s
5	3	0.63 – 0.69	Spatial Resolution	25 m at 20 km (65,000 ft)
6	C	0.69 – 0.75	Swath Width	15.6 km (8.4 nmi) at 20 km
7	4	0.76 – 0.90		
8	D	0.91 – 1.05		
9	5	1.55 – 1.75		
10	7	2.08 – 2.35		
11	6	8.5 – 14.0	High Gain	
12	6	8.5 – 14.0	Low Gain	

3.2.2 Satellite Data

3.2.2.1 Advanced Spaceborne Thermal Emission and Reflection Radiometer

The Advanced Spaceborne Thermal Emission and Reflection Radiometer (ASTER) is one of five instruments on board the Terra platform, which was launched in December 1999 (Abrams, 2000). The ASTER provides 14 spectral bands with 15- to 90-m resolution depending on the bands (see Tables 3.9 and 3.10). The ASTER does not acquire data continuously, and its sensors are activated only upon request to collect specific scenes. The instrument consists of three separate telescopes, each of which provides a different spectral range and resolution. The visible and near-infrared (VNIR) sensor provides four bands at 15-m resolution. The short-wave infrared (SWIR) sensor provides six bands at 30-m resolution. The thermal infrared (TIR) sensor provides five bands at 90-m resolution. The swath width for all sensors is 60 km. The ASTER data are archived, processed, and distributed by the LP DAAC at EDC.

The ASTER standard data products are provided in the Universal Transverse Mercator (UTM) projection. The data are referenced to the World Geodetic Survey (WGS) system of 1984 (WGS84) and distributed in the Hierarchical Data Format EOS (HDF-EOS) format. The ASTER standard products include all scenes that have been acquired to date, processed by the ground data system in Japan, and delivered to the LP DAAC.

Table 3.9 ASTER sensor characteristics and radiometric performance requirements

Subsystem	Band No.	Spectral Range (μm)	Spatial Resolution (m)	Signal Quantization Levels
VNIR	1	0.52 – 0.60	15	8 bits
	2	0.63 – 0.69		
	3N	0.78 – 0.86		
	3B	0.78 – 0.86		
SWIR	4	1.600 – 1.700	30	8 bits
	5	2.145 – 2.185		
	6	2.185 – 2.225		
	7	2.235 – 2.285		
	8	2.295 – 2.365		
	9	2.360 – 2.430		
TIR	10	8.125 – 8.475	90	12 bits
	11	8.475 – 8.825		
	12	8.925 – 9.275		
	13	10.25 – 10.95		
	14	10.95 – 11.65		

Table 3.10 ASTER sensor characteristics and geometric performance requirements

Swath Width	60 (kms)
Total Cross-Track Coverage	± 116 to ± 318 km
Stereo Base-to-Height Ratio	0.6 (along-track)
Modulation Transfer Frequency	0.25 (cross-track) 0.20 (along-track)
Band-to-Band Registration	0.2 pixels (intra-telescope) 0.3 pixels (inter-telescope)
Duty Cycle	8% (VNIR & SWIR) 16% (TIR)
Peak Data Rate	89.2 Mbps

The processing levels directly available from the archive include: raw uncorrected data (Level-1A), and data that have had radiometric and geometric corrections applied (Level-1B). The ASTER On-Demand Products are created by applying a specific processing algorithm to Level-1B data (Level-1A for the ASTER DEM). The following higher-level products are available upon request, as long as a Level-1B scene exists: on-demand decorrelation stretch, brightness temperature, surface emissivity, surface reflectance, surface kinetic temperature, surface radiance (VNIR and SWIR), surface radiance (TIR), polar surface and cloud classification, and ASTER DEM (created from Level-1A data).

ASTER Data Acquisition Requests (DARs) may be submitted because ASTER data are not acquired on a continuous basis. With the exception of the ASTER Science Team global mapping strategy, the sensors are only activated to collect specific scenes upon request. Many locations may not be available, but full global coverage should be available by 2005 with data that have been Updated. Users will need to run a search to determine if a particular location has been acquired. Authorized investigators can request the acquisition of a particular scene(s), although each request will be subject to approval by the ASTER Science Team. The standard Level-1A and Level-1B products are orderable through the EOS Data Gateway; Level-1B data are also searchable through GloVis. The higher-level on-demand products are orderable through the ASTER On-Demand Gateway.

3.2.2.2 Moderate Resolution Imaging Spectroradiometer

The Moderate Resolution Imaging Spectroradiometer (MODIS) is one of five instruments carried on board the Terra and Aqua satellites, which were launched in December 1999 and May 2002, respectively (Justice and Townshend, 2002). Terra has a 10:30am equatorial crossing time whereas Aqua is at 1:30am. MODIS provides continuous global coverage every one to two days, and collects data from 36 spectral bands (see Table 3.11).

Table 3.11 MODIS sensor system: spectral bands and applications

Band	Wavelength (nm) (reflected)	Wavelength (μm) (emitted)	Key Use
1	620 – 670		Absolute Land Cover Transformation, Vegetation Chlorophyll
2	841 – 876		Cloud Amount, Vegetation Land Cover Transformation
3	459 – 479		Soil/Vegetation Differences
4	545 – 565		Green Vegetation
5	1,230 – 1,250		Leaf/Canopy Differences
6	1,628 – 1,652		Snow/Cloud Differences
7	2,105 – 2,155		Cloud Properties, Land Properties
8	405 – 420		Chlorophyll
9	438 – 448		Chlorophyll

Continued

Band	Wavelength (nm) (reflected)	Wavelength (μm) (emitted)	Key Use
10	483 – 493		Chlorophyll
11	526 – 536		Chlorophyll
12	546 – 556		Sediments
13h	662 – 672		Atmosphere, Sediments
13l	662 – 672		Atmosphere, Sediments
14h	673 – 683		Chlorophyll Fluorescence
14l	673 – 683		Chlorophyll Fluorescence
15	743 – 753		Aerosol Properties
16	862 – 877		Aerosol Properties, Atmospheric Properties
17	890 – 920		Atmospheric Properties, Cloud Properties
18	931 – 941		Atmospheric Properties, Cloud Properties
19	915 – 965		Atmospheric Properties, Cloud Properties
20		3.660 – 3.840	Sea Surface Temperature
21		3.929 – 3.989	Forest Fires & Volcanoes
22		3.929 – 3.989	Cloud Temperature, Surface Temperature
23		4.020 – 4.080	Cloud Temperature, Surface Temperature
24		4.433 – 4.498	Cloud Fraction, Troposphere Temperature
25		4.482 – 4.549	Cloud Fraction, Troposphere Temperature
26	1.360 – 1.390		Cloud Fraction (Thin Cirrus), Troposphere Temperature
27		6.535 – 6.895	Mid Troposphere Humidity
28		7.175 – 7.475	Upper Troposphere Humidity
29		8.400 – 8.700	Surface Temperature

Continued

Band	Wavelength (nm) (reflected)	Wavelength (μm) (emitted)	Key Use
30		9.580 – 9.880	Total Ozone
31		10.780 – 11.280	Cloud Temperature, Forest Fires & Volcanoes, Surface Temperature
32		11.770 – 12.270	Cloud Height, Forest Fires & Volcanoes, Surface Temperature
33		13.185 – 13.485	Cloud Fraction, Cloud Height
34		13.485 – 13.785	Cloud Fraction, Cloud Height
35		13.785 – 14.085	Cloud Fraction, Cloud Height
36		14.085 – 14.385	Cloud Fraction, Cloud Height

Note: Bands 1 and 2 have nominal 250 m resolution, bands 3 – 7 are nominally 500 m, and bands 8 – 36 are nominally 1,000 m resolution.The Swath width for MODIS is 2,330 km.

MODIS low-level instrument data (Level-1A and -1B) are archived and distributed by the NASA Goddard Earth Sciences Distributed Active Archive Center (GES DAAC); whereas the higher-level land products (see Tables 3.12 – 3.14) are archived and distributed by the LP DAAC at the USGS/EDC. All LP DAAC MODIS products are provided in HDF-EOS format.

Table 3.12 MODIS land products – energy balance suite

Product LongName	Terra ShortName	Aqua ShortName	Terra/Aqua Combined ShortName
Surface Reflectance 8-Day L3 Global 500 m SIN Grid	MOD09A1	MYD09A1	
Surface Reflectance Daily L3 Global 0.05 Deg CMG	MOD09CMG	MYD09CMG	
Surface Reflectance Daily L2G Global 500 m SIN Grid	MOD09GHK	MYD09GHK	
Surface Reflectance Daily L2G Global 250 m SIN Grid	MOD09GQK	MYD09GQK	
Surface Reflectance Quality Daily L2G Global 1 km SIN Grid	MOD09GST	MYD09GST	
Surface Reflectance 8-Day L3 Global 250 m SIN Grid	MOD09Q1	MYD09Q1	

Continued

Product LongName	Terra ShortName	Aqua ShortName	Terra/Aqua Combined ShortName
Land Surface Temperature/Emissivity 5-min L2 Swath 1 km	MOD11_L2	MYD11_L2	MCD11_L2
Land Surface Temperature/Emissivity Daily L3 Global 1 km SIN Grid	MOD11A1	MYD11A1	MCD11A1
Land Surface Temperature/Emissivity 8-Day L3 Global 1 km SIN Grid	MOD11A2	MYD11A2	MCD11A2
Land Surface Temperature/Emissivity Daily L3 Global 5 km SIN Grid	MOD11B1	MYD11B1	MCD11B1
Land Surface Temperature/Emissivity Daily L3 Global 0.05 Deg CMG	MOD11C1	MYD11C1	
Land Surface Temperature/Emissivity 8-Day L3 Global 0.05 Deg CMG	MOD11C2	MYD11C2	
Land Surface Temperature/Emissivity Monthly L3 Global 0.05 Deg CMG	MOD11C3	MYD11C3	
BRDF/Albedo Model-1 16-Day L3 Global 1 km SIN Grid	MOD43B1		MCD43B1
BRDF/Albedo Model-2 16-Day L3 Global 1 km SIN Grid	MOD43B2		MCD43B2
Albedo 16-Day L3 Global 1 km SIN Grid	MOD43B3		MCD43B3
Nadir BRDF-Adjusted Reflectance 16-Day L3 Global 1 km SIN Grid	MOD43B4		MCD43B4
Albedo 16-Day L3 Global 0.05 Deg CMG	MOD43C1		MCD43C1
BRDF/Albedo Parameters 16-Day L3 Global 0.05 Deg CMG	MOD43C2		MCD43C2
Nadir BRDF-Adjusted Reflectance 16-Day L3 Global 0.05 Deg CMG	MOD43C3		MCD43C3
Geolocation Angles Daily L2G Global 1 km SIN Grid Day	MODMGGAD	MYDMGGAD	
Observation Pointers Daily L2G Global 1 km SIN Grid Day	MODPT1KD	MYDPT1KD	
Observation Pointers Daily L2G Global 500 m SIN Grid	MODPTHKM	MYDPTHKM	
Observation Pointers Daily L2G Global 250 m SIN Grid	MODPTQKM	MYDPTQKM	

Table 3.13 MODIS land products— vegetation parameter suite

Product LongName	Terra ShortName	Aqua ShortName	Terra/Aqua Combined ShortName
Land Cover Change 96-Day L3 Global 1 km SIN Grid	MOD12Q2	MYD12Q2	
Vegetation Indices 16-Day L3 Global 500 m SIN Grid	MOD13A1	MYD13A1	
Vegetation Indices 16-Day L3 Global 1 km SIN Grid	MOD13A2	MYD13A2	
Vegetation Indices Monthly L3 Global 1 km SIN Grid	MOD13A3	MYD13A3	
Vegetation Indices 16-Day L3 Global 0.05 Deg CMG	MOD13C1	MYD13C1	
Vegetation Indices Monthly L3 Global 0.05 Deg CMG	MOD13C3	MYD13C3	
Vegetation Indices 16-Day L3 Global 250 m SIN Grid	MOD13Q1	MYD13Q1	
Leaf Area Index/FPAR 8-Day L4 Global 1 km SIN Grid	MOD15A2	MYD15A2	
Leaf Area Index/FPAR 32-Day L4 Global 0.05 Deg CMG	MOD15CM	MYD15CM	
Evapotranspiration/Surface Resistance 8-Day L4 Global 1 km SIN Grid	MOD16A2	MYD16A2	
Net Photosynthesis 8-Day L4 Global 1 km SIN Grid	MOD17A2	MYD17A2	
Net Primary Production Yearly L4 Global 1 km SIN Grid	MOD17A3	MYD17A3	
Net Photosynthesis 32-Day L4 Global 0.05 Deg CMG	MOD17CM	MYD17CM	
Net Primary Production Yearly L4 Global 0.05 Deg CMG	MOD17CY	MYD17CY	
Burn Scars 16-Day L3 Global 500 m SIN Grid	MOD40A3	MYD40A3	
Vegetation Cover Conversion 32-Day L3 Global 250 m SIN Grid	MOD44A	MYD44A	
Vegetation Continuous Fields Yearly L3 Global 500 m SIN Grid	MOD44B	MYD44B	

Table 3.14 MODIS land products—land use/land cover suite

Product LongName	Terra ShortName	Aqua ShortName	Terra/Aqua Combined ShortName
Land Cover Type Yearly L3 Global 0.05 Deg CMG	MOD12C1	MYD12C1	MCD12C1
Land Cover Change 96-Day L3 Global 0.05 Deg CMG	MOD12C2	MYD12C2	
Land Cover Data Base 32-Day L3 Global 1 km SIN Grid	MOD12M	MYD12M	
Land Cover Type Yearly L3 Global 1 km SIN Grid	MOD12Q1	MYD12Q1	MCD12Q1
Thermal Anomalies/Fire 5-Min L2 Swath 1 km	MOD14	MYD14	
Thermal Anomalies/Fire Daily L3 Global 1 km SIN Grid	MOD14A1	MYD14A1	
Thermal Anomalies/Fire 8-Day L3 Global 1 km SIN Grid	MOD14A2	MYD14A2	
Thermal Anomalies/Fire 32-Day L3 Global 0.05 Deg CMG	MOD14C3	MYD14C3	

The Terra and Aqua satellites provide the capability to broadcast MODIS data in real-time during imaging to appropriately equipped ground stations. The USGS/EDC receives Terra and Aqua MODIS direct broadcast data and processes these data to calibrated radiances (Level-1B), calibrated radiances gridded to uniform time/space intervals, estimates of surface reflectance, and estimates of surface reflectance gridded to uniform time/space intervals. The 250-m data products contain bands 1 and 2 at their native 250-m resolution; the 500-m data products contain bands 3 through 7 at their native 500-m resolution, along with bands 1 and 2 aggregated to 500-m; and the 1,000-m data products contain bands 8 through 36 at their native 1,000-m resolution, along with bands 1 – 7 aggregated to 1,000 m. The MODIS direct broadcast data distributed by the EDC are available in GeoTIFF and HDF formats.

3.2.2.3 Earth Observing-1

The Earth Observing-1 (EO-1) satellite was launched November 2000 as a one-year technology demonstration/validation mission. The EO-1 satellite follows the Landsat WRS-2 path and row system with a 16-day repeat cycle (Ungar, 1997). The spacecraft follows Landsat-7 by approximately one minute, and is also capable of cross-track pointing to allow potential imaging within one adjacent WRS path in each direction from nadir. There are three instruments on board the EO-1 spacecraft: Hyperion, the Advanced Land Imager (ALI), and the Linear Etalon Imaging Spectrometer Array (LEISA) Atmospheric Corrector (LAC).

After the initial technology mission was completed, NASA and the USGS agreed to the continuation of the EO-1 program as an extended mission. The EO-1 extended mission is chartered to collect and distribute Hyperion hyperspectral and ALI multispectral data according to customer tasking requests.

Hyperion collects 220 unique spectral channels ranging from 357 to 2,576 nm with a 10-nm bandwidth. The instrument operates in a pushbroom fashion, with a spatial resolution of 30 m for all bands. A Hyperion scene has a standard width of 7.7 km and a standard length of 42 km, with an optional increased scene length of 185 km. Hyperion provides detailed spectral mapping across all 220 channels with high radiometric accuracy.

The ALI provides image data from 10 spectral bands (see Table 3.15). The instrument operates in a pushbroom fashion, with a spatial resolution of 30 m for the multispectral bands and 10 m for the panchromatic band. The scenes have a standard width of 37 km and a standard length is 42 km, with an optional increased scene length of 185 km.

Table 3.15 Advanced land imager spectral band characteristics

Band	Wavelength (μm)	Resolution (m)
1	0.048 – 0.690	10
2	0.433 – 0.453	30
3	0.45 – 0.515	30
4	0.525 – 0.605	30
5	0.630 – 0.690	30
6	0.775 – 0.805	30
7	0.845 – 0.89	30
8	1.2 – 1.3	30
9	1.55 – 1.75	30
10	2.08 – 2.35	30

The LAC provided the first space-based test of an on-board atmospheric corrector for increasing the accuracy of surface reflectance estimates. These data are no longer being acquired as part of the EO-1 extended mission.

Depending on the date of acquisition (past vs. future), archived or tasked products are available for both Hyperion and ALI. Tasked products consist of imagery to be collected in the future, based on a customer's data acquisition request (DAR). All newly collected scene(s) acquired by EO-1 as a result of the DAR will be provided exclusively to the requesting customer for 90 days. Following this 90-day period, the data will become available to the general public as an archived product. Archived products consist of imagery collected during the original one-year technology demonstration/validation mission, as well as more recent acquisitions from the subsequent EO-1 extended mission period. Archived data will include all scenes that have been acquired to date, except scenes that fall within the 90-day exclusion period. Hyperion and ALI data products are radiometrically corrected (Level-1R) with no geometric correction applied, and the image data are provided as 16-bit radiance values. The data are archived and distributed as HDF files.

3.2.2.4 Enhanced Thematic Mapper Plus

The Landsat series of earth observing satellites provides one of the longest records of land surface remotely sensed measurements (Goward and Williams, 1997; Lauer et al., 1997; Mika, 1997). The Enhanced Thematic Mapper Plus is a multispectral scanning radiometer that is carried on board the Landsat-7 satellite that was launched in April 1999. The sensor has provided continuous coverage since July 1999, with a 16-day repeat cycle. The ETM+ instrument provides image data from eight spectral bands (see Table 3.16). The spatial resolution is 30 m for the visible and near-infrared (bands 1 – 5 and 7). Resolution of the panchromatic (band 8) is 15 m, and the thermal infrared (band 6) is 60 m. The approximate scene size is 170 km× 183 km (106 miles × 115 miles).

Table 3.16 Enhanced thematic mapper plus: spectral band characteristics

Landsat 7	Wavelength (μm)	Resolution (m)
Band 1	0.45 – 0.52	30
Band 2	0.53 – 0.61	30
Band 3	0.63 – 0.69	30
Band 4	0.78 – 0.90	30
Band 5	1.55 – 1.75	30
Band 6	10.40 – 12.50	60
Band 7	2.09 – 2.35	30
Band 8	0.52 – 0.90	15

There are several product levels available to the general public:

(1) Level-0Rp—raw uncorrected data that have no radiometric or geometric corrections applied. Scan lines are reversed and nominally aligned. Image data are provided as 8-bit (DN) values. Specialized processing of these data is required to create higher-level products.

(2) Level-1R—radiometric corrections have been applied, but no geometric corrections. Scan lines are reversed and nominally aligned. Image data are provided as 16-bit (radiance) values. These data are of limited use without the geometric corrections, except for analyzing the radiometric performance of the instrument.

(3) Level-1G—systematic radiometric and geometric corrections have been applied. Image data are provided as rescaled 8-bit (DN) values. The scene will be rotated, aligned, and georeferenced to a user-defined map projection. Geometric accuracy of the systematically corrected product should be within 250 m (1 sigma) for low-relief areas at sea level.

The following products with further levels of correction are available to approved USGS researchers only:

(1) Level-1P—precision correction includes radiometric and geometric correction, as well as the use of ground control points (GCPs) to improve the accuracy of geographic referencing. For locations outside the United States, accuracy of the precision-corrected product will depend upon the availability and accuracy of the GCPs.

(2) Level-1T—terrain correction includes radiometric, geometric, and precision correction, as well as the use of a digital elevation model (DEM) to correct parallax error due to local topographic relief. For locations outside the United States, the accuracy of the terrain-corrected product will depend upon the availability and accuracy of GCPs and resolution of the available DEM data.

Raw and radiometrically corrected data will have no projection or datum applied. For systematic-, precision-, and terrain-corrected products, the user-defined parameters are given in Table 3.17.

Table 3.17 Processing parameter options for systematic-, precision-, and terrain-corrected Landsat ETM+ data products

Processing Parameters	Available Options
Map Projection	UTM, Space Oblique Mercator (SOM-B) Albers Equal-Area Conic Other*
Horizontal Datum	WGS84 NAD83 NAD27 Other*
Resampling Method	Cubic Convolution (CC) Nearest Neighbor (NN) Other*
Image Orientation	Map (north up) Path (satellite) Other*
Pixel Zize	30 m (30 m/15 m/60 m) 28.5 m (28.5 m/14.25 m/57 m) 25 m (25 m/12.5 m/50 m) Other*
Multi-Scene**	Available for up to 3 scenes (L1R, L1G, L1P, L1T) Available for up to 10 scenes (L0Rp) (CD limit = 2 scenes; 4.5 scenes for L0Rp)
Scene Shift	Available in 10% increments (along-path shift only)
SLC-off Scan Gap Interpolation	2 – 15 pixels
SLC-Off Gap-Filled	Missing data are replaced with histogram matched values from an alternative acquisition date.

* Other options may be available upon request.

** Multi-scenes are only available for scenes (rows) that were acquired and archived consecutively along the same path for the same date.

An ETM + instrument malfunction occurred onboard Landsat-7 on May 31, 2003. The problem was caused by failure of the Scan Line Corrector (SLC), which compensates for the forward motion of the satellite. All attempts to fix the SLC failed, but data are still being collected in SLC-off mode. To enhance the overall usability of SLC-off data, the USGS/EDC is implementing a series of improvements to SLC-off products and processing. Current USGS processing systems use a standard 2-pixel interpolation along the scanline boundary during Level-1G processing of all Landsat-7 SLC-off data. The user may select an alternate number of pixels to be interpolated across the scanline boundary. Allowable interpolation values range from 0 to 15 pixels, with each pixel representing 30 m. The default value for interpolation is 2 pixels. Selection of a high interpolation value will increase the number of replaced (interpolated) pixel values into the scan gap regions, and will therefore increase the overall population of pixel values throughout the image. Interpolated data values within affected scanlines will therefore not be correlative with actual features. Previously unaffected scanlines will be of similar geometric quality to previous Landsat-7 data acquired before the anomaly (SLC-on) data.

Another product that is available is one in which the scan gaps are filled by replacing all missing pixels with values derived from a coregistered, histogram-matched SLC-on scene. A localized linear transform will be performed in a moving window throughout the missing pixel region. The USGS and NSA are continuing to investigate the development of a gap-filled product, which would be produced from the merge of two or more SLC-off scenes.

3.2.2.5 Thematic Mapper

The Thematic Mapper is a multispectral scanning radiometer that was carried on board Landsats-4 and -5. The TM sensors have provided nearly continuous coverage from July 1982 to present, with a 16-day repeat cycle. TM image data consists of seven spectral bands (see Table 3.18) with a spatial resolution of 30 m for bands 1 – 5 and 7, and 120 m for the thermal infrared (band 6). The approximate scene size is 170 km × 183 km (106 miles × 115 miles).

Table 3.18 Thematic mapper: spectral band characteristics

Landsats 4 – 5	Wavelength (μm)	Resolution (m)
Band 1	0.45 – 0.52	30
Band 2	0.52 – 0.60	30
Band 3	0.63 – 0.69	30
Band 4	0.76 – 0.90	30
Band 5	1.55 – 1.75	30
Band 6	10.40 – 12.50	120*
Band 7	2.08 – 2.35	30

* TM band 6 was acquired at 120 m resolution, but the delivered product is resampled to 30 m.

Only systematically corrected TM data products, including radiometric and geometric corrections, are available to the general public. The scene will be rotated, aligned, and georeferenced to a user-defined map projection. Absolute geometric accuracy of the systematically corrected TM product can vary, depending upon the accuracy of the predicted ephemeris data that are used for processing.

Precision- and terrain-corrected products are only available to approved US Government and its Affiliated Users (USGAU). The user-defined processing parameters are given in Table 3.19.

Table 3.19 Processing parameter options for systematic-, precision-, and terrain-corrected Landsat TM data products

Processing Parameters	Available Options
Map Projection	UTM Space Oblique Mercator Albers Equal-Area Other*
Horizontal Datum	WGS84 NAD83 NAD27 Other*
Resampling Method	Cubic Convolution Nearest Neighbor Other*
Image Orientation	Map (north up) Path (satellite; not recommended for UTM projection)
Pixel Zize	30 m (30 m/120 m) 28.5 m (28.5 m/114 m) Other*
Multi-Scene**	Available for up to 3 scenes
Scene Shift	Available in 10% increments (north-to-south only)

* Other options may be available upon request.

** Multi-scenes are only available for scenes (rows) that were acquired and consecutively along the same path for the same date.

3.2.2.6 Multispectral Scanner

The Multispectral Scanner (MSS) is an imaging radiometer that was carried on board Landsats 1 through 5. The instrument provided nearly continuous coverage from July 1972 to October 1992, with an 18-day repeat cycle for Landsats-1 – -3 and a 16-day repeat cycle for Landsats-4 – -5. MSS data generally consist of four spectral bands, which include the visible green, visible red, and near-infrared wavelengths, although the specific band designations were changed from Landsats-1 – -3 to Landsats-4 – -5 (see Table 3.20). MSS data from Landsat-3 will also include a fifth band in the thermal infrared. The resolution for all bands is 79 m, and the approximate scene size is 185 km× 170 km (115 miles × 106 miles).

Table 3.20 Multispectral sensor spectral band characteristics

Landsats-1 – -3	Landsats-4 – -5	Wavelength (μm)	Resolution (m)
Band 4	Band 1	0.5 – 0.6	80
Band 5	Band 2	0.6 – 0.7	80
Band 6	Band 3	0.7 – 0.8	80
Band 7	Band 4	0.8 – 1.1	80

Only systematically corrected MSS data products are available to the general public, whereas precision- and terrain-corrected products are available to approved USGAU. The user-defined processing parameters for all levels of correction are given in Table 3.21.

Table 3.21 Processing parameter options for systematic-, precision-, and terrain-corrected Landsat MSS data products

Processing Parameters	Available Options
Map Projection	Universal Transverse Mercator (UTM) Space Oblique Mercator Albers Equal-Area Other*
Horizontal Datum	WGS84 NAD83 NAD27 Other*
Resampling Method	Cubic Convolution (CC) Nearest Neighbor (NN) Other*
Image Orientation	Map (north up) Path (satellite)
Pixel Size	Standard (57 m) Other*
Multi-Scene**	Available for up to 3 scenes (1981 – 1992 only)
Scene Shift	Available in 10% north-to-south increments (1981 – 1992 only)

* Other options may be available upon request.

** Multi-scenes are only available for scenes (rows) that were acquired and archived consecutively along the same path for the same date.

3.2.2.7 Multi-Resolution Land Characteristics 2001

The Multi-Resolution Land Characteristics 2001 (MRLC 2001) data consist of a collection of terrain-corrected Landsat-7 ETM+ and limited Landsat-5 TM scenes that have been acquired by the MRLC Consortium (EPA, NOAA, USGS, GAP Analysis Program, US Fish and Wildlife Service, National Park Service). The MRLC2001 data covers the conterminous United States, Alaska, and Hawaii. Most of the Landsat scenes have year-2000 acquisition dates, although individual dates may range from 1999 to present. Multi-temporal scenes may also be available, depending on location. Most of the images are very high quality, and cloud cover is generally less than 10%. The data also include a 30-m DEM for scenes that do not include international (Mexico or Canada) borders. Because of USGS restrictions on the distribution of terrain-corrected data, the MRLC 2001 data are available to MRLC Consortium members and approved USGS researchers only. All MRLC 2001 data products were created using a standardized set of processing parameters (see Table 3.22).

Table 3.22 MRLC2001 standard processing parameters

Number of Bands (Band Designations)	8 (Landsat-7 ETM +) 7 (Landsat-5 TM)
Resolution	30 m (most bands) 15 m (Landsat-7 band 8)
Projection	Albers Equal-Area Conic
Datum	NAD83
Resampling	Cubic Convolution
Format	NDF (NLAPS Data format)*
Media	CD-ROM, FTP-pull

* National Land Archive Production System (NLAPS)

3.2.2.8 Landsat Orthorectified Data

The Landsat orthorectified data collection consists of a global set of high-quality, relatively cloud-free orthorectified MSS, TM, and ETM+ imagery (Tucker et al., 2004). This data set was selected and generated through NASA's Commercial Remote Sensing Program, as part of a cooperative effort between NASA and the commercial remote sensing community to provide users with access to quality-screened, high-resolution satellite images with global coverage over the earth's land masses. The data collection was compiled through a NASA contract with the Earth Satellite Corporation (Rockville, MD) in association with NASA's Scientific Data Purchase program. The USGS contributed the Landsat-7 ETM + data.

The Landsat orthorectified data collection consists of approximately 7,500 MSS scenes, 7,600 TM images and approximately 8,500 ETM + images, which were selected to provide three full sets of global coverage over an approximate 10-year interval (circa 1980, circa 1990, and circa 2000). All selected images were either cloud-free or contained minimal cloud cover. In addition, only images with a high quality ranking in regards to the possible presence of errors such as missing scans or saturated bands were selected. Table 3.23 summarizes the characteristics of the following data products:

(1) Landsat Orthorectified TM: The average acquisition date for the Landsat orthorectified TM data is 1990 (± 3 years). However, individual scenes may range from 1985 to 1996. These scenes have been orthorectified, using geodetic and elevation control data to correct for positional accuracy and relief displacement. Large blocks of Landsat data were adjusted through a patented procedure using pixel correlation to acquire tie points within the overlap area between adjacent Landsat images.

(2) Landsat Orthorectified ETM + : The average acquisition date for the Landsat orthorectified ETM+ data is 2000 (± 1 year). All scenes were acquired between 1999 and 2003, and have been orthorectified for coregistration with the earlier (circa 1990) Landsat TM coverage.

(3) Landsat Orthorectified ETM + Pansharpened: These data products consist of pansharpened versions of the ETM+ scenes above. The pansharpening process allows the fusion of the higher-resolution panchromatic band (8) with selected lower-resolution multispectral bands (7, 4, 2) to create a higher-resolution ("pansharpened") color image.

(4) Landsat Orthorectified MSS: There are plans for a collection of Landsat orthorectified MSS data to be delivered to the USGS/EDC for archive and distribution, but the production schedule and delivery dates are not known at this time.

Table 3.23 Landsat orthorectified data product characteristics

	MSS	TM	ETM+	ETM + Pansharpened
Level of Processing	Terrain Corrected	Terrain Corrected	Terrain Corrected	Terrain Corrected
Number of Bands	4	7	9	3 (bands 7, 4, 2)
Resolution (m)	57	28.5	28.5	14.25
Projection	UTM	UTM	UTM	UTM
Datum	WGS84	WGS84	WGS84	WGS84
Resampling	Nearest Neighbor	Nearest Neighbor	Nearest Neighbor	Cubic Convolution
Format	GeoTIFF	GeoTIFF*	GeoTIFF	GeoTIFF

* Includes a separate NLAPS header file.

All Landsat orthorectified data products have been processed according to a standardized set of parameters that will vary slightly according to product type.

Five- by six-degree mosaics were created from both the TM and pansharpened

ETM + data. Bands 7, 4, and 2 were used for each of these epochs and combined as red, green, and blue, respectively, as color composites available in GeoTIFF format. An adaptive histogram matching technique was used to balance radiometric variations between adjacent scenes. The image mosaics are also in the UTM projection.

3.2.2.9 NASA Landsat Data Collection

The NASA Landsat Data Collection (NLDC) consists of selected MSS and TM scenes that have been acquired by NASA investigators for use in a variety of projects. The Landsat scenes date from 1975 to the present. All scenes have been preselected for minimum cloud cover and optimum band quality, and are processed to Level-1 (systematic geometric and radiometric corrections applied). Specific processing parameters, such as projection, resampling method, pixel and cell size, will vary according to investigator specifications at the time of the original data purchase. The format of the NLDC data will vary depending on where the data were purchased (EOSAT, USGS, or foreign ground stations).

3.2.2.10 North American Landscape Characterization

North American Landscape Characterization (NALC) data consist of Landsat MSS triplicates that were acquired in 1973, 1986, and 1991 (± one year). The data cover selected locations throughout the conterminous United States and Mexico (Sohl and Dwyer, 1998). Each NALC triplicate will consist of at least one MSS scene or scene composite for each of the selected years. The specific temporal windows will vary for geographical regions, based on the seasonal characteristics of the vegetation. The data will also include a 60-m DEM for all scenes that do not include Mexico or its borders.

Most of the images for each year are single-scene acquisitions, although some 1973 images may be mosaics of two scenes because the path-row footprint for Landsats-1 – -3 (Worldwide Reference System-1) is different than the path-row footprint of Landsats-4, -5 and -7 (Worldwide Reference System-2). In addition, some images may be multi-temporal composites for years in which a high-quality cloud-free single scene was not available. In most cases, the 1980s image has been precision corrected; the 1970s and 1990s scenes have been radiometrically and geometrically corrected, then coregistered to the 1980s image. All images have been resampled to 60-m pixels and referenced to the UTM projection referenced to the North American Datum of 1927 (NAD 27).

3.2.2.11 Advanced Very High Resolution Radiometer

The Advanced Very High Resolution Radiometer (AVHRR) provides four- to six-band multispectral data from the NOAA polar-orbiting satellite series. There is fairly continuous global coverage since June 1979, with morning and afternoon acquisitions available. The resolution is 1.1 km at nadir, and the number of available bands will depend on the satellite (see Tables 3.24 and 3.25).

Table 3.24 Current and historical NOAA AVHRR missions

Satellite	Dates (Month/year)	Overpass Time	Number of Bands
TIROS-N	10/78 – 1/80	pm	5
NOAA-6	06/79 – 11/86	am	4
NOAA-7	08/81 – 06/86	pm	5
NOAA-8	05/83 – 10/85	am	4
NOAA-9	02/85 – 11/88	pm	5
NOAA-10	11/86 – 09/91	am	4
NOAA-11	11/88 – 09/94	pm	5
NOAA-12	05/91 – 12/94	am	5
NOAA-13	failed	pm	n/a
NOAA-14	12/94 – present	pm	5
NOAA-15	05/98 – present	am	6
NOAA-16	09/00 – present	pm	6
NOAA-17	06/02 – present	am	6

Table 3.25 NOAA satellite platform and the corresponding AVHRR spectral bands

Band	TIROS – *N*	NOAA-6, -8, -10	NOAA-7, -9,-11,-12,-14	NOAA-13	NOAA-15, -16, -17
1	0.55 – 0.90	0.58 – 0.68	0.58 – 0.68	0.58 – 0.68	0.58 – 0.68
2	0.725 – 1.10	0.725 – 1.10	0.725 – 1.10	0.725 – 1.0	0.725 – 1.0
3	3.55 – 3.93	3.55 – 3.93	3.55 – 3.93	3.55 – 3.93	1.58 – 1.64
4	10.5 – 11.5	10.5 – 11.5	10.3 – 11.3	10.3 – 11.3	3.55 – 3.93
5	Channel 4 repeated	Channel 4 repeated	11.5 – 12.5	11.4 – 12.4	10.3 – 11.3
6					11.5 – 12.5

Several AVHRR data sets are available: Level-1B, stitched orbital segments, and georegistered Level-1B. In addition to the AVHRR data listed below, the USGS/EDC also produces a series of derived AVHRR NDVI composites and Global Land Cover Characterization (GLCC) data sets (see Section 3.2.3.3). AVHRR Level-1B data have radiometric calibration coefficients and earth location data appended but not applied. Level-1B data are available from the USGS/EDC as three product types:

Raw (Level-1B) single-scene AVHRR data consist of a single scene that was collected by a single AVHRR satellite on a specific date and time. The approximate scene size is 2,400 km × 6,400 km (1,491 miles × 4,000 miles). The data are available to approved USGS researchers only, and are provided in binary

(10-bit packed) format. These data are also publicly available from the NOAA Comprehensive Large Array-data Stewardship System (CLASS) located in Suitland, Maryland and Asheville, North Carolina.

Raw (Level-1B) stitched orbital segments consist of single-scene AVHRR Level-1B data stitched together within an orbit, and are available for a limited number of acquisition periods. Each segment is composed of five channels with a resolution of 1.1 km at nadir. The data are provided in binary (10-bit packed) format.

Georegistered (Level-1B) single-scene AVHRR data consist of atsensor radiances with latitude/longitude. The data are available in binary (8- or 10-bit) format.

3.2.2.12 Declassified Satellite Imagery-1

Declassified Satellite Imagery-1 consists of approximately 880,000 photographic images that were taken between 1959 and 1972. These images were primarily collected by the CORONA satellite series, but the data set also includes photos taken from the ARGON and LANYARD satellites. The images used to be distributed only as photographic products (see Tables 3.26 – 3.28), but like the aerial photographic data, these will be available as scanned or digitized images in the future. Browse images are available for preview. The images have multiple scales, the image quality can be variable, and substantial cloud cover extent is common. The original film and technical mission-related documents are maintained by the National Archives and Records Administration (NARA) in College Park, Maryland.

Table 3.26 Product specifications for declassified satellite imagery-1 (CORONA)

Mission Designators	Image Size	Media
KH-1 KH-2 KH-3 KH-4 KH-4A KH-4B	2.25 in × 29.8 in	Paper Print Film Negative Film Positive

Table 3.27 Product specifications for declassified satellite imagery-1 (ARGON)

Mission Designator	Image Size	Media
KH-5	4.5 in × 4.5 in 10 in × 10 in 20 in × 20 in	Paper Print
	4.5 in × 4.5 in	Film Negative Film Positive

Table 3.28 Product specifications for declassified satellite imagery-1 (LANYARD)

Mission Designator	Image Size	Media
KH-6	4.5 in × 4.5 in	Paper Print Film Negative Film Positive

3.2.2.13 Declassified Satellite Imagery-2

Declassified satellite imagery-2 consists of approximately 50,000 photographic images that were collected from 1963 to 1980 by the KH-7 Surveillance System and the KH-9 Mapping System. The images have variable scales, the image quality can be variable, and substantial cloud cover extent is common. Declassified satellite imagery-2 should always be previewed on line prior to purchase. Browse images of all scenes are available by searching on Earth Explorer. The images used to be distributed only as photographic products (see Table 3.29), but like the aerial photographic data, these will be available as scanned or digitized images in the future. The original film and technical mission-related documents are maintained by the NARA.

Table 3.29 Product specifications for declassified satellite imagery-2

Mission Designators	Image Size	Media
KH-7*	9 in × variable lengths*	Black & White Paper Print Black & White Film Negative Black & White Film Positive Color Paper Print Color Positive Film
KH-9	9 in × 18 in	Black & White Paper Print Black & White Negative Film Black & White Positive Film Color Paper Print Color Positive Film

* Products from KH-7 will vary in length, ranging from 4 in to 500 ft. Prices are determined according to the number of 30-in segments that occur within a particular image. The number of segments per image will be identified when searching and ordering on Earth Explorer. (Note: Customers will need to purchase the entire image, regardless of the area/segment of interest. In addition to the price of the enlarged paper prints, a setup fee of $ 25 per frame will be charged for each standard and custom enlargement.)

3.2.2.14 Spaceborne Imaging Radar-C

Spaceborne Imaging Radar-C (SIR-C) is an imaging radar system that was flown on board two Space Shuttle flights (Apr. 9 – 20, 1994 and Sep. 30 – Oct. 11, 1994). The USGS/EDC distributes the C-band (5.8 cm) and L-band (23.5 cm) data. All X-band (3 cm) data are distributed by the Deutsches Zentrum für Luft- and Raumfahrt e.V. (DLR). The SIR-C ground data processing system was built as a one-of-a-kind system. If the processing software is not ported to more contemporary hardware architecture and the system becomes irreparable, then the data may not be made available.

Several types of products that are derived from the SIR-C data: survey data, precision (standard) data, single-look complex (SLC) data, multi-look complex (MLC) data, reformatted signal data (RSD), interferometric single-look complex (iSLC) data, raw interferogram product (RIn), and SIR-C educational CD-ROMs.

1. Survey Data

Survey Data are intended as "quick look" browse images for viewing the areas that were imaged by the SIR-C system. The data consist of a strip image of an entire data swath. Resolution is approximately 100 m, processed to 50-m pixel spacing.

2. Precision (Standard) Data

Precision (Standard) Data consist of a frame image of a data segment, which represents a processed subset of the data swath. It contains high-resolution multi-frequency (C-band, L-band) and multi-polarization (HH, HV, VH, VV) data. The following types of precision data products are available:

(1) Single-Look Complex data consist of one single-look file for each scene, per frequency. Each data segment will cover 50 km along the flight track, and is broken into four processing runs (two L-band, two C-band). Resolution and polarization will depend on the mode in which the data were collected, and are available as calibrated or uncalibrated data.

(2) Multi-Look Complex data are based on an averaging of multiple looks, and consist of one file for each scene per frequency. Each data segment will cover 100 km along the flight track, and is broken into two processing runs (one L-band and one C-band). Polarization will depend on the modes in which the looks were collected. The data are available in 12.5- or 25-m pixel spacing.

(3) Reformatted Signal Data consist of the raw radar signal data only. Each data segment will cover 100 km along the flight track, and the segment will be broken into two processing runs (L-band and C-band).

3. Interferometry Data

Interferometry Data consist of experimental multi-temporal data that cover the same area, and are available only upon special request. Most of the data were collected during repeat passes within the second flight (days 7, 8, 9, and/or 10). In addition, nine data takes were collected during the second flight that was repeat passes of the first flight. Most of these data takes were single polarization, although dual and quad polarization data were also collected on some passes. Estimates of the baseline (difference in the location of the radar signal propagation and receipt positions from one date to the next) will not be available prior to processing, but the baseline for all of the processed scenes thus far have been less than 400 m and most are less than 200 m. A DEM will not be included with any of the following SIR-C interferometry products that are available:

(1) Interferometric Single-Look Complex data consist of two or more uncalibrated SLC images that have been processed with the same Doppler centroid to allow interferometric processing. Each image will cover 50 kilometers along the flight track.

(2) Raw Interferogram product data are the combination of two data takes over the same area to produce an interferogram for each frequency (L-band and C-band).

(3) Reformatted Signal Data data consist of radar signal data that have been processed from two or more data takes over the same area, but the data have not been combined. Although this is not technically an interferometric product, the RSD can be used to generate an interferogram. Each frame will cover 100 km along the flight track.

There are approximately 18,300+ previously processed data products (precision, single look complex, multi-look complex, reformatted signal data, or interferometric data) that are available through the EOS Data Gateway. The usefulness of these products may be constrained by the processing parameters specified for the original customer order.

The SIR-C educational CD-ROMs are products designed for use by educators and students in middle schools, high schools, and universities. The CD-ROM contains numerous SIR-C images from around the world, and also includes lesson plans, presentation materials, and a teacher's guide.

3.2.3 Derived Satellite Data Products

3.2.3.1 Shuttle Radar Topography Mission

The Shuttle Radar Topography Mission (SRTM) is a joint project between NASA and the National Geospatial-Intelligence Agency (NGA) to map the world in three dimensions (Farr and Kobrick, 2000). The SRTM utilized dual SIR-C and dual X-band Synthetic Aperture Radar (X-SAR) configured as a baseline interferometer, acquiring two images simultaneously. These images, when combined, can produce a single 3-D image. Flown aboard the NASA Space Shuttle Endeavour Feb. 11 – 22, 2000, SRTM successfully collected data over 80% of the earth's land surface, for most of the area between 60 degrees N and 56 degrees S latitude.

NASA's JPL performed preliminary processing of SRTM data and forwarded partially finished data directly to the NGA for finishing by their contractors and subsequent monthly delivery to the NGA's Digital Products Data Warehouse (DPDW). All data products delivered by the contractors conform to the NGA SRTM data products and the NGA Digital Terrain Elevation Data (DTED) specifications. The DPDW ingests the SRTM data products, checks them for formatting errors, loads the public SRTM DTED into the NGA data distribution system, and ships them to the USGS/EDC. In addition to NGA's SRTM DTED format, the USGS/EDC has reformatted the data into a non- proprietary generic raster binary SRTM format that is readable by most remote sensing software packages.

The SRTM DTED is a uniform matrix of elevation values indexed to specific points on the ground. The horizontal reference datum is the WGS84 and the vertical datum is mean sea level as determined by the WGS84 Earth Gravitational Model (EGM 96) geoid. The DTED Level-2 elevation values are spaced 1 arc second apart between 0 and 50 degrees latitude, and spaced 1 arc second apart in latitude and 2 arc seconds apart in longitude between 50 and 60 degrees latitude. SRTM DTED Level-1 values are derived from the SRTM DTED Level-2 values such that the DTED-1 values are identical to the DTED-2 values at coincident points. However, the SRTM DTED-1 values are spaced 3 arc seconds apart between 0 and 50 degrees latitude, and 3 arc seconds apart in latitude and 6 arc seconds apart in longitude between 50 and 60 degrees latitude.

NGA performs quality assurance checks on the JPL SRTM data, and their contractors perform several additional finishing steps. Spikes and wells in the data are detected and voided out if they exceed 100 m compared to surrounding elevations. Small voids (16 contiguous posts or less) are filled by interpolation of the surrounding elevations. Large voids are left in the data. Water bodies are depicted in the SRTM DTED. The ocean elevation is set to 0 m. Lakes of 600 m or more in length are flattened and set to a constant height. Rivers that exceed 183 m in width are delineated and monotonically stepped down in height. Islands are depicted if they have a major axis exceeding 300 m or the relief is greater than 15 m. The data are processed in one-degree-by-one-degree "cells." The edges of each cell are matched with the edges of adjacent cells to assure continuity.

3.2.3.2 National Land Cover Data 1992

The National Land Cover Data Set 1992 (NLCD 92) is a 21-category land-cover classification scheme that has been applied consistently over the conterminous United States and that is based primarily on the unsupervised classification of Landsat TM 1992 imagery (Vogelmann et al., 2001). Ancillary data sources included topography, census, agricultural statistics, soil characteristics, other land-cover maps, and wetlands data. The NLCD 92 classification is provided as raster data with a spatial resolution of 30 m. The NLCD 92 data are available as state or seamless products.

The NLCD 92 by State data are referenced to the Albers equal-area conic projection and are referenced to the North American Datum of 1983 (NAD83). Accuracy assessment has not been completed for every location, so there are two separate categories of the NLCD 92 product to reflect this status. The "preliminary" product consists of the NLCD 92 data for which accuracy assessment is not complete; the "final" product consists of the NLCD 92 data for which accuracy assessment has been completed.

The NLCD 92 Seamless product is accessible through The National Map and consists of NLCD 92 for the conterminous United States, which is provided according to a user-specified area of coverage. The data are in geographic coordinates (latitude/longitude), and are referenced to the NAD83.

3.2.3.3 Global Land Cover Characterization

The Global Land Cover Characterization (GLCC) data consist of a series of global land cover classification data sets that are based primarily on the unsupervised classification of 1-km AVHRR 10-day NDVI composites (Brown et al., 1999; Loveland et al., 1999; 2000). The AVHRR source imagery dates range from April 1992 through March 1993. Ancillary data sources included digital elevation data, ecoregions interpretation, and country- or regional-level vegetation and land cover maps. The GLCC database includes the following classifications: global ecosystems, International Geosphere-Biosphere Programme and cover classification, USGS land use/land cover system, simple biosphere model, simple biosphere 2 model, biosphere atmosphere transfer scheme, and vegetation lifeform. All GLCC classifications consist of raster images, which contain class number values for each pixel that correspond to the appropriate classification scheme legend.

The GLCC data are available in two forms: as continental or as a global data sets. GLCC (by continent) provides the GLCC classifications, as well as monthly Normalized Difference Vegetation Index (NDVI) composites, on a continent-by-continent basis (Africa, North America, South America, Eurasia, and Australia/Pacific). These data are available in the Interrupted Goode Homolosine or in the Lambert azimuthal equal-area projection with a nominal spatial resolution of 1 km. The GLCC (global) data consist of the continent GLCC data sets that have been combined into a set of global GLCC composites. The global data are available in either the Interrupted Goode Homolosine or in geographic (latitude/longitude) coordinates with a nominal spatial resolution of 1 km. For each GLCC data set there are two versions available: Version 1 consists of GLCC data that have undergone complete accuracy assessment; Version 2 is updated GLCC which reflect additional data and sources, but accuracy assessment has not been completed.

3.2.3.4 Normalized Difference Vegetation Index Composites

The Normalized Difference Vegetation Index (NDVI) Composites are produced from multiple AVHRR daily observations that have been combined to create a nearly cloud-free image showing maximum greenness. An NDVI ratio is produced from the difference between the near-infrared and reflected red, normalized by the sum of the near-infrared and reflected red bands (see Eq. (3.1)). The AVHRR composite is created by selecting pixels having the maximum NDVI value from two or more dates. Several NDVI products are available.

$$\frac{\text{NearIR} - \text{Red}}{\text{NearIR} + \text{Red}} \tag{3.1}$$

The Conterminous United States and Alaska weekly and biweekly NDVI composites date from January 1989 to present. US composites are referenced to the Lambert Azimuthal Equal-Area projection whereas Alaska composites are cast to the Albers Equal-Area Conic projection.

The 7- and 14-day (current) NDVI composites are weekly and biweekly updates that have been produced within the most recent four weeks of data acquisitions. The data include a composite for each of the acquired AVHRR bands, as well as a derived NDVI band.

The 7- and 14-day (recent) NDVI composites consist of weekly and biweekly updates that were produced from the end of the previous year until four weeks prior to current acquisition date. The data include a composite for each of the acquired AVHRR bands, as well as a derived NDVI band. Satellite zenith, solar zenith, relative azimuth, and acquisition dates are also provided.

The 14-day (historical) NDVI composites consist of biweekly composites that were produced from January 1989 to the end of the year previous to the current acquisition year. The data include a composite for each of the acquired AVHRR bands, as well as a derived NDVI band. Satellite zenith, solar zenith, relative azimuth, and acquisition dates are also provided.

The North American 10-day composites include a composite for each of the acquired AVHRR bands, a derived NDVI band, satellite zenith, solar zenith, relative azimuth, and acquisition dates. These data are available for limited dates between April 1992 and May 1996. The data are referenced to the Interrupted Goode's Homolosine projection.

Global 10-day composites consist of a 10-day composited AVHRR data and a derived NDVI band for limited dates between April 1992 and May 1996. The data are referenced to the Interrupted Goode's Homolosine projection. Subsampled resolution and geographic subsets are also available.

Conterminous United States Land Cover Characteristics CD-ROM contains a land cover characteristics database with seasonal land cover classification, based on the analysis of AVHRR images collected from March to October 1990. The data are referenced to the Lambert Azimuthal Equal-Area projection, with a resolution of 1.1 km. Files on the CD-ROM include the original 1990 AVHRR data used for analysis, initial and final interpretations of land cover regions, descriptive and quantitative attributes of each land-cover region, thematic interpretations of land regions, and display software.

Northern Great Plains CD-ROM contains 33 weekly AVHRR composite images for the northern Great Plains. The composites were produced in June and August of 1987, and March through October 1988. The data are referenced to the Albers Equal Area projection, with a resolution of 1 km. Files on the disc include weekly AVHRR composites (bands 1-5, afternoon coverage), weekly NDVI composites, satellite and solar azimuth and zenith data (by pixel), and state, county, and ecological region information.

3.3 Conclusions

The USGS/EDC manages the largest archive of satellite remotely sensed data over the earth's landmasses, and this archive continues to grow. Access to these data is provided on a non-discriminatory basis. On site capabilities enable direct reception and processing of data acquired by Landsat-5 TM, Landsat-7 ETM + , EO-1 ALI and Hyperion, NOAA-17 AVHRR, and Terra and Aqua MODIS. In addition, full global coverage of ASTER Level-1 data is received from the ASTER GDS in Japan; requests for higher-level products are fulfilled on demand. Global coverage of higher-level land products derived from Terra and Aqua MODIS data are received from the MODIS Adaptive Processing System (MODAPS) at NASA GSFC.

The science quality of the ASTER and MODIS land products continues to improve as the processing algorithms are refined in response to results from product validation studies. The USGS and NASA continue to investigate methods by which to fill data gaps in the ETM + imagery resulting from the failure of the SLC. A number of derived products are available as well, including information on land-cover, vegetation condition, and hydrologic features. Aerial photography and aircraft scanner data are often used to assist with the validation of products derived from satellite data.

The community of users accessing USGS/EDC data products consists of government, academic, and commercial organizations, as well as private citizens from the United States and numerous other countries around the world. The USGS/EDC provides comprehensive customer support services, including data searches, order processing, answering questions, and providing technical assistance. Open source software tools have been developed and made freely available in order to facilitate maximum use of the data products. The USGS is in the process of changing its business model from the reproduction and distribution of aerial photography and declassified satellite imagery as photographic products to scanning and converting these data into digital format.

In addition to the breadth of data capture, processing, archive, distribution, and support services that are the foundation of the USGS/EDC mission, there are also numerous research and applications projects that are conducted by the staff. These projects are undertaken in collaboration with other United States government agencies, universities, and international organizations, the results of which are summarized in peer reviewed publications. These projects include the development of national land-cover data bases, mapping and quantifying land-cover change and surface deformation, investigating the phenology of natural vegetation and its response to drought and climate variability, wildfire mapping and ecosystem recovery, and emergency response to natural hazards and disasters.

Acknowledgements

This work was made possible, in part, by Science Applications International Corporation under U.S. Geological Survey contract 1434-CR-97-CN-40274.

References

Abrams M (2000) The Advanced Spaceborne Thermal Emission and Reflection Radiometer (ASTER): data products for the high spatial resolution imager on NASA's Terra platform. Int J Remote Sensing 1(5): 847 – 859

Brown JF, Loveland TR, Ohlen DO, Zhu Z (1999) The global land cover characteristics database: the users' perspective. Photogrammetric Engineering and Remote Sensing 65(9): 1,069 – 1,074

Farr TG, Kobrick M (2000) Shuttle radar topography mission produces a wealth of data. Eos Transactions American Geophysical Union 81(48): 583 – 585

Goward SN, Williams DL (1997) Landsat and Earth systems science: development of terrestrial monitoring. Photogrammetric Engineering and Remote Sensing 63(7): 887 – 900

Justice C, Townshend J (2002) Special issue on the Moderate Resolution Imaging Spectroradiometer. Remote Sensing of Environment Issues 83(1 – 2): 1 – 359

Lauer DT, Morain SA, Salomonson VV (1997) The Landsat program: its origins, evolution, and impacts. Photogrammetric Engineering and Remote Sensing 63 (7): 831 – 838

Loveland TR, Reed BC, Brown JF, Ohlen DO, Zhu J, Yang L, Merchant JW (2000) Development of a global land cover characteristics database and IGBP DISCover from 1-km AVHRR data. International Journal of Remote Sensing 21(6/7): 1,303 – 1,330

Loveland TR, Zhu Z, Ohlen DO, Brown JF, Reed BC, Yang L (1999) An analysis of the IGBP global land-cover characterization process. Photogrammetric Engineering and Remote Sensing 65(9): 1,021 – 1,032

Mika AM (1997) Three decades of Landsat instruments. Photogrammetric Engineering and Remote Sensing 63 (7): 839 – 852

Sohl T, Dwyer JL (1998) North American landscape characterization project: the production of a continental scale three-decade Landsat data set. Geocarto International 13 (3): 43 – 51

Tucker CJ, Grant DM, Dykstra JD (2004) NASA's global orthorectified Landsat data set. Photogrammetric Engineering and Remote Sensing, 70(3): 313 – 322

Ungar SG (1997) Technologies for future Landsat missions. Photogrammetric Engineering and Remote Sensing 63(7): 901 – 905

Vogelmann JE, Howard SM, Yang L, Larson CR, Wylie BK, Van Driel JN (2001) Completion of the 1990's national land cover data set for the conterminous United States. Photogrammetric Engineering and Remote Sensing 67: 650 – 662

4 NASA Direct Readout for Its Polar Orbiting Satellites

Patrick L. Coronado and Kelvin W. Brentzel

4.1 Introduction

The two overarching concepts that encompass the content of this Chapter are Direct Broadcast and Direct Readout. Fortunately, the conceptual separation is simple; one refers to the space segment, the other to everything on the ground.

Direct Broadcast (DB) is the real-time transmission of satellite data to the ground. As the Earth is observed by satellite instruments, data is formatted and broadcast omni-directionally in a hemispheric pattern to Earth in real-time. Users who have compatible ground receiving equipment and are in direct line of sight to the satellite may receive these transmissions.

Direct Readout (DR) is the process of acquiring freely transmitted live satellite data. As DR technologies have become more affordable and accessible—such as with the onset of the Internet—tools have been developed by the remote sensing community to make satellite data easier to acquire, process, and utilize. As a member of this community, NASA supplies many of these tools to foster global data exchange and scientific collaboration. Live local and regional environmental data, in turn, benefits environmental, commercial, and public interest decision making.

With that stated, when referring to this topic, the term DR is not used, but instead, DB is the overarching descriptor. This was not done by design, but by the human need to simplify terminology. In this chapter, DB will be described briefly to differentiate it from the primary method of acquiring instrument data from Earth remote sensing satellites, such as POES, EOS series, NPP, and future NPOESS satellites. Instead, DR will be the focus where history, methods of acquisition, supporting technologies, and its roadmap will be described. Further, this chapter will address only NASA's satellites including its involvement in the POES DR system's development and how it has, and still is providing a bridge between NASA DB Earth science missions and the broader user community.

4.2 Context in History

In the late 1970s the stage was set for DB of satellite data to come into the forefront. But this stage was pre-defined by the first of ten application satellites named Television Infra Red Observation Satellite (TIROS, 1960 – 1965) as depicted in Fig. 4.1. TIROS was launched primarily to provide images of cloud formations. The second satellite had a five-channel radiometer and an infra-red camera which could provide both day and night cloud images. The fourth satellite in this series, launched in February 1962, had a low data rate direct broadcast capability called Automatic Picture Transmission (APT).

Normal satellite operations involved recording the instrument measurements onto an onboard tape recorder and the data would later be dumped to a dedicated ground receiving station. This data was then operationally distributed to the broad user community by the National Oceanographic and Atmospheric Administration (NOAA). It was shortly thereafter realized that the satellites making the measurements would make excellent distribution vehicles, and thus the concept of direct broadcast of satellite data was developed. In this mode of operation, as the satellite measurements were made, the data was both recorded onboard the satellite and directed to a separate transmitter to be sent directly to the ground whereby any receiving station within the line of sight with the satellite could receive the data in real-time, such as APT.

These first series of satellites lead to the development of the ESSA/TOS (TIROS Operational Satellites, 1966 – 1969) series of weather satellites which had advanced video cameras and APT low resolution DB capability. In late 1960's and early 1970's the spectacular Gemini and Sky-lab Photographs had sparked the imagination of the world, but more importantly, the entire science community, which was now able to clearly see the potential that satellite measurements could have on all aspects of Earth science.

During this early stage of satellite remote sensing, scientists and application engineers realized that the varied ways the electromagnetic radiation interacts with matter could provide quantitative measurements of the parameters of interest to each Earth science discipline. Detailed studies of absorption and scatter of the atmosphere were undertaken, and the interaction of radiation with various surfaces was investigated, which in turn lead to the development of unique new satellite instruments that were available via direct broadcast, all-be-it using very expensive DR systems.

The first images of the Earth used visible light-based instruments to show clouds and cloud patterns, ocean features such as boundaries, eddies and currents, and land features such as forest, grasslands and urban centers. The meteorological instruments concentrated on using infrared and microwave measurements which were designed to measure sea surface temperature, cloud top temperature and detailed atmospheric temperature profiles.

An important technology breakthrough occurred with the NIMBUS series of experimental weather satellites where many of these new and innovative suites of instruments allowed experimenters to explore the use of the entire spectrum to monitor and measure a significant range of phenomena not only of atmospheric interest, but of oceans, land and Earth's vegetation.

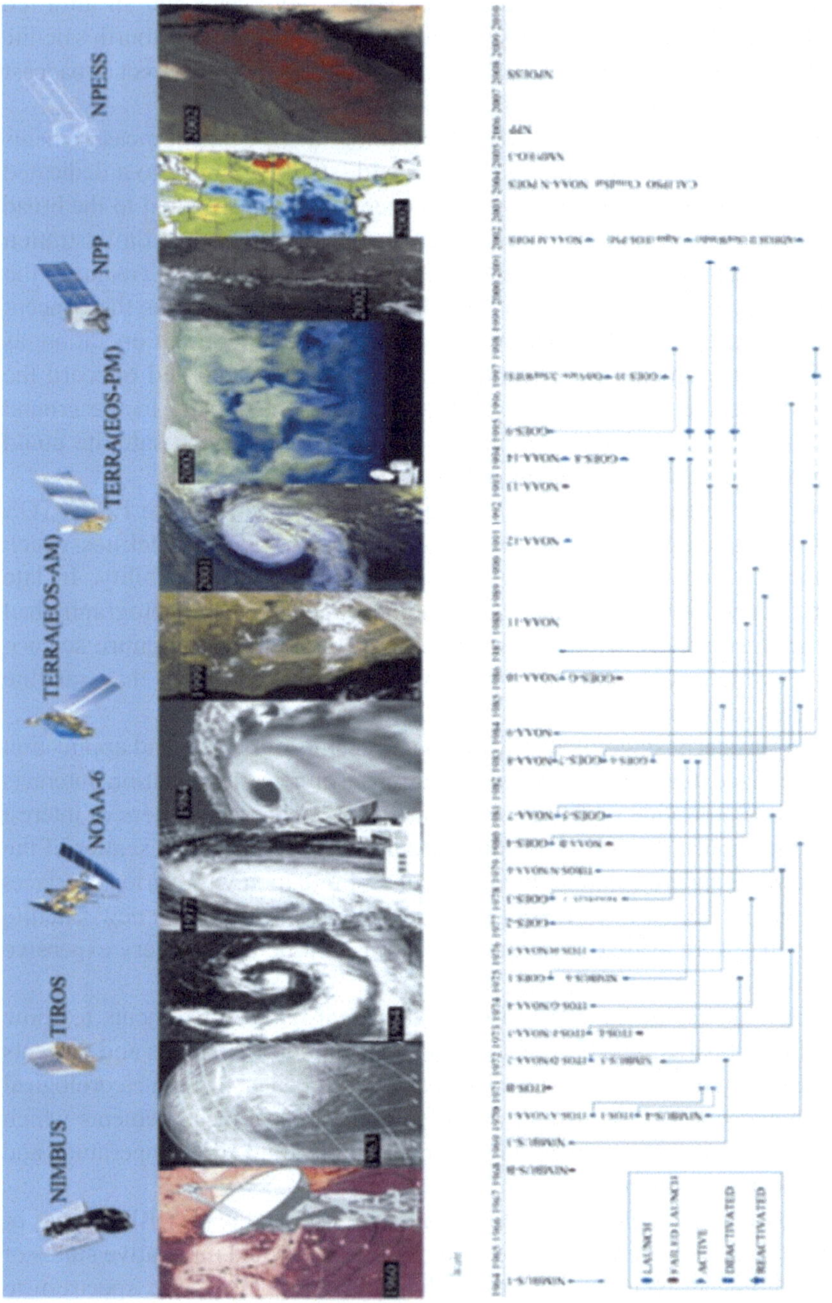

Figure 4.1 NASA/NOAA Direct Broadcast Satellite history timeline

In 1972 the first High Resolution Picture Transmission (HRPT) DR system was developed by the NASA Direct Readout Laboratory which was established at Goddard Space Flight Center (GSFC). This HRPT capability enabled the Very High Resolution Radiometer (VHRR) and Vertical Temperature Profiling Radiometer (VTPR) data to be sent directly to DR ground receiving stations. The laboratory made investments in low noise amplifiers and down converters, inexpensive receivers, demodulators and the design and development of a multi-satellite frame formatter and ingest board that resided directly on Digital Equipment Corp. VAX's computer bus. Also at this time, the costly hardware-based image processing systems were replaced with software.

With the advent of Microvax computers, NASA was able to significantly reduce the cost of DR ground stations. A single Microvax computer could now ingest the HRPT data stream at 665 Kbps from the new ITOS/NOAA. At this time, the cost of a ground receiving stations was still high, on the order of 1 million to 1.5 million US dollars, but the expectation of a long continuous series of satellites producing significant data sets with multiple applications made the decision to invest in the new ground systems acceptable for large organizations and institutions.

In 1983, NASA GSFC and NOAA joined forces to develop a complete satellite ground system based entirely on the use of direct broadcast data from the ITOS.

By 1989 NASA GSFC had made significant improvements to the hardware and software of DR ground receiving systems such that multiple polar orbiting satellites (DMSP—Defense Meteorological Satellite Program & TIROS) and Geo-synchronous satellite data from GOES and GMS could now be received, processed and analyzed in near real time in an automated fashion. The DMSP satellites were encrypted, but these satellites had unique microwave instruments which were based on the NIMBUS instruments. In the late 1980's to early 1990's the data products from DMSP satellites were made available to the research community. These products would eventually influence the design of the future National Polar Orbiting Environmental Satellite System (NPOESS).

During this period, many of the advanced DR ground system technologies were transferred to industry. NASA GSFC was able to demonstrate that previous ground systems costing about 1 million US dollars could now be procured for less than 300 thousand US dollars. Shortly thereafter, the industry would further reduce the cost and complexity such that, at the time of this writing, many basic TIROS and geosynchronous satellite-based ground systems can be purchased for less than 15 thousand US dollars as further described in Fig. 4.2.

Beginning in 1987 NASA GSFC's primary interest in DR was the demonstration of practical applications of DR ground system technologies. This included data mining, distributed processing computers, data archiving and data management technologies. These activities lead to important studies such as the relationship between sea surface temperature anomalies and climate changes (El-Nino Southern Oscillation), Global deforestation, sea surface rise, volcano

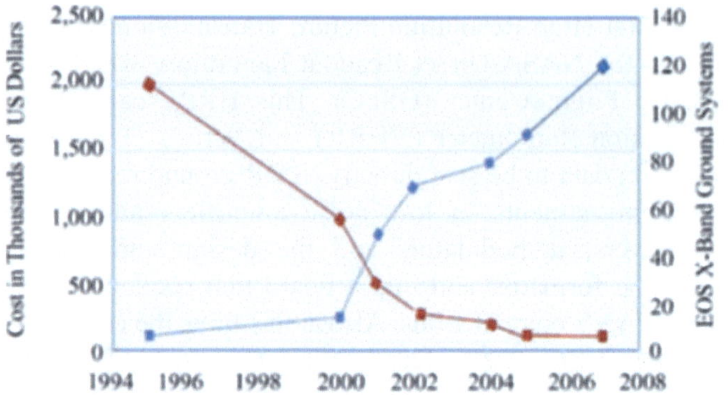

Figure 4.2 Correlation of X-band ground system costs to their number increase

eruptions and sand storm dust distribution, snow and ice distributions and ozone anomalies to name a few of the scientific applications which will lead to the need for the next generation of both research (TERRA, AQUA) and operational (NPOESS) and Geostationary (GOES) satellites.

In the early 1990s, the next technology project involving DR ground system development was in support of the Orbview-2 ocean color commercial satellite, formally known as SeaStar carrying the SeaWiFS ocean color instrument. In this project the development of new antenna feed horns allowed both L-band DB data and S-band satellite tape dumps to be received using the same ground system.

At this juncture, the development cycle began to repeat itself; NASA's new research DB satellites; TERRA and AQUA have state of the art instruments, such as, MODIS, HIRS & AMSU. The data rates have increased significantly to an average of 13 Mbps to 15 Mbps, and the frequency band has changed from L-band to X-Band (7.8 GHz – 8.5 GHz). Aqua and Terra are sister spacecraft with planned mission lives of 6 years. Terra was launched on Dec. 18, 1999, followed by Aqua on May 04, 2002. Each spacecraft is in a 98-degree inclination near-polar orbit so that its motion is largely in a north or south direction when passing over the Equator. Aqua and Terra each circle the Earth 16 times daily at a nearly constant altitude of about 705 km or 438 miles.

Terra follows a sun-synchronous orbit across the Equator heading southward at about 10:30 am, local time, when cloud cover tends to be minimal and visibility greatest. Aqua crosses the Equator northbound at about 1:30 pm local time. The crossing times were selected to gain insight into the daily cycling of precipitation, ocean circulation, and other key scientific parameters.

4.3 The Next Step

These new NASA experimental research spacecraft such as Terra and Aqua are continuing to make the long-term measurements which have been defined by the

NASA and collaborating communities. The NOAA operational instruments have been partially defined by these prototype NASA instruments and plans are underway to launch a new generation of satellites, the National Polar-orbiting Observational Earth Satellite System (NPOESS) Preparatory Project (mission) in 2009, and NPOESS series in 2012.

The Net Primary Productivity (NPP) mission was formulated by NASA, GSFC and the National Polar-orbiting Operational Environmental Satellite System (NPOESS) Integrated Program Office (IPO).

For NASA, the NPP mission is part of the Earth Observing System (EOS) program that provides measurements of environmental phenomena identified in the EOS Science Plan. Scheduled for launch in Oct. 2006, the NPP mission furnishes remotely-sensed land, ocean, and atmospheric data to the meteorological and global climate change communities as the responsibility for these measurements transitions from existing Earth-observing missions, e.g. Aqua, Terra, to the NPOESS.

For the IPO, the NPP mission presents an opportunity to demonstrate and validate new instruments and science data processing algorithms, as well as to demonstrate and validate aspects of the NPOESS command, control, communications and ground processing capabilities prior to the launch of the first NPOESS spacecraft. The NPP does not officially supply operational data to military users; but for all practical purposes it does so to its civilian support structure, and the general world community.

NPOESS is a tri-agency, Department of Defense (DoD), NOAA and NASA, program that merges two environmental satellite systems into one. Currently, NOAA and the Department of Defense each operate their own systems—the Polar Operational Environmental Satellites (POES) program and the Defense Meteorological Satellite Program (DMSP), respectively.

The plan is for the NPOESS constellation to consist of three polar-orbiting satellites. Each satellite will be designed for a seven-year lifetime and will host 10 to 12 sensor payloads. These sensors will provide a wide range of weather and environmental data to be disseminated worldwide, such as improved three- to five-day, seasonal and interannual weather forecasts, storm tracking, and agricultural crop management. The first launch is scheduled to coincide with the end of the expected lifetimes of the POES and DMSP constellations in 2009.

4.4 DB Community

There is significant reference to the DB community, but for many, they are difficult to identify and quantify, especially since the term DB can mean many things to many people. Within this context, a DB community member is any person or group that uses, benefits from, or contributes to the acquisition of remotely-sensed, DB data. The DB community can be classified into two categories: DR stations and DB data users. The DR station category can be further segmented into low, mid and high data rate acquisition systems. The low rate (2.4 Kbps) DR systems include the APT stations, of which there are over 4,900 world-wide. The mid data rate (665 Kbps – 3 Mbps) DR stations consist of the HRPT systems of which there are approximately 3,000. The last segment constitute the next generation of remote sensing DR stations. These high data rate (13 Mbps – 25 Mbps), X-band DR stations, consist of approximately 110 world-wide at the time of this writing.

The second DB community category is the derived product user. This category includes the ground station but has an additional estimated 10,000 users that make direct use of the information derived from DB data. Members of this category range from the commercial sector that provides processing components to the scientists that evaluate the most recent data acquisition results. Other members of this community include universities, the USDA forest service, fisheries, weather forecasters, DoD, NOAA, Dept of Transportation, Commerce and the Interior, NASA, amateurs and foreign meteorological offices, to name a few.

4.5 Technologies and Data Flows in Direct Broadcast and Direct Readout

Satellite direct broadcast (DB) is supported by technologies that are ever evolving. Nevertheless, over the last ten years and for the next twenty, these technologies will evolve within two well defined categories—data packaging and Radio Frequency (RF) communications packaging.

At NASA GSFC, the Direct Readout Program promotes continuity and compatibility among evolving EOS direct broadcast satellite downlink configurations and direct readout acquisition and processing systems. This Program bridges the EOS missions with the global direct readout community by establishing a clear path and foundation for the continued use of NASA's Earth science DB data.

From Fig. 4.3, the primary components of DB are depicted. The Consultative Committee for Space Data Systems (CCSDS) is the primary method for packaging on-board instrument data. This standard incorporates robust error-correction methods such as Reed Solomon, as well as a structure for ordering multiple data streams from one or more instruments.

The second category from Fig. 4.3 constitutes the RF packaging and transmission. Most DB data users identify DB by the link characteristics. As described in the "Context in History" section of this chapter, DB has been perceptually defined by broadcasting data in the L-band frequencies at a rate of 665 Kbps, which is synonymous with HRPT. But as technologies evolve, data requirements increase and new utility for the instrument data are developed, therefore new methods of DB must follow. These advancements are predominantly in the area of data encoding and in the methods for data correction.

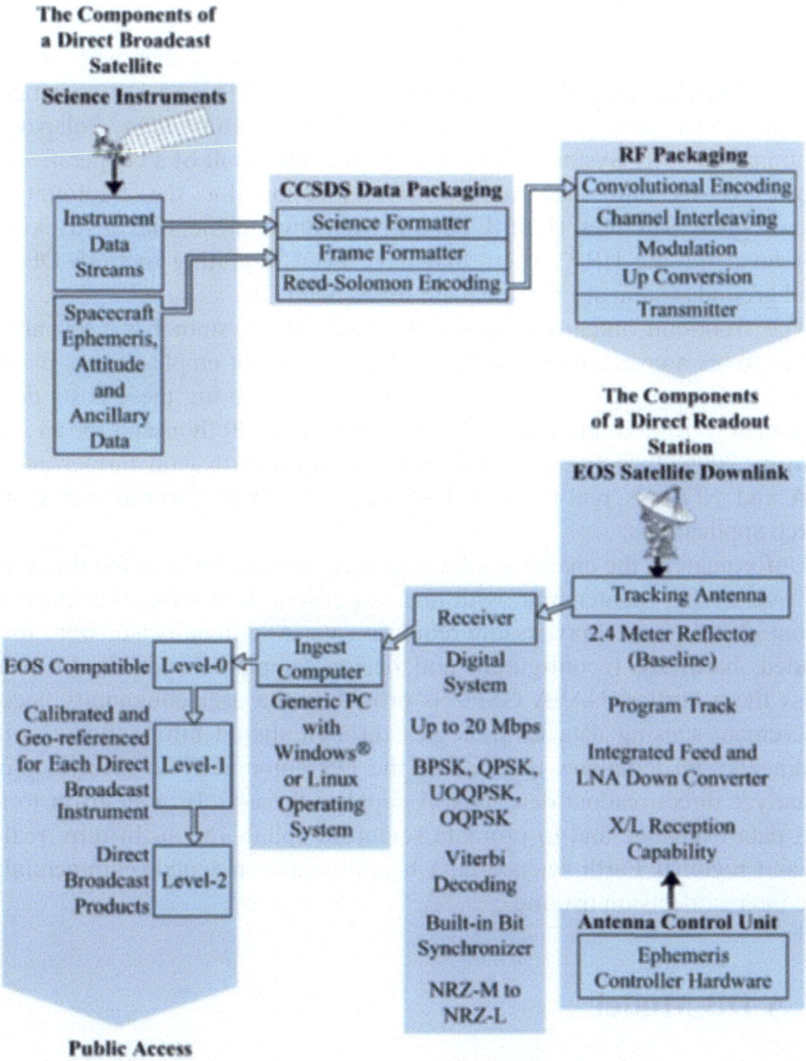

Figure 4.3 Functional components of DB through DR

Current and future designs have settled on multi-phase encoding (QPSK, SQPSK, etc.) with convolutional (1/2 and 3/4 rate viterbi) forward error correction. Additional variations of these are coming online, but most importantly, they represent a standard which enables easier transition for the direct readout (DR) end user on the ground from one spacecraft to the next.

Based on the improved DB transmission methods and data packet assembly schemes, the end-user has to adapt to and comply with these changes. And even though the spacecraft changes have been intentionally designed to ease the transition to new satellites and instruments, there is still a significant expense involved with compliance. This is most evident in the 2006 and 2007 time frame when HRPT users are transitioning to Metop and subsequently to NPP and NPOESS Low Rate Data (LRD) DB. The Metop and NPOESS DR transition will be made easier with the establishment of the Initial Joint Polar-orbiting Operational Satellite System (IJPS). The IJPS is the result of a cooperative effort between NOAA and the European Organization for the Exploitation of Meteorological Satellites (EUMETSAT). Transition difficulties also exist at a higher level for the HRPT users that have been migrating to EOS DB, since L-band ground system are not adequate for X-band DB.

The front-end antenna component of the DR system has been and will continue to be, a cost-driver, and therefore a significant emphasis was placed by NASA GSFC to work with the commercial sector to bring these costs down to approximately 15thousand for an HRPT system and 120 thousand for an X-band system by 2006. In doing so, it allowed the end-user to gain further access to NASA and NOAA's real-time satellite data for entrepreneurial and real-time research applications.

Unfortunately, the challenges for end-users to keep up with the times are far from over. After contending with the expensive front-end, the end-user is confronted with several processing requirements. Additionally, software must be upgraded, hardware reconfigured, and design templates updated. In order to address these hurtles, NASA GSFC is providing the user community access to Earth remote sensing data technologies through shared information resources, including design templates that enable the DB community to receive, process, and analyze direct readout data. NASA supplies these tools in an effort to foster global data exchange and to promote scientific collaboration. In turn, real-time local and regional Earth science data benefits environmental, commercial, and public interest decision making.

4.6 A DB Model

The Direct Readout Program (DRP) at NASA GSFC plays a vital role as the intermediary between remote sensing missions, such as Terra, Aqua, and the NPP, and the DB community. The project encourages communication and maintains an open-door policy with the commercial and research and development sectors. This two-way information exchange is part of NASA DR model, depicted in Fig. 4.4. This model provides the DRP with critical information on the equipment currently in use, the technology being developed commercially, and the needs of the direct broadcast community. As a result, the Direct Readout Program is able to report the "state of the community" to the NASA mission while providing the DB community with a two-way dialogue between the mission objectives and the user needs.

Figure 4.4 NASA Direct Readout model applied to the NPP mission

An example of an application of the DB model, as depicted in Fig. 4.4, is the NPP In-Situ Ground System (NISGS) development. Using this process, NISGS inherits the Direct Readout Program's knowledge, experience and "lessons learned" from prior missions, bringing continuity from legacy missions to NPP and NPOESS. This experience enables the DRP to predict problem areas in the design and development of NISGS and mitigate them before problems develop. The NISGS model is both, system architecture, and a methodology. In this methodology, the direct broadcast community is the driving force for DB service-type requirements, which often translates to system design requirements.

The DRP also developed the Mobile NISGS, a mobile DB system, as an accessible prototype and evaluation platform. The Mobile NISGS represents the state-of-the-art technologies used by the commercial and end-user DB communities. This development system allows for immediate compatibility testing of the Direct Readout Program technologies and provides demonstrations to the end-user community and commercialization partners. The Mobile NISGS also interfaces with development DB components of the spacecraft for compatibility and validation of the RF system, protocol and data processing, and ground network interfaces. It also serves as a tool for gathering real-time Earth science data in remote areas for instrument calibration and validation campaigns.

4.7 Technology Roadmap

One of the primary results of the DR model, as described in Fig. 4.5, is the identification of key technology categories that the DR end-user would have to contend with in order to be compliant with a multi-satellite environment. These categories include: a multi-mission autonomous task manager and dispatcher, Consultative Committee for Space Data Systems (CCSDS) packet re-assembly and standard data reformatting tools, instrument-specific calibration and geo-registration algorithms, and real-time data distribution mechanisms. NASA GSFC through the DRP and the NPP mission have addressed these key technology categories in the form of specific technologies which are generic in nature and can be integrated into existing or developing DR systems.

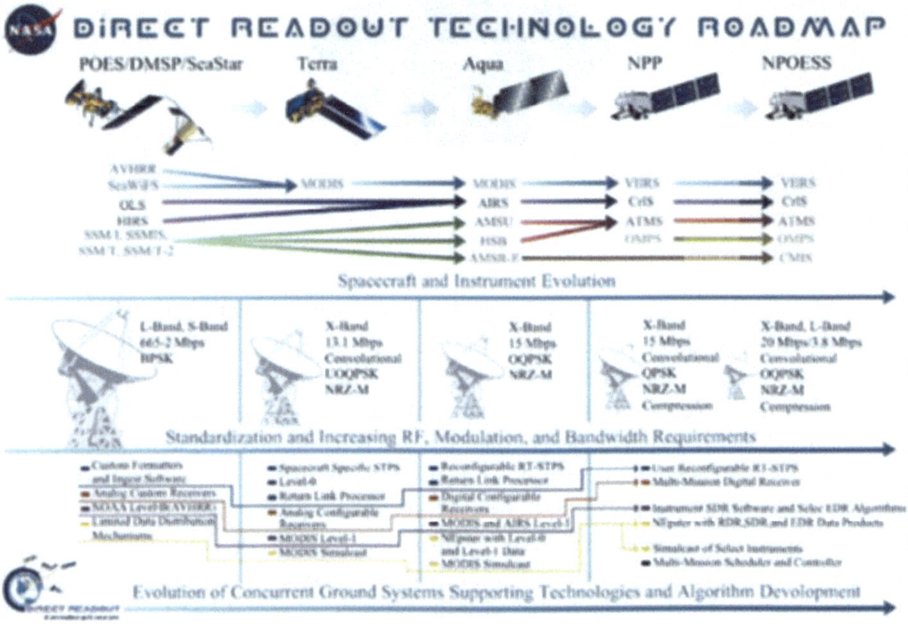

Figure 4.5 Direct Readout technology roadmap

As previously described, the DRP serves as the intermediary between missions and the user community, but more so, the project's history of involvement with direct broadcast missions ensures that the legacy of information and technology is incorporated into future mission efforts. In this way, the DRP brings continuity, cohesiveness, and standardization to these technologies, providing the DR community with a technology roadmap as described in Fig. 4.5.

4.7.1 Multi-Mission Scheduler

Within the NISGS example, the DR Multi-Mission Scheduler (MMS) is used as a baseline technology. The MMS is a configurable, client-server control software environment that automates ground system data acquisition, data processing, and data distribution, see Fig. 4.6. MMS also provides a documented Application Programming Interface (API) so that it can be easily integrated into any ground system environment.

MMS is a client-server software package where the client translator sits on each piece of equipment to be controlled. A client template is part of the MMS

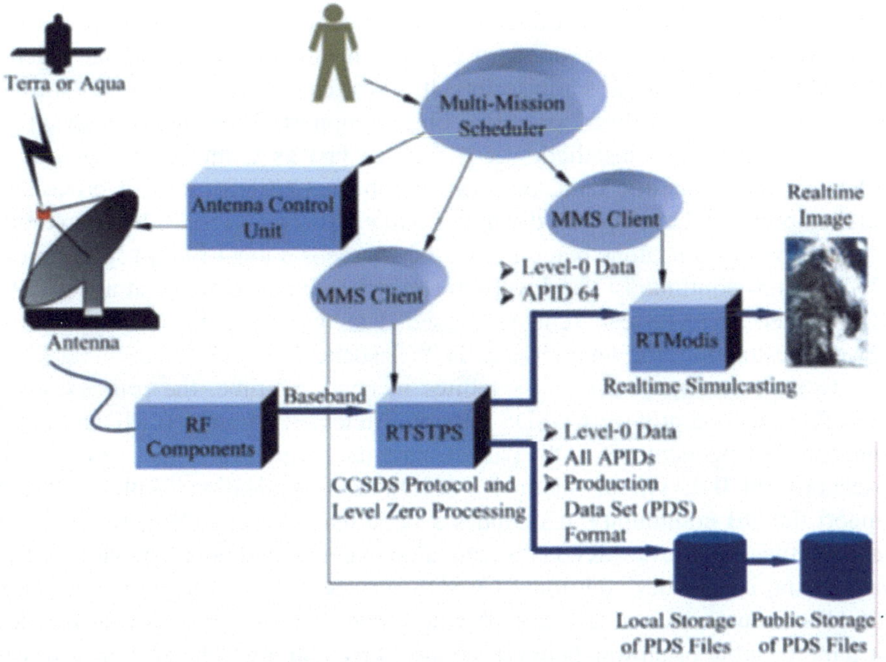

Figure 4.6 Multi-Mission Scheduler functional diagram

package. Scheduler is the MMS server. Scheduler provides a central control point for all the ground system software and hardware. The client template is customized, and then installed on each ground system component. Scheduler communicates with each client via a User Datagram Protocol (UDP) socket, and Scheduler uses data and events provided by each client to control data acquisition and processing. Based on Scheduler input, the client starts and stops software tasks local to the device on which it is installed.

Scheduler provides a graphical user interface (GUI), which allows the operator to completely configure and control MMS activities. Data from multiple satellites can be acquired, processed, and distributed by defining parameters provided via the GUI.

4.7.2 Real-Time Software Telemetry Processing System

Real-Time Software Telemetry Processing System (RTSTPS) provides the baseline technology for NASA standard's Protocol Processing and Level-0 Data Production elements. RTSTPS is a software package that provides front-end processing in support of Earth remote sensing missions. This technology is a pure software implementation of a Consultative Committee for Space Data Systems protocol and level-zero processing system data gateway. It performs protocol and Level Zero processing for legacy and upcoming missions. RTSTPS is platform independent. It is designed as a collection of independent processing nodes, so a user can create standard or custom pipelines by plugging together selected nodes using XML setup scripts. It is written entirely in Java and does not require any special hardware or native library support. Therefore, it runs on any system where Java is installed, yet it runs as fast as C and C++ equivalents. RTSTPS ingests satellite data, performs complete CCSDS protocol processing, and then provides Level-0 data distribution to end-users. RTSTPS has two modes of operation. In standalone batch mode, the source is a local file. In server mode, RTSTPS runs continuously and waits for raw data input to arrive at a TCP/IP or UDP socket. RTSTPS can send processed data units to a file or directly to a network destination, for which it uses TCP/IP sockets.

To support remote sensing satellites in near real-time, the primary job of RTSTPS is to synchronize CCSDS version 2 telemetry frames (CADUs) and to compose independent CCSDS packet streams, which are distinguished by spacecraft (SCID), virtual channel (VCID), and application (ApID). Also to support the DB community, it includes a node that creates EOS-compliant PDS Level-0 files. However, RTSTPS can also receive and identify non-CCSDS frames, and it provides additional CCSDS services to those listed above. Other RTSTPS features include a rate-buffering client, which regulates transmissions for unpredictable or slower network connections, and a graphical viewer to ease configuration and observe processing status. RTSTPS processing has been successfully verified with live Terra and Aqua MODIS data. The most CPU intensive RTSTPS processing, which includes Reed-Solomon decoding, has been benchmarked at rates in excess of 20 Mbps.

Since RTSTPS has a modular design, it has distinct components that work together to create the telemetry processor. In addition, several utility programs can provide additional functions. Figure 4.7 shows the layers that create the RTSTPS program. Utility programs exist outside this diagram.

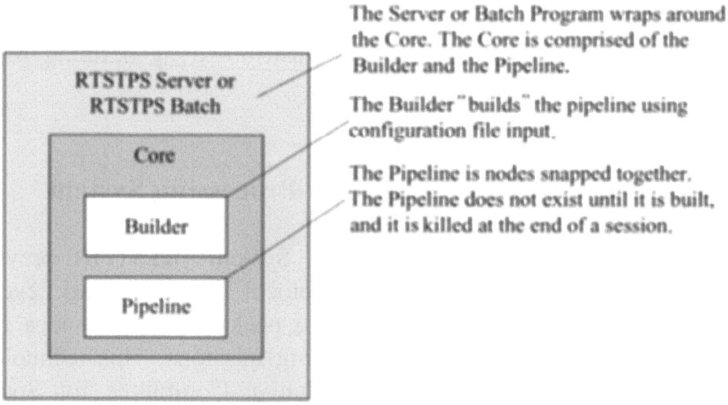

Figure 4.7 RTSTPS architectural layers

The RTSTPS components and the supporting utility programs can be summarized as follows (Fig. 4.8):

(1) RTSTPS Server or RTSTPS Batch Program: The processing component of the RTSTPS is either a standalone batch version, which ingests a single raw data file, or a server (TCP/IP or UDP versions), which runs continually and waits for raw data to arrive at an input socket. To satisfy EOS and NPP requirements, both the server and batch program include a node that creates EOS-compliant PDS files. The following diagram illustrates the Server, which could be replaced by the batch program, and its relationship to two other internal components: Builder and Pipeline. The server loads the XML Configuration file, which is defined below, via the "configuration and commands" input path. It receives telemetry data input through the "data" path.

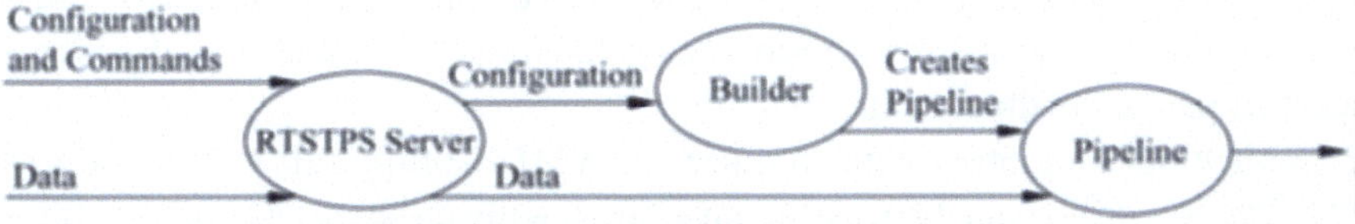

Figure 4.8 RTSTPS Data Flow

(2) Nodes: RTSTPS Nodes define data processing functions. Each node is a Java object, which performs one primary function. Once connected, nodes pass along data units. Each node is independent of all other nodes, and nodes can be modified or new nodes created without having to understand system-wide dependencies or relationships. Nodes implement connection interfaces to create the connectors that allow nodes to be linked together. The interfaces empower a node to be either a sender or a receiver; Sender and Receiver are the two categories of interfaces.

(3) Builder: Each time a user loads an XML Configuration file into the server or the batch program, the Builder interprets it to create a pipeline of linked nodes that process raw telemetry into products such as packets.

(4) XML Configuration File: The XML Configuration File is a script that is input to the server/batch program. The script tells it which nodes to create and how to link them together. The user can create or modify the configuration file with any text editor.

(5) Pipeline: The pipeline is a collection of independent nodes that are linked together to define the data processing path. The XML configuration script defines the pipeline. It tells the server (the Builder specifically) which nodes to create and how to link them together.

(6) RTSTPS Utility Programs: The RTSTPS system has several other utilities besides the batch program and the UDP and TCP/IP servers:

- Viewer: The Viewer is a graphics tool that allows the user to load a configuration file into the server. It also shows status that the Viewer receives from the Server.

- Editor: The Editor is part of future development. It is a graphics editor that allows users to edit RSTPS configuration files.
- Sender: The Sender is a small graphics program that sends raw data files to the server. Mainly used for test purposes, it substitutes for the front-end antenna system.
- RAT: This program is a rate-buffering output client, which regulates transmissions to slow receivers. It can be attached on the output side of the server to spool the server's socket output.

4.7.2.1 Modes of Operation

The RTSTPS can run in two independent modes. It can run as a stand-alone program that processes a single raw data file (batch mode), or it can run as a server, in which case it runs continuously, listening to a socket and processing spacecraft passes as they occur.

Both modes of processing require an XML Configuration file for system setup. The specific data processing provided with either mode of operation is dependent on the pipeline (consisting of nodes) defined in the configuration file.

1. Batch Mode

Batch Mode is a run-once program. It reads all data in a source data file, and then produces output units as defined in the configuration file. It terminates when finished. Batch Mode is invoked with a command line that includes the program name "STPSBATCH," a configuration file, and the name of a file that contains the unsynchronized source data.

Once invoked, STPSBATCH creates a data pipeline as seen in Fig. 4.9. It then opens the data file and reads chunks of it, which it stuffs into the frame synchronizer node at the head of the pipeline. When it exhausts the file, it terminates.

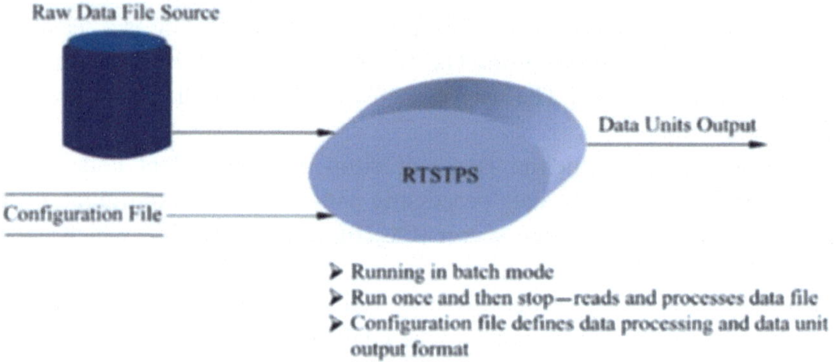

Figure 4.9 RTSTPS Batch Mode

2. Server Mode

The second mode, Server Mode, runs continuously. It listens for spacecraft data on an input port, and it also listens for commands on a command port, which can be used to load a specific configuration. The server accepts data from a TCP/IP socket, but there is also a second version of the server that connects to a UDP socket (Fig. 4.10).

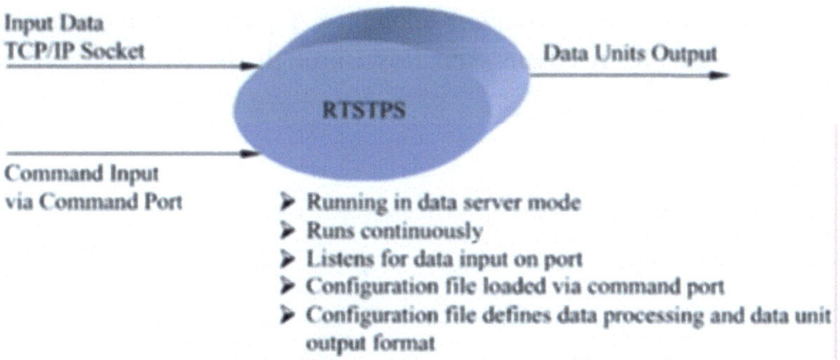

Figure 4.10 RTSTPS Server Mode

The server runs continuously. It listens on a TCP/IP socket for connections and spacecraft data. It will accept only one connection at a time. (There is a second server version that uses UDP sockets for data input, but most use the TCP/IP version.)

The server is multi-threaded. One thread represents the pipeline. It receives raw data from the socket and processes it through the pipeline. Another thread listens for commands through a second socket. This is a special java interface called Remote Method Invocation (RMI) through which the RTSTPS viewer sends load and shutdown commands to the server. The server also uses this interface to return status information back to the viewer. Finally, there is a third thread called the proxy thread that also allows users to send commands to the server. Unlike the RMI interface, which requires a Java sender, the proxy thread is a pure TCP/IP socket connection, and the commands are text strings. This interface was added so that programmers could send commands to the server from programs that were not written in Java.

When operating in Server Mode, RTSTPS runs in sessions. A session usually corresponds to a spacecraft pass. Just prior to a pass, either a user or an automated scheduler sends a setup configuration to the RTSTPS server. The server will prepare itself to receive data, which involves creating a pipeline and waiting for data on a TCP/IP socket. RTSTPS begins to process data as soon at it arrives at the socket connection.

If the server receives data before it has been configured, it automatically configures it itself. It either uses the last loaded configuration or a default configuration if one has never been loaded. If the user shuts down the server while it still has data to process, it will discard all subsequent buffers.

A shutdown ends a session. Either the user or automated scheduler sends the shutdown command to the server. The server shuts down automatically if the data sender breaks its connection. When the server shuts down, it closes all output files and all output sockets, and then it discards the current pipeline.

A node is a Java object, which performs primarily one function. For example, the Reed Solomon node performs Reed Solomon decoding and correction. When configuring the RTSTPS, the first and only required node is the frame synchronizer. It accepts blocks of unsynchronized telemetry. Figure 4.11 provides an node implementation and the resultant data flow through the RTSTPS Server/Batch program.

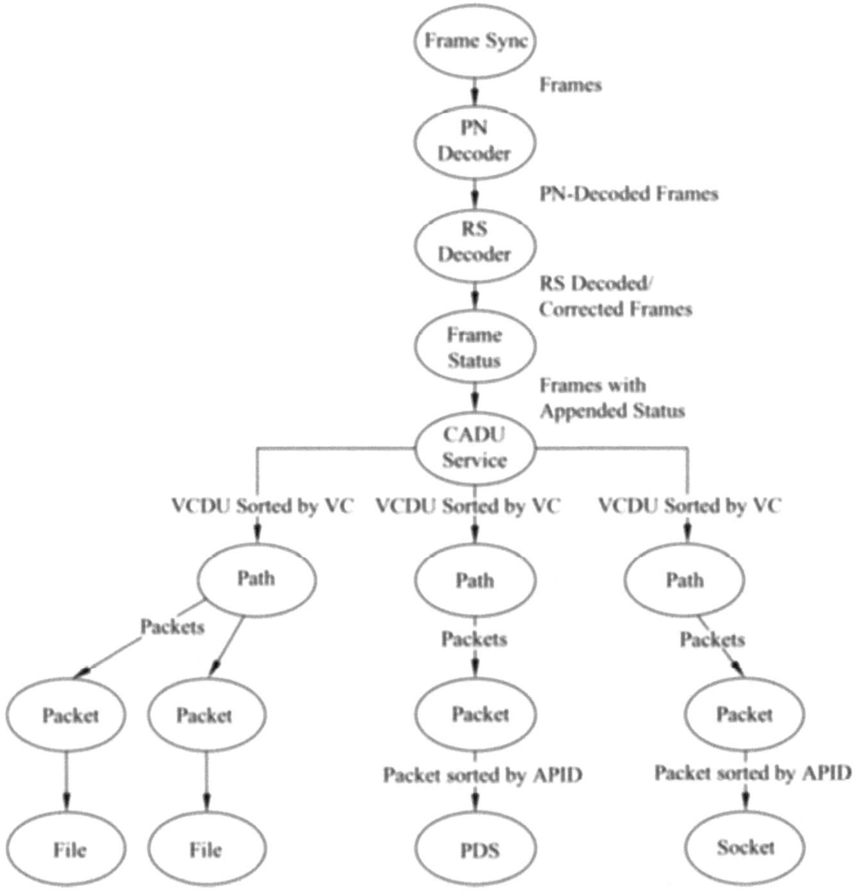

Figure 4.11 RTSTPS Server/Batch node implementation

The behavior of any one node is independent of all other nodes. A node has one input "port" and zero or more output ports. With multiple output ports, a node broadcasts. For data input, a node implements a frame, packet, or unit Receiver Interface. For data output, a node implements a Sender Interface. Since a node is isolated from all other nodes and can be defined to use any interface, a node can be plugged together with little regard to what is being plugged. Once connected, nodes pass along units. The term unit is used loosely to define all items such as frames, packets, VCDUs, and MPDUs.

4.7.3 Simulcast

Simulcast constitutes a component of a satellite DR system that enables real-time Earth science data utility and sharing. The system is designed to read the full multi-instrument CCSDS packet stream from a front-end satellite receiving station, create reduced data volume intermediate products, distribute the products simultaneously to multiple remotely networked workstations, and provide an appropriate interactive near-real-time display of instrument data.

The design and development of the simulcast system is based largely on interest and lessons derived from a similar software program known as RTModis (for Real-time Modis) depicted in Fig. 4.6 in the MMS section. The RTModis system was developed in 1999 to support the direct readout sites for the then, soon to be launched, TERRA satellite. The RTModis system successfully processed data beginning with the very first DB transmission from the TERRA satellite and is in operational use today providing real-time display of MODIS data for both the TERRA and AQUA satellites at both Mission Operations' Centers.

To support multiple remote displays and to minimize the network bandwidth requirement, the system is designed around a client/server model. A data processing component reads, decodes, decommutates, and performs any other process necessary to create intermediate data products. These products are designed to provide a reduced bandwidth representation of the instrument data as well as performance and state of health information for the spacecraft, instrument, and ground station.

The remote clients connect to the data processing server and receive the intermediate products as they are generated. The clients will render the intermediate products and provide an interactive interface to the end user. The remote display capabilities range from updating textual data about the spacecraft or other system components, to enhanced image products generated from instrument data.

The system is implemented in Java to provide cross-platform support and maintainability. The system will receive data from the front-end system. The data processing server functionality can be broken up across several processes or computers to spread out the necessary workload of generating intermediate products defined by the end-user; single band and image composites are the available public baseline.

The current EOS satellites, Terra and Aqua, as well as the next generation programs, NPP and NPOESS, will all transmit data using standards accepted by the Consultative Committee for Space Data Systems. There are several hardware and software (including a Java implementation) systems that process raw serial data from an antenna system and produce CCSDS packet streams. These streams can be filtered to include a subset of the instrument data, including the possibility of different streams each for different instruments. The filtering/grouping field for the CCSDS packet is the Application Process Identifier (APID).

The primary input for the Simulcast system is a CCSDS packet stream. The overall goal of the Simulcast Server is to produce intermediate products for each instrument. The processing required for this task may overwhelm one physical computer, or other considerations may make it desirable to handle different instruments on different machines or different processes. Hence, the first Simulcast component, a CCSDS Router, filters and groups the input stream to multiple streams and routes them to different instances of product Extractors described in Fig. 4.12. The same Router functionality will be present within the product Extractors to direct the packets from different instruments to the proper processing routines.

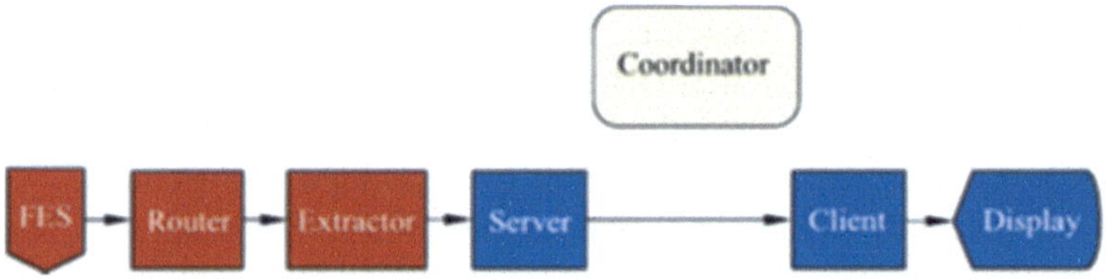

Figure 4.12 Simulcast block diagram (simple configuration)

At one extreme there can be a Server for each APID producing an intermediate product for that instrument. Similar operational configurations lead to the need for some coordination regarding the delivery of the intermediate products to the connecting Clients. To accomplish this, one additional component is added to the client/server system, the Client Coordinator. This component provides a single point of contact for remote clients and data processing servers. Information will be exchanged to direct the clients to the proper server for the desired product. This also provides the added benefit of allowing the data processing to be moved to backup machines or even alternate antenna sites without burdening the client with reconfiguration.

One additional feature is derived from the need to protect the operational environment from denial of service and other network attacks. The most efficient way to do this is to isolate the majority of resources behind tight network security and present only the needed resources to the wider Internet community. To accomplish this, a distinct element will reflect the intermediate products to the multitude of remote clients; in effect being a single client to the data processing servers and reflecting the intermediate products to the end user display clients. This design also addresses the concerns of providing the resources to handle many different legitimate display clients.

The end system results in a programming paradigm that is a grown trend: a set of general processing components forming a complete system via a coordinating configuration. This paradigm relates very well with the general principles of object oriented programming and distributed processing. In fact, using Java and TCP/IP client/server technology as a backbone of development, pulls a system design in this general direction; a happy synergy within the design cycle.

In its simplest form (see Fig. 4.12) some of the components only add processing overhead and do not serve a useful purpose.

However, the key to this system is designed extensibility. The operational system, as described in Fig. 4.13, includes a backup Front-End System (FES) that is transparent to the end user, split product generations processors (Extractors), an isolated server to reflect products generated behind a firewall, multiple remote clients, and a remote client which processes and reflects products via a Web Server.

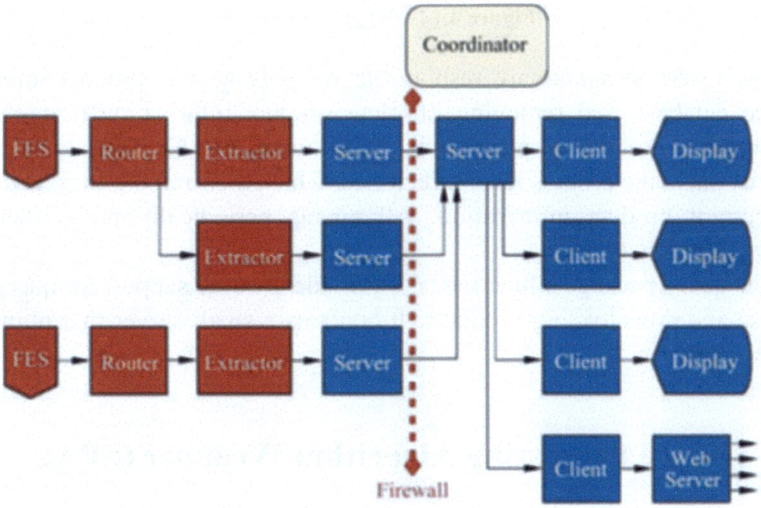

Figure 4.13 Simulcast block diagram (operational configuration)

4.7.4 NEpster

NEpster is last in a set of technologies needed to support a DR multi-mission environment as described in the Technology Roadmap section. NEpste is a software system that stores information about instrument products. Internet users, through their browsers, can search the system and download for specific products. NEpster also includes an assortment of agents who periodically probe remote data sites to collect product information. NEpster is not designed to store the actual products. Instead it stores the information that describes the products (the "metadata").

The core element of the NEpster system is a MySql database as described in Fig. 4.14, although any commercial database would work. The database stores information about products, product types, and collection stations. It is not meant to be a complete instrument archive but is intended to hold only recent information, typically 1 to 30 days of current data which is dependent on how long the source-data-keeper retains the data.

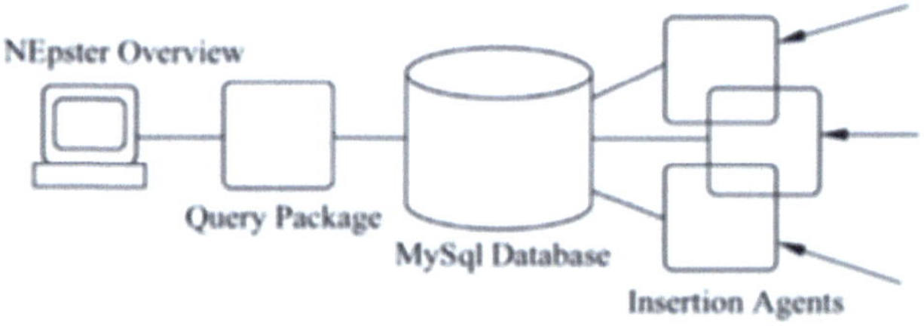

Figure 4.14 NEpster Overview

The insertion agents are responsible for putting new product information into the database and removing obsolete product information. Each agent is customized to probe one specific site. Most agents use FTP to examine a site's products, but some process local directories, which their remote sites have loaded with current product information. All agents periodically probe their sites, usually every few hours.

The query package's function is to provide browser support for querying the database and downloading products. It contains a small server that plugs into a larger web server.

4.8 Science Processing Algorithm Wrapper (SPA)

One of the final components in the DR processing chain that requires standardization is the SPA wrapper which is a computer science approach to encapsulate multi-mission science processing algorithms and allow them to run and function in a standalone environment. Functionality within this environment is required by the Direct Readout Laboratory (DRL) for dissemination of science algorithm to the broad user community; hence, SPAs wrapped with this tool would provide both, a common interface to algorithms, and a structured procedure for future algorithm updates which would require minimum effort.

The wrapper module assumes that the algorithm code:

(1) Consists of one or more executable programs that can be run from a command-line.

(2) Takes in one or more input data files, processes them, and emits one or more output data files. The files are labeled with a simple symbolic type.

(3) May also use other ancillary data files (e.g. digital elevation model) or dynamically generated control files (e.g. lists of input files and parameters) which can be hidden inside the wrapper.

(4) Makes use of a scratch directory while running (for intermediate temporary files).

The wrapper code can accept a single executable, multiple executables, or source code to be compiled and linked as long as those programs can run from the command line. The wrapper code takes in one or more input data files (including ancillary data), processes them, and emits one or more output data files. The wrapper code uses a scratch directory while running to store intermediate temporary files.

Wrapper creation

Before wrapping an algorithm, it should be decided how this code should appear in terms of the SPA wrapper model. In its simplest form, a wrapped algorithm is a black box that is given a list of (label, filename) pairs that specifies its input and output files—when it runs, it reads data from the input files, transforms it, and writes the results to the output files as shown in Fig. 4.15.

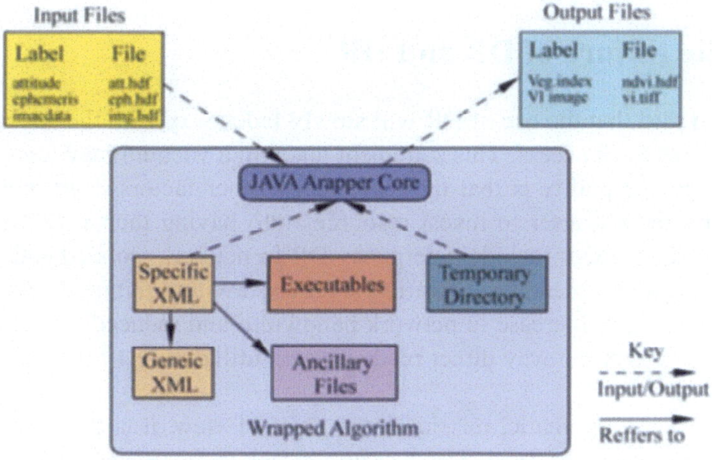

Figure 4.15 SPA Algorithm Wrapper Concept

SPA wrappers can actually have slightly more complex behavior than this. Normally each label must have exactly one file associated with it—when a wrapped algorithm is run, it checks its list of input and output labels against the (label, filename) list submitted to it, and it will complain unless all the input files exist and are readable, and all the output files are writeable.

To go beyond this, a label can be marked as *optional*, which means the wrapper will not complain if no file is submitted (input) or created (output) for that label. A label can also be marked as a *collection*, which means the label must be associated with either a directory (and all the files in the directory are part of the collection) or a space- delimited list of files. The exact mechanics of these specifications will be explained below.

A label can be any string that describes the file(s) associated with it. They don't have to be short—we have been using a platform.sensor.filetype convention (e.g., aqua.modis.pds). The most important thing to be aware of when choosing labels is the existing labels of any other SPA wrapped algorithms you may wish to pass data into or out of—consistency is a virtue.

Once labels and file associations have been decided, this information will be recorded in two XML files. One file is called the installation file, and contains information about the specific installation of the algorithm (mostly paths to executables, static data, and parameter files). The other file is called the generic file, and contains instructions on how to run the algorithm code, based on the information in the installation file.

As new algorithms and/or updates become available for public access they will be wrapped so to provide a consistent and transparent interface to the end-user, so as to maximize algorithm and data utility and minimize the possibility of algorithm corruption.

4.9 The Future of DB and DR

It has been said that the use of DR will slowly fade away with the advent of high bandwidth network access. This statement taken in a vacuum may logically seem accurate, but the reality is that there are many other factors contributing to the decision by the end user to invest resources into having their own DR system. Documented instances include emergency DR in network-isolated and/or remote areas where environmental information is required within minutes—currently not possible otherwise. Increase in network bandwidth and reduced access costs will nevertheless affect the way direct readout users utilize direct broadcast data and derived products.

From a programmatic, mission level point-of-view, it is the author's belief that DB will never go away for the simple fact that DB has proven to be an indispensable backup to the primary store and forward method of obtaining high resolution global data. Because of the innate and built-in robustness of the DB system design on-board the spacecraft, DB data has been used as a gap filler when satellite data relays fail and access to near-real-time global data is delayed due to data processing backlogs. This last scenario can be counter-argued by the fact that computational systems and network bandwidths will increase, but as described in Fig. 4.16, data volumes are commensurate with computational power. In other words, we're barely keeping up. And when one considers expected data volumes and computational processing requirements for the next 20 years, the rate of increase will not be decreasing.

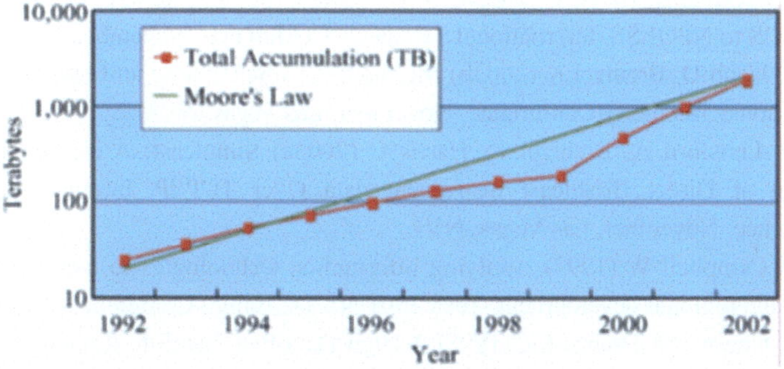

Figure 4.16 Archived data in Earth Observing System Data & Information System. By permission from Chris Lynnes at the NASA Distributed Active Archive Center

There is also an economic factor when considering the longevity of DR. When analyzing the DR system growth trend, it is clear that not only is cost significantly decreasing but there is a direct correlation between a step decrease in cost and a step increase in DR systems as described in Fig. 4.2.

On the other hand, there will be a significant change in how the majority of DB users will gain access to near-real-time products; after all, the point of DB is to get a data product and/or product to the end user as soon as possible over their area. With the on-set of the NPOESS, a comprehensive, all-be-it not new, method of acquiring, processing and distributing near-real-time data products will be established. This concept will also be mirrored by many additional profit and non-profit organizations which will take advantage of real-time raw data access to generate and distribute near-real-time value added products. This "new" concept, fueled by increases in network bandwidth and global access, is fundamentally a regression to the centralized, large data processing system's model with a new distribution twist, which includes broadcasting of data products, of much lower volume, to the end user via ubiquitous communication methods.

Acknowledgements

Research and development discussed in this chapter have been supported by the NPOESS Preparatory Project (NPP) and Dr. James Dodge at NASA former Code

YS Earth Sciences Division. Special recognition is also given to Mr. Charles Vermillion and Mr. Fran Stetina, former NASA civil servants for their pioneering work in DB and DR systems.

References

Coronado P, Brentzel K, Harris C (2003a) NPP In-Situ Ground System: Bridging Technologies from EOS to NPOESS. International Telemetry Conference. November, Las Vegas, NV

Coronado P, Jacob D, Brentzel K (2003b) Multi-Mission Scheduling and Dispatching System. International Telemetry Conference. November, Las Vegas, NV

Coronado P, Lunsford A, Brentzel K, Harris C (2003c) Simulcast: A method for real-time viewing of Direct Broadcast instrument data Over TCP/IP. International Telemetry Conference. November, Las Vegas, NV

Coronado P, Campbell W (1997) Applying information technologies to facilitate information access and regional development. Tech 2007 NASA Annual Conference. March

Coronado P, Ghuman P, Brentzel K (1998) A High Data Rate Satellite Remote Sensing Ingest System. International Telemetry Conference. October

Wallach J (1997) National Oceanographic and Atmospheric Administration User's Guide for Building and Operating Environmental Satellite Receiving Stations. DOC Publication. July

5 MODIS Calibration and Characterization

Xiaoxiong Xiong and William Barnes

5.1 Instrument Background

The Moderate Resolution Imaging Spectroradiometer (MODIS) is one of the key instruments for the NASA's Earth Observing System (EOS). It was developed based on the desire of the science community for frequent global observations of the Earth's land, oceans, and atmosphere at a moderated spatial resolution with carefully selected spectral bands at wavelengths ranging from the visible (VIS) through the long-wave infrared (LWIR). The MODIS is a major advance over its heritage sensors in terms of its spectral, spatial, and temporal resolutions and its broad range of science applications. A detailed description of the MODIS development history, science data products, and data processing is provided in Volume Ⅰ Chapter 2 "Introduction to MODIS and an Overview of Associated Activities", and Table 2.1 (MODIS design parameters) provides an overview of key MODIS design parameters. Two nearly identical copies of the MODIS instrument, the Protoflight Model (PFM) and the Flight Model 1 (FM1), were built by the instrument contractor—Raytheon/Santa Barbara Remote Sensing (SBRS) previously known as Hughes/Santa Barbara Research Center (SBRC). The MODIS PFM was launched in December 1999 on-board the EOS Terra (EOS-AM) spacecraft and the FM1 in May 2002 on-board the EOS Aqua (EOS-PM) spacecraft.

The MODIS has 36 spectral bands with center wavelengths from 0.41 to 14.5 μm. It is a passive cross-track scanning (whisk-broom) radiometer with a double-sided paddle wheel scan mirror that continuously rotates at 20.3 rpm. Each scan of 1.478 s produces a swath of 10 km (nadir) in the along track direction and 2,330 km in the along scan direction, thus providing complete global coverage of the Earth every two days. MODIS makes measurements at three different nadir spatial resolutions: 250 m for bands 1 – 2 (40 detectors/band), 500 m for bands 3 – 7 (20 detectors/band), and 1,000 m for bands 8 – 36 (10 detectors/band). The 36 spectral bands are located on four focal plane assemblies (FPAs): VIS, NIR, short-and mid-wave infrared (SMIR), and long-wave infrared (LWIR) as shown in Fig. 5.1. Two of the spectral bands (bands 13 and 14), each with two columns of 10 detectors, output data at low and high gains through the use of time-delay and integration (TDI). Thus, there are a total

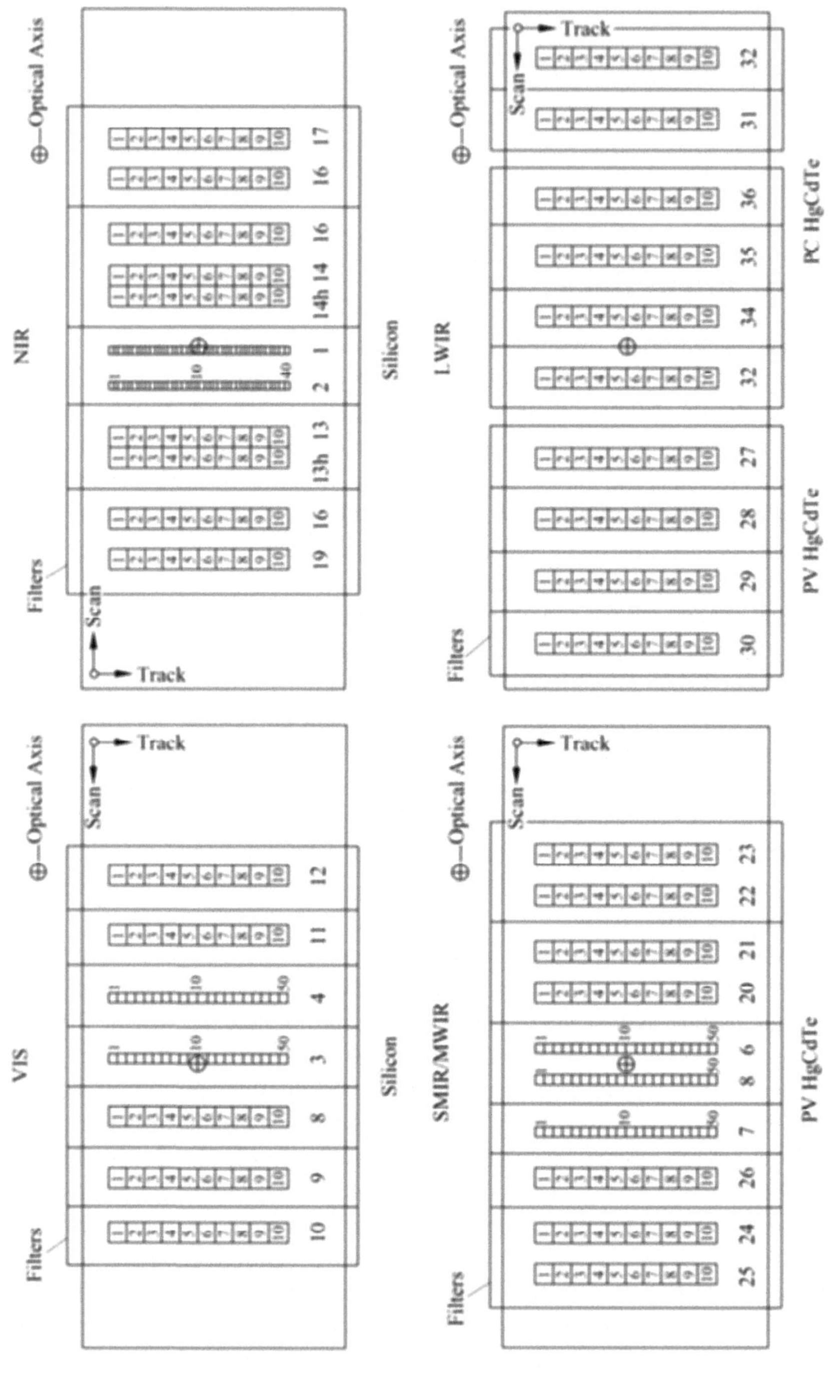

Figure 5.1 MODIS focal plane assemblies: VIS, NIR, SMIR, and LWIR

of 490 detectors. For each 1 km cross-track integration period (frame), there are 4 sub-frames for the 250 m resolution bands and 2 sub-frames for the 500 m resolution bands.

The scene radiant flux reflects from the scan mirror and then impinges upon a fold mirror followed by an off-axis telescope with a primary and secondary mirror. The aft optics include a dichroic beam splitter assembly consisting of 3 beam splitters, four objective assemblies for the VIS, NIR, SMIR, and LWIR FPAs, fold mirrors, and blocking and spectral band-pass filters. The VIS and NIR detector arrays are photovoltaic (PV) silicon hybrids that are operated at instrument ambient temperature. The SMIR FPA uses PV HgCdTe hybrid arrays. The LWIR FPA consists of PV HgCdTe detector arrays for bands with wavelengths less than 10 μm and photoconductive (PC) HgCdTe detectors for bands beyond 10 μm. Both the SMIR and LWIR FPAs are operated at 83 K on-orbit via a three stage passive radiative cooler.

The analog output signals produced by the PV FPAs are buffered and digitized by the space-view analog module (SAM). The signals produced by the PC detectors on the LWIR FPA are pre-amplified by the cooler located amplifier module (CLAM) and then post-amplified and digitized by the forward-viewing analog module (FAM). The digital outputs from SAM and FAM are formatted into science data packets by a formatter processor in the main electronics module (MEM). They are then buffered and sent to the spacecraft through a first-in first-out (FIFO) buffer and fiber distributed data interface (FDDI) circuits.

As shown in Fig. 2.1 (Vol. Ⅰ: Chapter 2), the MODIS OBCs include a solar diffuser (SD), a blackbody (BB), and a spectro-radiometric calibration assembly (SRCA). In addition, there is a space view (SV) port that is used to provide a zero reference. The on-orbit degradation of the SD is monitored by the solar diffuser stability monitor (SDSM).

Utilizing the rotating scan mirror, the sensor takes data each scan from the OBC and Earth view (EV) sectors: 50 frames of data from the SD sector, 10 from the SRCA, 50 from the BB, 50 from the SV, and 1,354 from the EV. The SD/SDSM system is used for the on-orbit radiometric calibration of the sensor's 20 reflective solar bands (RSB) with wavelengths from 0.41 μm to 2.1 μm and the BB for the radiometric calibration of the 16 thermal emissive bands (TEB) with wavelengths above 3.7 μm. The SDSM only takes data during scheduled SD calibrations. The SRCA is used primarily for spatial (all bands) and spectral (RSB only) characterization. It also provides limited radiometric monitoring capability for the RSB.

Overall, the two MODIS instruments (PFM and FM1) have been performing very well, exceeding the majority of the design specifications and science requirements (Barnes et al., 2002; Salomonson et al., 2002). The OBCs have been functioning as designed, providing continuous on-orbit instrument calibration and characterization in support of the production of high quality data products for the studies of short- and long-term global and regional Earth processes and trends (Xiong et al., 2003a; 2003b).

5.2 MODIS Pre-Launch Calibration and Characterization

The MODIS Level-1B (L1B) radiometric calibration algorithm converts the digital Earth view output of the 36 spectral bands to top of the atmosphere (TOA) reflectance factors for the RSB, and in-band spectral radiances for both the RSB and TEB. The absolute radiometric accuracy requirements (1σ) at the typical spectral radiance levels are $\pm 2\%$ for the RSB reflectance factors and $\pm 5\%$ for the RSB radiance products. The TEB requirements are $\pm 1\%$ except for $\pm 0.75\%$ for band 20 at 3.75 μm and $\pm 0.5\%$ for bands 31 and 32 at 11 μm and 12 μm. At radiance levels other than the typical radiance between 0.3 of the specified typical radiance and 0.9 of the specified maximum radiance, an extra 1% is added to the radiometric accuracy requirements. Band 21 is used for fire detection with a very low gain setting. Its calibration requirement is less demanding ($\pm$ 10%). Other requirements include spectral band center wavelength and band-width, band-to-band registration (BBR), modulation transfer function (MTF), and polarization sensitivity for the reflective solar bands (NASA, 1993).

To verify that the instruments met their specified design requirements both PFM and FM1 underwent extensive pre-launch calibration and characterization. Detailed data analyses were performed to evaluate each sensor's radiometric, spatial, and spectral performance. Most of the radiometric calibration and characterization measurements were performed in a thermal vacuum (TV) environment that was configured to closely simulate in-flight conditions. Three instrument temperature plateaus (cold, nominal, and hot) were used during the TV calibration and characterization. Each band, detector, and sub-sample (for the sub-kilometer resolution bands 1 – 7) is fully characterized. Both mirror sides are also treated individually.

5.2.1 Pre-Launch Calibration of the Reflective Solar Bands

Pre-launch, the MODIS reflective solar bands, 1-19 and 26, were calibrated using a 1-m diameter spectral integrating sphere source (SIS-100). The combination of different lamps of the SIS-100 provided numerous radiance levels, allowing the measurement of each RSB detector's response, dynamic range, nonlinearity, and signal-to-noise ratio (SNR). A linear algorithm was used for the RSB pre-launch radiometric calibration (Guenther et al., 1998; Xiong et al., 2002a). The radiometric calibration was performed at the three TV instrument temperature plateaus using both the primary and redundant on-board signal control and data acquisition electronics sub-systems. This data was used to characterize system response to changes in instrument and focal plane temperature. Because the VIS and NIR FPAs are not actively controlled, their temperatures are closely coupled with (or correlated to) the instrument temperature.

Calibration of the RSB reflectance factor was achieved by characterization of the solar diffuser's (SD) bi-directional reflectance factor (BRF). The SD panel, shown in Fig. 5.2, is made from space-grade Spectralon with a near Lambertian reflectance profile in the VIS/NIR/SWIR regions. The MODIS SD BRF was carefully characterized pre-launch by the instrument vendor using a scattering goniometer in a comparison mode with reference samples traceable to reflectance standards at the National Institute of Standards and Technology (NIST). The SD BRF measurements were made at 400, 500, 600, 700, 900, and 1,700 nm over a two-dimensional grid of nine directions of incidence, a combination of 3 elevation and 3 azimuth angles. The directions of incidence were chosen to cover the anticipated range that would be observed during on-orbit calibration. The viewing direction of the scan mirror to the SD was fixed during these tests. Additionally the BRF at 2,100 nm was indirectly derived from the SD BRF at other wavelengths and total integrated scattering (TIS) measurements of reference samples. A set of second order polynomial functions (of incident directions) was determined from these pre-launch measured BRF values. An interpolation from the polynomial values was made for each reflective solar band (RSB) to all possible incident directions (solar zenith and azimuth angles: θ and ϕ).

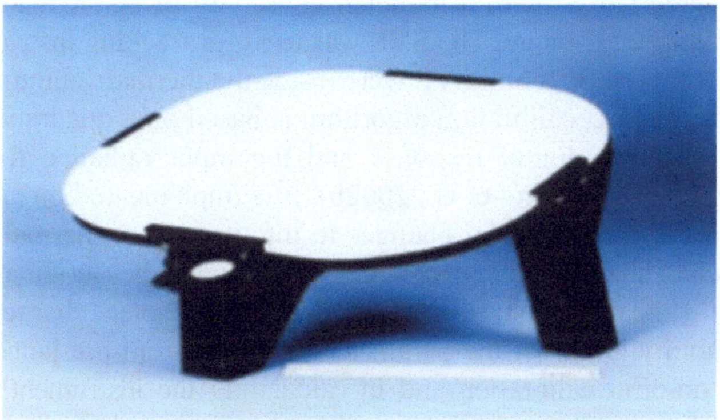

Figure 5.2 MODIS onboard SD panel used for the reflective solar bands calibration

MODIS uses a double-sided scan mirror to make observations of the Earth scenes in a $\pm 55°$ scan angle range from the instrument nadir. For each mirror side, this (scan angle) range corresponds to 1,354 EV frames of data with angles of incidence (AOI) on the scan mirror ranging from 10.5° to 65.5°. Since the calibration at each frame is performed at a different scan mirror AOI and since the reflectance of the protected silver surface varies with the AOI, it is essential to accurately characterize the sensor's response versus scan-angle (RVS). The RVS of the reflective solar bands was characterized using the SIS-100 at a system level with MODIS on a rotary table such that the radiant flux from the SIS can illuminate the scan mirror at different angles of incidence. In this setup, the SIS was fixed and the MODIS was rotated around the scan mirror's rotation axis. The RSB RVS test was performed in an ambient environment. Several SIS lamp levels were used to achieve mid-range signals for most of the spectral bands. A space view source (SVS), effectively a blackbody, was operated at a very low temperature to simulate deep space. The results from all detectors within a spectral band were averaged to produce a band dependent RVS. The RVS was characterized for each mirror side. The final RVS was determined by fitting the sensor's response to a quadratic function of AOI to the scan mirror, or equivalently to the data frame number.

5.2.2 Pre-Launch Calibration of the Thermal Emissive Bands

Pre-launch TEB radiometric calibration was performed in a thermal vacuum environment at cold, nominal, and hot instrument temperature plateaus. Like the RSB calibration, both primary and redundant electronics configurations were used. In addition, different SMIR and LWIR FPA temperatures (83 K, 85 K, and 88 K) were used to characterize the TEB detectors' temperature sensitivity. The pre-launch calibration source used for the TEB was a large aperture blackbody calibration source (BCS) with a modeled emissivity of better than 0.9999 over the entire TEB spectral range. The SVS was used for tracking instrument thermal background. Both the BCS and SVS were inside the thermal vacuum chamber.

The MODIS TEB calibration algorithm is based on a quadratic relationship between the sensor's digital response and the input radiance from the BCS (Guenther et al., 1998; Xiong et al., 2002b). It is implemented on a scan-by-scan basis to minimize the effects of changes in the instrument thermal background. The pre-launch calibration and characterization included measuring and evaluating system gains (linear response term), non-linearity, noise characteristics, and short-term stability. Some of the parameters obtained from pre-launch calibration are used for on-orbit calibration and in calculating the instrument's radiometric uncertainty. The temperature of the on-board calibrator blackbody was set at 290 K (PFM) and 285 K (FM1) during most of the thermal vacuum calibration tests while the BCS temperature was varied from 170 K to 340 K providing multiple radiance levels.

One of the key objectives of the TEB pre-launch calibration was to determine the OBC BB (see Fig. 5.3) emissivity since it is the primary on-orbit radiometric calibration source for the thermal emissive bands. The OBC BB is a v-grooved panel with 12 embedded thermistors. The thermistors are calibrated to the NIST temperature scale. The OBC emissivity was determined using the response of the MODIS detectors to the BCS and to the OBC BB at the same time. This was performed under different operational configurations and at a number of radiance levels as part of the pre-launch to on-orbit calibration reference and scale transfer activities.

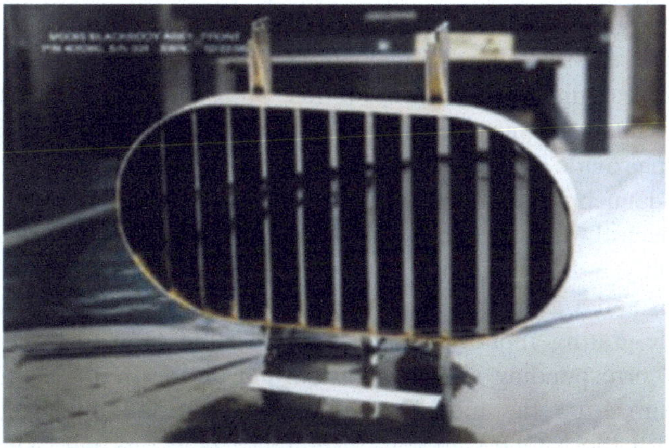

Figure 5.3 MODIS on-board blackbody used for the thermal emissive bands calibration

The TEB RVS characterization was performed in an ambient environment using a similar setup to the RSB RVS measurements. The BCS was used as the input source with MODIS rotating to produce the desired range of AOIs to the scan mirror. Both the BCS and OBC BB temperatures were set above ambient. The space view source was attached to the rotary table to enable the space view measurement of the instrument's thermal background. A bench test cooler was used to cool the SMIR and LWIR FPAs and a charge subtraction technique was used to prevent saturation. This is a very difficult and challenging characterization. The PFM TEB pre-launch RVS characterization did not produce valid data sets that could be used for the RVS analysis. Consequently, its at-launch RVS was derived from scan mirror witness sample reflectance measurements and aft optics polarization parameters from the FM1 RVS measurements. It was later improved via on-orbit characterization (Xiong et al., 2003c).

5.2.3 Pre-Launch Spatial and Spectral Characterization

System level spatial characterization was performed pre-launch using an integration and alignment collimator (IAC). This characterization included evaluation of the line spread function (LSF), modulation transfer function (MTF), instantaneous field of view (IFOV) size, and band-to-band registration (BBR) for each detector. The IAC BBR along-scan measurement was made using a narrow slit with a nadir-equivalent width of 1/20 km to 1/5 km to scan across the detectors. A series of phase-delays of the data sampling time, with an equivalent step size of 0.1 IFOV, were applied to enable the profiling of detector responses. The centroid value of each detector's response curve determines the detector's IFOV position along-scan. The along-track position was measured using a reticle with three inclined slits (0.15 IFOV in width) scanned across each band. In order to assure high measurement precision, the slit inclination angle was chosen so that the slit moves 1 IFOV along-track when the reticle scans 12 IFOVs in the along-scan direction. Because of this, only three 1 km IFOV detector positions can be determined to provide an along-track band position.

The pre-launch spectral characterization of MODIS' 36 bands was measured using the spectral measurement assembly (SpMA). The SpMA consists of a source assembly, a double-grating monochromator, and a collimator. The sources include lamps and a glow bar mounted on a rotating mechanism. The slit width of the double-grating monochromator is adjustable to match the wavelength selected, its corresponding signal intensity, and the desired spectral resolution. The exit beam from the monochromator is collimated by a set of mirrors designed to fill the MODIS aperture. Reference detectors with known spectral response are used to monitor the source spectrum of the SpMA during spectral characterization. In addition to removing the SpMA source spectrum from the relative spectral response (RSR) measurements, the spectral reflectivity of the mirrors after the reference detector and the transmittance of the atmosphere along the optical path between the reference detector and the sensor were also corrected. Pre-launch spectral characterization included measurements of each detector's system level RSR, spectral band center wavelength, bandwidth, and out-of-band (OOB) response.

The on-board Spectro-radiometric Calibration Assembly (SRCA), shown in Fig. 5.4, is a calibration device that can be operated in three different modes (spatial, spectral, and radiometric). The spatial mode is used to monitor the band-to-band registration of all 36 spectral bands, including FPA to FPA relative

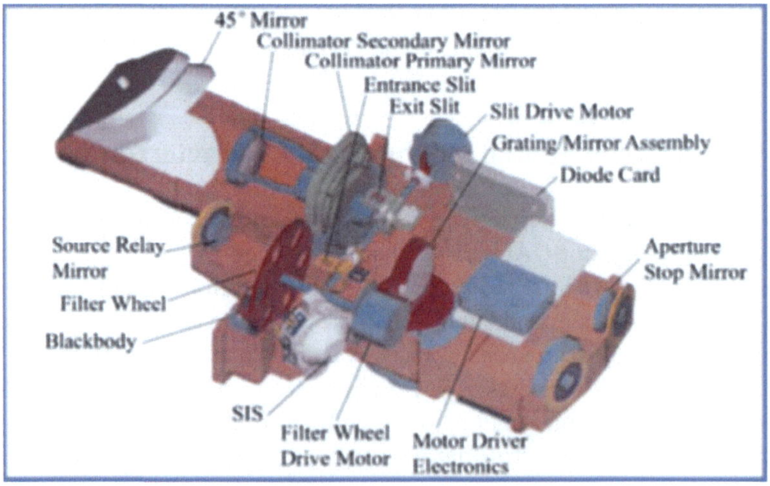

Figure 5.4 MODIS on-board Spectro-radiometric Calibration Assembly used for spectral and spatial characterization

shifts, while the spectral mode is used to track wavelength stability. Due to internal source limitations, the SRCA spectral characterization is only performed for the (RSB). For the same reason, the SRCA radiometric mode also applies only to the RSB. During pre-launch spatial and spectral characterization, the SRCA was operated under the same test conditions as the IAC and SpMA and at nearly the same time. The SRCA data was compared with that from the IAC and SpMA so that the high quality ground measurements could be transferred to the SRCA for later on-orbit spatial and spectral monitoring (Barnes et al., 1998; Xiong et al., 2002c; Che et al., 2003).

5.2.4 Pre-Launch Calibration and Characterization Summary

Table 5.1 summarizes some of the key pre-launch calibration and characterization activities described in Sections 5.2.1 to 5.2.3. The MODIS Characterization Support Team (MCST), in support of the MODIS science team leader and team members, performed most of the pre-launch data analyses and instrument performance evaluation. The instrument calibration and characterization results were presented at MODIS Science Team meetings and several calibration workshops. They were also used to support the L1B algorithm design and development. For the reflective solar bands, a linear approach was chosen for the radiometric calibration. For the thermal emissive bands, a quadratic algorithm was used. MCST is also responsible for the design and development of the MODIS Level-1B algorithms (Volume, Chapter 3). In addition to those listed in Table 5.1, many other tests were also performed pre-launch at either the component, sub-system, or system level to fully characterize the MODIS sensors. We cannot emphasize enough the importance of instrument pre-launch calibration

Table 5.1 MODIS pre-launch calibration and characterization activities

Radiometric	Detectors' gains (or responses), non-linearity, and noise characterization at cold, nominal, and hot instrument temperature plateaus	TEB and RSB
	Blackbody emissivity	TEB
	Solar diffuser Bi-directional Reflectance Factor	RSB
Spectral	Relative Spectral Response (RSR)	TEB and RSB
	Out-of-Band Response	TEB and RSB
Spatial	Band-to-Band Registration	TEB and RSB
	Modulation Transfer Function	TEB and RSB
	Instantaneous Field of View	TEB and RSB
Others	Polarization sensitivity	RSB
	Response Versus Scan angle	TEB and RSB

and characterization as some of the on-orbit calibration and characterization activities depend heavily on input from pre-launch measurements (Barnes et al., 1998; Guenther et al., 1998).

5.3 MODIS On-Orbit Calibration and Characterization

The MODIS on-orbit calibration and characterization activities are performed using its on-board calibrators shown in Figs. 5.2 – 5.4. The SD/SDSM system is used for RSB radiometric calibration, the OBC BB for TEB radiometric calibration, and the SRCA for spatial (RSB and TEB) and spectral (RSB only) characterization.

The RSB SD/SDSM calibration was performed weekly during the instruments' first year of on-orbit operation and then reduced to a biweekly schedule. This activity is controlled through uploaded ground commands provided by the MCST instrument operation team (IOT) with input from MCST calibration analysts. On the other hand, the TEB calibration is performed on a scan-by-scan basis. Table 5.2 lists key MODIS on-orbit radiometric, spatial, and spectral calibration and characterization activities and their execution frequencies.

Table 5.2 MODIS on-orbit calibration and characterization activities and schedules

SD and SDSM Observations	Bi-weekly (weekly during the first year)	RSB
Blackbody warm-up and cool-down (TEB)	Quarterly	TEB
SRCA radiometric mode	Monthly	RSB
SRCA spatial mode	Bi-monthly	TEB and RSB
SRCA spectral mode	Quarterly	RSB
Lunar observations (with spacecraft roll maneuvers)	Monthly	RSB and TEB
Yaw maneuvers to characterize BRF and SD screen vignetting function	Special request	RSB
Pitch maneuvers to characterize RVS	Special request	TEB

5.3.1 Reflective Solar Bands Calibration Algorithm and Performance

The top of the atmosphere scene reflectance factor is the primary L1B data product for the MODIS reflective solar bands. Its on-orbit calibration is based on solar observations via the instrument's on-board solar diffuser. The SD bi-directional reflectance factor was characterized pre-launch using NIST traceable reference samples. The Earth view reflectance factor, $\rho_{EV}\cos(\theta_{EV})$, is related to the detector's response and given by (Xiong et al., 2002a, 2003a):

$$\rho_{EV}\cos(\theta_{EV}) = m_1 \cdot \mathrm{d}n_{EV}^{*} \cdot d_{ES}^{2} \tag{5.1}$$

where θ_{EV} is the solar zenith angle of the EV pixel, m_1 is a calibration coefficient determined from the SD calibration, d_{ES} is the Earth-sun distance (in AU) at the time of the EV scene observation, and $\mathrm{d}n_{EV}^{*}$ is the sensor's digital response to the EV scene with the instrument background subtracted, the effect of the sensor's viewing angle normalized, and the instrumental temperature effect corrected. It is evaluated by the following expression:

$$\mathrm{d}n_{EV}^{*} = \mathrm{d}n_{EV} \cdot (1 + k_{inst} \cdot \Delta T_{inst}) / \mathrm{RVS}_{EV} \tag{5.2}$$

where k_{inst} is the instrument temperature correction coefficient, ΔT_{inst} is the difference between the instrument temperature and its reference value, and RVS_{EV} is the scan mirror's response versus scan angle at the EV pixel's angle of incidence. The T_{inst} reference value was chosen pre-launch and used for deriving the temperature correction coefficient k_{inst}.

The dark background subtracted response, $\mathrm{d}n_{EV}$, is computed by

$$\mathrm{d}n_{EV} = \mathrm{DN}_{EV} - \langle \mathrm{DN}_{SV} \rangle \tag{5.3}$$

where DN_{EV} is the EV digital response (raw data) and $\langle \mathrm{DN}_{SV} \rangle$ is the average of the SV digital response.

The SD calibration coefficient m_1 is determined by

$$m_1 = \frac{\rho_{SD}\cos(\theta_{SD})}{\mathrm{d}n_{SD}^{*} \cdot d_{ES}^{2}} \cdot \varGamma_{SDS} \cdot \Delta_{SD} \tag{5.4}$$

where ρ_{SD} is the SD pre-launch BRF, θ_{SD} is the solar zenith angle at SD view, $\mathrm{d}n_{SD}^{*}$ is the corrected detector response to the SD using Eq. (5.2) with the EV subscript replaced by SD. The d_{ES} is the Earth-sun distance in AU at the time of the SD measurements. Except for the SD degradation factor, $\varDelta_{SD}$, and SD screen (SDS) vignetting function, $\varGamma_{SDS}$, Eq. (5.4) is exactly the same as Eq. (5.1) when it is applied to the SD observations. The SD reflectance deteriorates on-orbit due to its exposure to the sunlight. Its degradation, $\varDelta_{SD}$, is tracked by the Solar Diffuser Stability Monitor (SDSM) during each SD calibration. The SDSM alternately measures the response of the SD view and the response of the direct sun view through an attenuation screen of 1.44% nominal transmittance so that the responses from both views are closely matched. The SD degradation is monitored by 9 individually filtered detectors in the SDSM (Xiong et al., 2001). For high gain bands (bands 8 – 16), a retractable solar diffuser screen (SDS) is placed in front of the SD to attenuate the direct sunlight and to avoid saturation. The SDS attenuation is represented by the vignetting function Γ_{SDS}. For other bands (bands 1 – 7, 17 – 19, and 26) which are calibrated without the SDS, the vignetting function is set to unity. Generally two SD calibration events, one with the SDS open and one with the SDS closed, are performed to calibrate the reflective solar bands.

MODIS L1B also produces a radiance product for the RSB. From the reflectance factor, the Earth view radiance can be calculated by

$$L_{\text{EV}} = e_{\text{Sun}} \cdot \rho_{\text{EV}} \cos(\theta_{\text{EV}}) / d_{\text{ES}}^2 \tag{5.5}$$

where e_{Sun} is the solar irradiance normalized with a factor of π at $d_{\text{ES}} = 1$ AU and is written as a global attribute in the L1B product so that the users can readily convert the RSB reflectance product to the radiance product.

Figure 5.5 shows examples of RSB response (gain) trending derived from Terra MODIS using three years of on-orbit SD calibration data sets. The quantities plotted are inversely proportional to the m_1 given in Eq. (5.4) and are corrected for SD degradation and normalized to the beginning of the mission. The major discontinuities in the response trending are due to on-orbit changes of the operational configuration. The nadir aperture door (NAD) of Terra MODIS was first opened on day 55 (of year 2000). On day 304, its configuration was switched from the A-side electronics to B-side and on day 532 it was switched

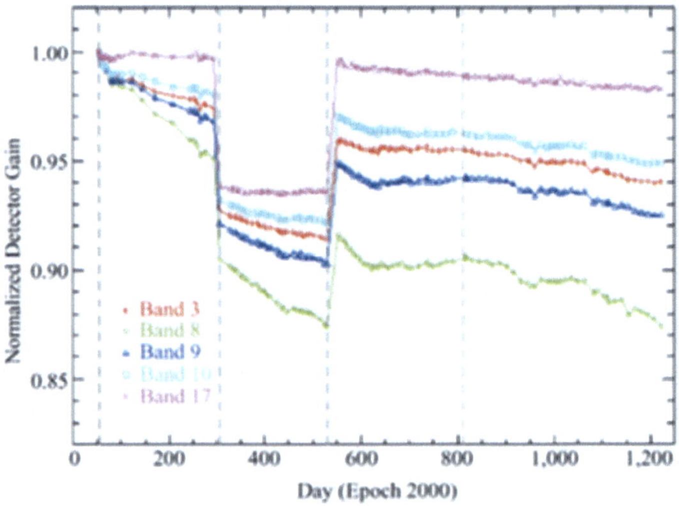

Figure 5.5 Terra MODIS RSB response trending for bands 8 (412 nm), 9 (442 nm), 3 (466 nm), 10 (487 nm), and 17 (904 nm)

back to the A-side. Starting from day 813, the A-side electronics was used together with Formatter B. The trending results show a clear wavelength dependent degradation for the Terra MODIS RSB; the shorter the wavelength the larger its degradation.

5.3.2 Thermal Emissive Bands Calibration Algorithm and Performance

The 16 thermal emissive bands, with a total of 160 individual detectors, are calibrated on-orbit using a quadratic fit of digital responses to the OBC BB radiances. Each scan the sensor views the OBC BB and deep space through the SV port for calibration, and then the EV for scene observations. The OBC BB view provides a known input radiance and the SV provides instrument thermal background. The calibration equation is given by (Xiong et al., 2002b)

$$\mathrm{RVS_{BB}} \cdot \varepsilon_{\mathrm{BB}} \cdot L_{\mathrm{BB}} + (\mathrm{RVS_{SV}} - \mathrm{RVS_{BB}}) \cdot L_{\mathrm{SM}} + \mathrm{RVS_{BB}} \cdot (1 - \varepsilon_{\mathrm{BB}}) \cdot \varepsilon_{\mathrm{CAV}} \cdot L_{\mathrm{CAV}} = a_0 + b_1 \cdot dn_{\mathrm{BB}} + a_2 \cdot dn_{\mathrm{BB}}^2 \quad (5.6)$$

where the first term on the LHS accounts for the radiance contribution of the BB (L_{BB}), the second term is from the scan mirror (L_{SM}), and the third term is from the instrument scan cavity (CAV). The response versus scan angle (RVS) is considered in all the source terms. The emissivities of the BB and the scan cavity, $\varepsilon_{\mathrm{BB}}$ and $\varepsilon_{\mathrm{CAV}}$, were determined during pre-launch calibration and characterization. The offset term a_0 and nonlinear term a_2 are updated periodically using on-orbit BB warm-up and cool-down cycles. The BB cycle provides the TEB responses over BB temperatures (or radiances) from 270 K to 315 K. The dominant linear coefficient b_1 of each detector is determined every scan. The instrument background subtracted response dn_{BB} is computed each scan from the average of 50 frames of BB and SV sector data,

$$dn_{\mathrm{BB}} = \langle DN_{\mathrm{BB}} \rangle - \langle DN_{\mathrm{SV}} \rangle. \quad (5.7)$$

The source radiance, L_{S} (S = BB, SM, CAV), is determined from the source temperature, T_{S}, using Planck's equation averaged over the relative spectral response of each detector,

$$L_{\mathrm{S}}(\lambda, T_{\mathrm{S}}) = \frac{\sum \mathrm{RSR}(\lambda) \cdot \mathrm{Planck}(\lambda, T_{\mathrm{S}})}{\sum \mathrm{RSR}(\lambda)} \quad (5.8)$$

Similarly the EV scene radiance retrieval equation can be written as,

$$\mathrm{RVS}_{\mathrm{EV}} \cdot L_{\mathrm{EV}} + (\mathrm{RVS}_{\mathrm{SV}} - \mathrm{RVS}_{\mathrm{EV}}) \cdot L_{\mathrm{SM}} = a_0 + b_1 \cdot dn_{\mathrm{EV}} + a_2 \cdot dn_{\mathrm{EV}}^2 \qquad (5.9)$$

notice the scan mirror contribution varies with the Earth scene AOI. Since the dominant linear term b_1 is computed for each scan, there is no need for additional temperature correction. Figure 5.6 illustrates the sensor's short-term stability using Terra MODIS detectors' scan-by-scan responses over an entire orbit of about 100 minutes, including both day and night. The long-term stability is shown in Fig. 5.7 using on-orbit response trending where each point is averaged over a 5-min granule. All the thermal emissive bands are located on the cold FPAs (SMIR and LWIR) controlled by a radiative cooler.

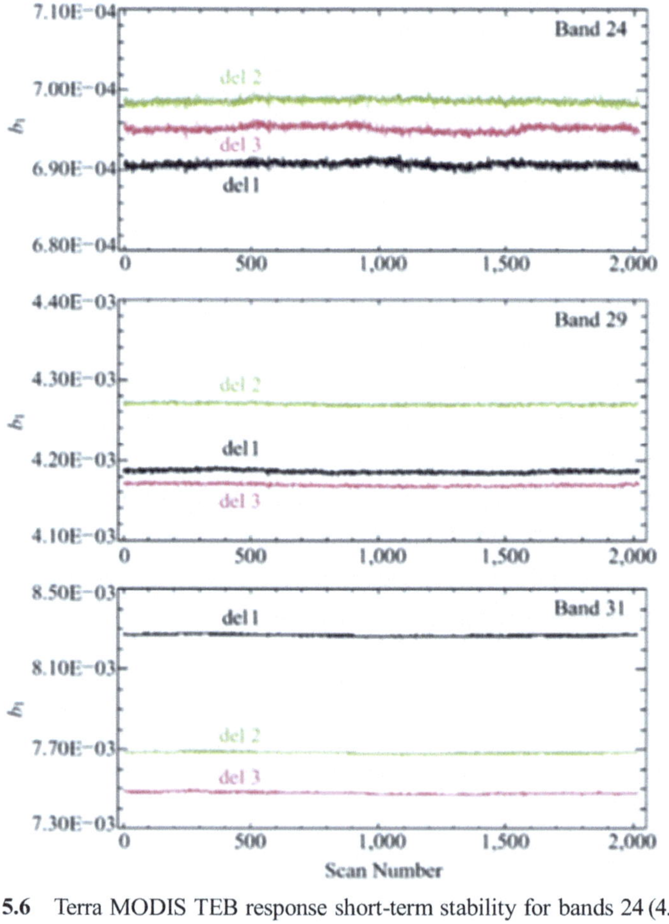

Figure 5.6 Terra MODIS TEB response short-term stability for bands 24 (4.47 μm), 29 (8.55 μm), and 31 (11.03 μm): one orbit (2003204 from 00:00 to 01:40 GMT) scan-by-scan linear response b_1

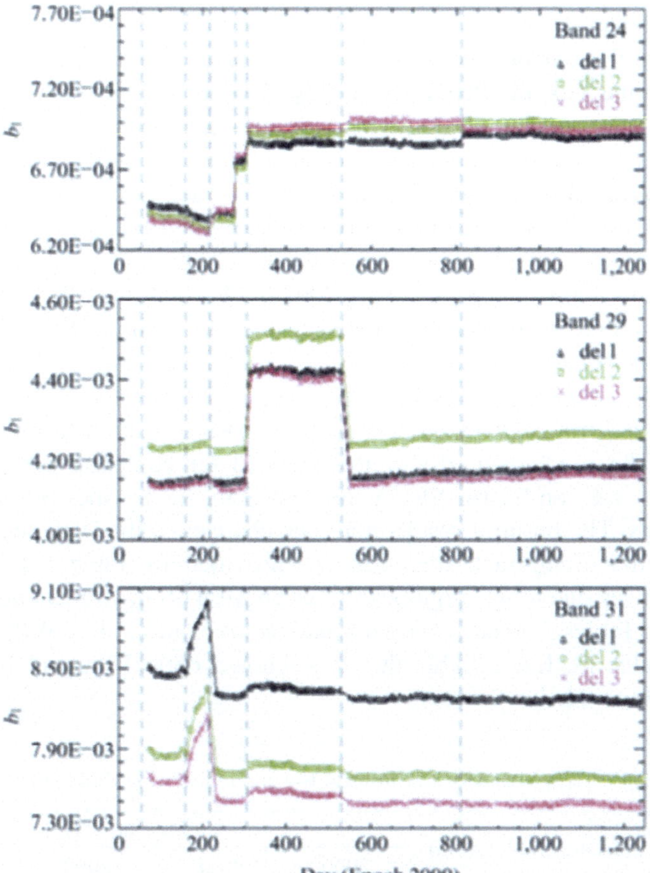

Figure 5.7 Terra MODIS TEB response long-term stability for bands 24 (4.47 μm), 29 (8.55 μm), and 31 (11.03 μm)

Shortly after its launch, the Terra MODIS' cooler margin dropped substantially which led to marginal control of the cold FPAs. During this period (day 160 – 218 of 2000), there were large variations in the thermal emissive bands responses, especially the PC bands as they are more sensitive to the FPA temperature. The configuration changes discussed earlier in the RSB examples also impact the TEB detectors' responses. For the TEB, the L1B data product is the EV top-of-the-atmosphere radiance.

5.3.3 On-Orbit Spatial and Spectral Characterization

The instrument on-orbit spatial (RSB and TEB) and spectral (RSB only) characterization is provided by the SRCA. In spectral mode, the SRCA subsystems include a light source, a monochromator, and a collimator. While in the spatial mode, the monochromator becomes a relay system with the grating replaced by a flat mirror. The light sources consist of an integration sphere with embedded lamps (for the RSB) and a glow bar (for the TEB). The SRCA uses two reticles in the spatial mode: one to measure detector to detector shift along-scan and the other to measure the band centroid shift along-track. Since there is no physical mirror scanning in the along-track direction, the reticle used for the along-track centroid shift has two offset along-track openings. A special algorithm was developed to determine the band centroid position along-track when using this approach (Xiong et al., 2002c). On-orbit, the SRCA measures the BBR that is corrected for the bias between the IAC and the SRCA determined from pre-launch characterization.

Table 5.3 presents the yearly averaged BBR results in the along-scan direction from Terra MODIS on-orbit spatial characterization from year 2000 to year 2003. The pre-launch results are included for comparison purposes. The results from the on-board SRCA are normalized to the pre-launch IAC measurements. The on-orbit results track the changes of the BBR over time. The changes in the along-track direction are provided in Table 5.4. To remove possible changes due to misalignment between MODIS instrument and its SRCA, the results in Tables 5.3 and 5.4 are normalized to band 1. The SWIR band (5 – 7 and 26) results are less reliable due to a known optical thermal leak and are therefore not included in the tables.

Table 5.3 Terra MODIS on-orbit band-to-band registration shifts (along scan)

Band	Pre-Launch Shifts (m)	On-Orbit Shifts (m)			
		2000	2001	2002	2003
1	0	0	0	0	0
2	18	– 1	– 1	– 1	– 1
3	– 35	– 6	0	– 5	– 5
4	– 47	6	6	3	0
8	– 30	– 4	1	– 2	– 1
9	– 30	– 18	– 13	– 15	– 15
10	– 45	– 44	– 43	– 45	– 47
11	– 44	21	21	18	17
12	2	64	66	64	95
13	– 8	– 28	– 28	– 28	– 16
14	17	27	26	26	25
15	16	20	19	20	16
16	33	27	26	27	23
17	32	40	40	40	34
18	– 33	– 37	– 36	– 37	– 36
19	– 51	– 58	– 57	– 57	– 56
20	– 22	– 60	– 59	– 61	– 67

Continued

Band	Pre-Launch Shifts (m)	On-Orbit Shifts (m)			
		2000	2001	2002	2003
21	– 101	– 7	– 5	– 7	– 8
22	– 59	– 74	– 67	– 65	– 70
23	– 96	– 48	– 27	– 27	– 32
24	10	– 17	– 4	– 6	– 14
25	23	– 6	4	3	– 5
27	– 66	– 103	– 108	– 111	– 115
28	– 55	– 107	– 110	– 111	– 117
29	– 65	– 70	– 78	– 79	– 84
30	– 55	– 169	– 177	– 176	– 178
31	– 46	– 7	– 1	– 1	– 6
32	24	99	105	104	98
33	– 46	– 50	– 48	– 47	– 53
34	– 64	– 38	– 30	– 42	– 40
35	– 57	– 26	– 19	– 22	– 26
36	– 41	– 66	– 58	– 63	– 62

Table 5.4 Terra MODIS on-orbit band-to-band registration (BBR) shifts (along track)

Band	Pre-Launch Shifts (m)	On-Orbit Shifts (m)			
		2000	2001	2002	2003
1	0	0	0	0	0
2	– 1	– 8	– 8	– 8	– 8
3	56	– 2	0	– 1	2
4	67	– 29	– 27	– 28	– 26
8	35	3	5	5	7
9	57	18	20	20	22
10	47	27	30	29	30
11	69	– 56	– 50	– 53	– 52
12	57	– 69	– 65	– 66	– 69
13	– 6	– 19	– 20	– 20	– 20
14	– 5	16	16	17	17
15	– 12	24	24	25	25
16	7	31	31	32	31
17	14	38	38	38	38
18	14	– 40	– 41	– 41	– 41
19	27	– 49	– 50	– 50	– 50
20	18	– 116	– 117	– 114	– 112
21	– 28	– 133	– 130	– 131	– 129
22	– 53	– 133	– 130	– 127	– 125

Continued

Band	Pre-Launch Shifts (m)	On-Orbit Shifts (m)			
		2000	2001	2002	2003
23	– 12	– 135	– 134	– 130	– 128
24	– 5	– 90	– 89	– 86	– 84
25	24	– 71	– 74	– 72	– 71
27	– 87	– 7	– 6	– 3	– 1
28	– 116	– 1	1	4	6
29	– 129	6	8	11	12
30	– 129	25	30	36	38
31	– 70	– 59	– 56	– 53	– 52
32	– 81	– 63	– 61	– 58	– 57
33	– 48	23	25	27	28
34	– 45	31	33	36	38
35	1	5	6	8	9
36	– 31	– 3	– 2	2	3

With the grating in place, the SRCA is used for spectral characterization. In addition to normalizing the results to the pre-launch SpMA measurements, the SRCA is capable of on-orbit wavelength self-calibration. This is achieved through the use of a didymium glass filter with sharp peaks at known wavelengths. Two silicon photodiodes (SiPD) are used in a ratioing mode for the wavelength calibration by comparing the observed SRCA profiles in the spectral calibration with the known didymium profiles.

Although the SRCA spectral resolution is lower than that of the ground measurements due to the use of a slit wider than that of the SpMA, it is still adequate for monitoring the RSB detectors' spectral response changes on-orbit. For Terra MODIS, the center wavelength shifts (band averaged) from existing on-orbit measurements are reported in Table 5.5. These shifts are relative to the pre-launch SRCA results. The center wavelengths listed in Table 5.5 were determined from pre-launch SpMA characterization. The center wavelength values determined from on-orbit SRCA measurements must include the corrections to transfer the SpMA reference to the SRCA. Due to SRCA source limits, on-orbit spectral characterization is performed only for the RSB.

5.3.4 Special Considerations and Activities

In addition to the on-orbit radiometric calibration and spatial and spectral characterization discussed above, many special activities were performed to enhance the instruments' calibration (Xiong et al., 2002d, 2003c; 2003d; 200e; Sun et al., 2003). These on-orbit activities include but are not limited to

(1) SD BRF validation and SDS vignetting function characterization using SD observations during spacecraft yaw maneuvers (RSB);

Table 5.5 On-orbit center wavelength shifts for Terra MODIS reflective solar bands

Band	Center Wavelength (nm)	Center Wavelength Shift (nm)			
	(Pre-Launch)	2000	2001	2002	2003
1	646.3	0.3	0.2	0.2	0.2
2	856.4	0.2	0.1	0.1	0.0
3	465.7	0.1	0.0	− 0.1	− 0.1
4	553.7	0.1	0.0	0.0	0.0
8	411.9	− 0.6	− 0.6	− 0.6	− 0.6
9	442.2	− 0.3	− 0.4	− 0.5	− 0.5
10	487.0	− 0.3	− 0.3	− 0.4	− 0.4
11	529.8	− 0.1	− 0.2	− 0.3	− 0.3
12	546.9	0.0	− 0.1	− 0.1	− 0.2
13	665.7	0.0	− 0.1	− 0.3	− 0.3
14	676.9	0.2	0.1	0.0	− 0.1
15	746.5	− 0.1	− 0.1	− 0.3	− 0.2
16	866.4	− 0.2	− 0.3	− 0.4	− 0.4
17	904.2	0.0	0.0	− 0.1	0.0
18	935.7	− 0.2	− 0.3	− 0.5	− 0.5
19	936.3	− 0.2	− 0.2	− 0.3	− 0.3

(2) Radiometric stability monitoring using monthly lunar observations through the SV port with spacecraft roll maneuvers (RSB);

(3) RVS characterization using data sets from scanning the closed nadir aperture door (TEB);

(4) RVS characterization using deep space view observations during spacecraft pitch maneuvers (Terra only, TEB);

(5) PC optical leak and SWIR crosstalk characterization using lunar observations (RSB and TEB);

(6) BBR characterization using lunar observations (RSB);

(7) SWIR bands data collected during spacecraft night to characterize thermal optical leak.

In addition to the general algorithms described above, a number of correction algorithms designed to deal with the sensors' special features have been implemented into the MODIS L1B (Guenther et al., 1998; 2002; Xiong et al., 2002a; 2002b), such as

(1) B21 (fire band with extremely lower gain) calibration algorithm with fixed calibration coefficients (Terra and Aqua);

(2) PC bands optical leak correction algorithm (Terra);

(3) SWIR bands thermal leak correction algorithm (Terra and Aqua);

(4) Bands 33, 35, and 36 calibration algorithm using pre-determined calibration coefficients when the BB temperature (during its warm-up and cool down cycle) is above their saturation limits (Aqua).

5.4 Summary

The MODIS sensors provide a broad range of radiometric data for the science community to study global and regional environmental and climate changes with both morning and afternoon observations. Because of its broad spectral coverage and wide range of viewing angles, MODIS is a very complicated instrument compared to its heritage sensors. It is equipped with a number of on-board calibrators for its on-orbit calibration and characterization. Both Terra MODIS (PFM) and Aqua MODIS (FM1) went through a series of comprehensive calibration and characterization activities pre-launch. There were a number of improvements made in the FM1 design as well as in its pre-launch thermal vacuum radiometric calibration procedures and processes based on the lessons learned from the PFM characterization that have led to better on-orbit performance. These improvements include a reduction of PC optical leak from band 31 to bands 32 – 36, a reduction of electronic cross-talk in the SWIR bands, a reduction of scattered light for the bands on the NIR PFA, and a better characterization of TEB response versus scan-angle. This chapter provides an overview of the MODIS instrument calibration and characterization, pre-launch and on-orbit. It describes the L1B algorithms used for the reflective solar bands and the thermal emissive bands and describes overall sensor performance. The MODIS Characterization Support Team (MCST), working closely with the instrument vendor and science team members, has provided major pre-launch calibration and characterization data analyses, designed and implemented the L1B algorithms, and is continuously monitoring the instruments' on-orbit operation and performance characteristics. Major efforts have been made to assure the quality of the on-orbit calibration and characterization, including many special activities designed and implemented to improve and enhance on-orbit calibration capability. The experience gained during the MODIS design, calibration and characterization is providing invaluable input for the development of its follow-on, the Visible Infrared Imaging Radiometer Suite (VIIRS).

References

Barnes WL, Xiong X, Salomonson VV (2002) Status of Terra MODIS and Aqua MODIS. In: Proceedings of IGARSS

Barnes WL, Pagano TS, Salomonson VV (1998) Pre-launch characteristics of the Moderate Resolution Imaging Spectroradiometer (MODIS) on EOS AM-1. IEEE Transactions on Geoscience and Remote Sensing 36: 1,088 – 1,100

Che N, Xiong X, Barnes WL (2003) On-orbit spectral characterization results for Terra MODIS reflective solar bands. Proceedings of SPIE—Earth Observing Systems Ⅷ 5151: 367 – 374

Guenther B, Xiong X, Salomonson VV, Barnes WL, Young J, (2002) On-orbit Performance of the Earth Observing System Moderate Resolution Imaging Spectroradiometer; first year of data. Remote Sensing of the Environment 83: 16 – 30

Guenther B, Godden GD, Xiong X, Knight EJ, Montgomery H, Hopkins MM, Khayat MG, Hao Z (1998) Pre-launch algorithm and data format for the Level 1 calibration products for the EOS AM-1 Moderate Resolution Imaging Spectroradiometer (MODIS). IEEE Transactions on Geoscience and Remote Sensing 36: 1,142 – 1,151

NASA (1993) Specifications for the Moderate Resolution Imaging Spectroradiometer (MODIS). GSFC 422-20-02, Revision A

Salomonson VV, Barnes WL, Xiong X, Kempler S, Masuoka E (2002) An overview of the Earth Observing System MODIS instrument and associated data systems performance. In: Proceedings of IGARSS

Sun J, Xiong X, Guenther B, Barnes WL (2003) Radiometric stability monitoring of the MODIS reflective solar bands using the Moon. Metrologia 40: 85 – 88

Xiong X, Chiang K, Esposito J, Guenther B, Barnes WL (2003a) MODIS on-orbit calibration and characterization. Metrologia 40: 89 – 92

Xiong X, Chiang K, Sun J, Che N, Barnes WL (2003b) Aqua first year on-orbit calibration and performance. In: Proceedings of SPIE—Sensors, Systems, and Next Generation of Satellites IX 5,234

Xiong X, Salomonson VV, Chiang K, Wu A, Xiong S, Chen N, Barnes WL, Guenther B (2003c) On-orbit characterization of Terra MODIS thermal emissive bands response versus scan angle. In: Proceedings of CALCON

Xiong X, Sun J, Esposito J, Liu X, Barnes WL, Guenther B (2003d) On-orbit characterization of a solar diffuser's bi-directional reflectance factor using spacecraft maneuvers. In: Proceedings of SPIE—Earth Observing Systems Ⅷ 5151: 375 – 383

Xiong X, Chiang K, Adimi F, Li W, Yatagai H, Barnes WL (2003e) MODIS correction algorithm for out-of-band response in the short-wave IR bands. In: Proceedings of SPIE—Sensors, Systems, and Next Generation of Satellites Ⅸ, 5,234

Xiong X, Sun J, Esposito J, Guenther B, Barnes WL (2002a) MODIS reflective solar bands calibration algorithm and on-orbit performance. In: Proceedings of SPIE—Optical Remote Sensing of the Atmosphere and Clouds Ⅲ 4891: 95 – 104

Xiong X, Chiang K, Guenther B, Barnes WL (2002b) MODIS thermal emissive bands calibration algorithm and on-orbit performance. In: Proceedings of SPIE—Optical Remote Sensing of the Atmosphere and Clouds Ⅲ 4,891: 392 – 401

Xiong X, Che N, Adimi F, Barnes WL (2002c) On-orbit spatial characterizations for Terra MODIS. In: Proceedings of SPIE—Earth Observing Systems Ⅶ 4,814: 347 – 357

Xiong X, Sun J, Chiang K, Xiong S, Barnes WL (2002d) MODIS on-orbit characterization using the Moon. In: Proceedings of SPIE—Sensors, Systems, and Next Generation Satellites Ⅵ 4,881: 299 – 307

Xiong X, Esposito J, Sun J, Pan C, Guenther B, Barnes WL (2001) Degradation of MODIS optics and its reflective solar bands calibration. In: Proceedings of SPIE—Sensors, Systems, and Next Generation Satellite Ⅴ 4,540: 62 – 70

6 Use of the Moon for Calibration and Characterization of MODIS, SeaWiFS, and VIRS

William Barnes, Xiaoxiong Xiong, Robert Eplee, Junqiang Sun and Cheng-Hsuan Lyu

6.1 Introduction

The need for long term global data sets to enable quantitative studies of changes in the Earth's climate leads directly to a requirement for a radiometric calibration source that is stable over several decades (and possibly several centuries) to levels of better than one part in a thousand, that can be readily used on-orbit and that is suitable for inter-comparison of a wide variety of imagng radiometers. For the visible through shortwave-infrared portion of the electromagnetic spectrum the moon meets this requirement. It's surface is stable over thousands of years, the sun's illumination is well characterized, and, if care is taken to account for changes in phase and libration, it can be used for radiometric calibration, calibration transfer and intercomparison of sensors.

During the early phases of NASA's Earth Observing System (EOS) program in the mid to late 1980's, the need for long-term, on-orbit radiometric calibration was recognized. Plans were developed to acquire a lunar spectral irradiance data set over a period of several years with enough detail to enable the development of a radiometric model suitable for the on-orbit calibration of remote sensing systems operating in the visible, near and shortwave-infrared portions of the solar spectrum. At the same time, sensor developers were encouraged to include diffusers as the primary method for solar calibration together with the ability to periodically view the moon as a means for monitoring the diffuser's stability.

This chapter will examine how lunar data is being used by three different imaging radiometers, the results from each sensor and how these techniques will be used by the next generation of operational environmental satellites being developed by the joint DOD/NOAA/NASA Integrated Program Office (IPO). The three sensors that will be examined are the Moderate Resolution Imaging Spectroradiometer (MODIS) onboard both the EOS Terra and Aqua satellites, the Sea-viewing Wide-Field-of-view Sensor (SeaWiFS) onboard the Orbview 1 satellite and the Visible and Infrared Scanner (VIRS) onboard the Tropical Rainfall Measuring Mission (TRMM).

6.1.1 The Lunar Radiometric Model

A program to characterize the brightness of the moon for the on-orbit calibration of Earth remote sensing imaging instruments has been established by the US Geological Survey (USGS) (Stone and Kieffer, 2002; Kieffer et al., 2003). The basis for this program is the Robotic Lunar Observatory (ROLO), an automated observatory dedicated to the radiometry of the Moon located in Flagstaff, Arizona. ROLO has been observing the moon in the visible and near-infrared wavelengths (347 – 945 nm) since 1995 and in the short-wave infrared wavelengths (945 – 2,390 nm) since 1997. The USGS has developed a spectral irradiance model of the moon that accounts for variations with lunar phase, lunar librations, and the location of the Earth-orbiting spacecraft. The model uses the star Vega for absolute scale and uses measurements of returned Apollo lunar samples as spectral reference standards.

Because of the strong dependence of lunar irradiance on the observing geometry, measurements by two spacecraft cannot be directly compared unless acquired at the same time and location. However, the intrinsic stability of the lunar surface photometric properties means that the lunar irradiance model can be applied to observations made at any time. This has the potential of interrelating the calibration scales of all spacecraft that view the moon, particularly if the instruments make nearly simultaneous observations (Kieffer et al., 2003). In the cases where lunar observations by single instruments have extended over several years, the instrument response degradations have been determined with precision of about 0.1% per year (Barnes et al., 2004; Eplee et al., 2004).

6.1.2 MODIS

The Moderate Resolution Imaging Spectroradiometer (MODIS) is one of the key instruments for the NASA's Earth Observing System missions. The MODIS design was based on the heritage of a number of earlier sensors and the need for improved spectral, spatial, and temporal resolutions. The Protoflight Model (PFM) was launched on Dec. 18, 1999 on-board the EOS Terra spacecraft and the Flight Model 1 (FM1) was launched on May 4, 2002 on-board the EOS Aqua spacecraft. Together they have produced more than six and a half years (four and a half years from Terra MODIS and two years from Aqua MODIS) of calibrated data for scienctific studies of the Earth's climate and emvironmental changes.

Terra and Aqua MODIS (PFM and FM1) are nearly identical sensors consisting of 36 spectral bands with center wavelengths from 0.41 to 14.4 μm located on four focal plane assemblies (FPAs): visible (VIS), near infrared (NIR), short- and mid-wave infrared (SMIR), and long-wave infrared (LWIR). Table 6.1 provides a summary of key design paramters for these bands, including their primary applications, wavelengths, number of along-track detectors per band,

Table 6.1 MODIS bands with principal uses, bandwidths, number of detectors along track, nadir instantaneous field of view (IFOV) and FPA

Primary Use	Band	λ (μm)	degree(°)	IFOV(m)	FPA
Land/Cloud/Aerosols Boundaries	1	0.620 – 0.670	40	250	NIR
	2	0.841 – 0.876	40	250	NIR
Land/Cloud/Aerosols Properties	3	0.459 – 0.479	20	500	VIS
	4	0.545 – 0.565	20	500	VIS
	5	1.230 – 1.250	20	500	SMIR
	6	1.628 – 1.652	20	500	SMIR
	7	2.105 – 2.155	20	500	SMIR
Ocean Color/ Phytoplankton/ Biogeochemistry	8	0.405 – 0.420	10	1,000	VIS
	9	0.438 – 0.448	10	1,000	VIS
	10	0.483 – 0.493	10	1,000	VIS
	11	0.526 – 0.536	10	1,000	VIS
	12	0.546 – 0.556	10	1,000	VIS
	13	0.662 – 0.672	10	1,000	NIR
	14	0.673 – 0.683	10	1,000	NIR
	15	0.743 – 0.753	10	1,000	NIR
	16	0.862 – 0.877	10	1,000	NIR
Atmospheric Water Vapor	17	0.890 – 0.920	10	1,000	NIR
	18	0.931 – 0.941	10	1,000	NIR
	19	0.915 – 0.965	10	1,000	NIR
Surface/Cloud Temperature	20	3.660 – 3.840	10	1,000	SMIR
	21	3.929 – 3.989	10	1,000	SMIR
	22	3.929 – 3.989	10	1,000	SMIR
	23	4.020 – 4.080	10	1,000	SMIR
Atmospheric Temperature	24	4.433 – 4.498	10	1,000	SMIR
	25	4.482 – 4.549	10	1,000	SMIR
Cirrus Clouds Water Vapro	26	1.360 – 1.390	10	1,000	SMIR
	27	6.535 – 6.895	10	1,000	LWIR
	28	7.175 – 7.475	10	1,000	LWIR
Cloud Properties	29	8.400 – 8.700	10	1,000	LWIR
Ozone	30	9.580 – 9.880	10	1,000	LWIR
Surface/Cloud Temperature	31	10.780 – 11.280	10	1,000	LWIR
	32	11.770 – 12.270	10	1,000	LWIR
Cloud Top Altitude	33	13.185 – 13.485	10	1,000	LWIR
	34	13.485 – 13.785	10	1,000	LWIR
	35	13.785 – 14.085	10	1,000	LWIR
	36	14.085 – 14.385	10	1,000	LWIR

nadir spatial resoultions, and locations on the focal plane assemblies (FPAs). MODIS bands 1 – 19 and 26 are the reflective solar bands (RSBs) collecting only daytime Earth scene data and bands 20 – 25 and 27 – 36 are the thermal emissive bands (TEBs) making measurements both day and night. A more detailed description of the MODIS sensor can be found in Volume Ⅰ: Sections 2.4 and 4.1. Figures 2.1 and 4.1 – 4.4 illustrate the MODIS design.

The MODIS is a sophisticated instrument that accommodates of a broad range of science applications. For on-orbit calibration and characterization purposes, each MODIS instrument is equipped with a number of calibration sub-systems, including a solar diffuser (SD), a solar diffuser stability monitor (SDSM), a blackbody (BB), and a spectro-radiometric calibration assembly (SRCA). In general, the SD/SDSM system is used for the RSB calibration on a weekly (first year) or bi-weekly (subsequent years) schedule. The SDSM is primarily used to monitor any SD degradation. The BB is used for the TEB calibration on a scan-by-scan basis. The SRCA is used to track instrument spectral (quarterly) and spatial (bi-monthly) performance. In addition to using the instrument's on-board calibrators (OBCs), both Terra and Aqua MODIS have aquired near-monthly lunar observations in support of on-orbit calibraion and characterization. In order to simplify monitoring RSB radiometric calibration stability, the phase angle of each lunar observation is set to be nearly the same (range of 55°-56°), thereby minimizing the corrections to the lunar data.

6.1.3 SeaWiFS

The Sea-viewing Wide Field-of-view Sensor (SeaWiFS) has been providing daily global imagery of the world's oceans since September 1997. The goal of the SeaWiFS Project is to produce an ocean-color data set with 5% absolute and 1% relative accuracies on the water-leaving radiances and 35% accuracy on the chlorophyll a concentrations in open-ocean regions. SeaWiFS is an eight-band, visible and near-infrared scanning radiometer designed for high radiometric sensitivity over the ocean without saturating over bright clouds. The SeaWiFS bands are provided in Table 6.2. Each band is comprised of three high sensitivity ocean detectors and one lower sensitivity cloud detector. For normal operations, the output from the four detectors in each band is averaged together on orbit, resulting in a bilinear gain for the instrument. The bilinear gain allows measurements of clouds or land to be made without saturating the instrument, thus allowing stray light corrections for the ocean data and allowing land measurements at the radiometric sensitivity of the cloud detectors.

The SeaWiFS Project uses monthly lunar calibrations to monitor the on-orbit radiometric stability of SeaWiFS over the course of its mission. Ongoing analyses of the steadily increasing lunar calibration data set have led to improvements in the calibration methodology over time.

Table 6.2 SeaWiFS bands. The nominal center wavelengths and bandwidths are in nanometers

Band	1	2	3	4	5	6	7	8
Wavelength	412	443	490	510	555	670	765	865
Bandwidth	20	20	20	20	20	20	40	40

6.1.4 VIRS

The Visible and Infrared Scanner (VIRS) was launched onboard the Tropical Rainfall Measuring Mission (TRMM) on Nov. 28 1997. After 6.5 years the sensor is performing well and has shown only minor changes from it's at-launch configuration. VIRS is a 5-band cross-track scanning radiometer. Its design has several features similar to those of the MODIS including a two-sided scan mirror, reflective telescope, passive cooler and solar diffuser. The bands are centered at 0.62, 1.61, 3.78, 10.83 μm and 12.03 μm. The two reflected solar bands at 0.62 μm and 1.61 μm are calibrated via periodic views of the solar illuminated diffuser. Performance is further verified by observing the moon through a space view port several times a year.

The three sensors, MODIS, SeaWiFS, and VIRS, have a number of characteristics and design features in common. They were all designed and built by the same instrument vendor, Raytheon Santa Barbara Remote Sensing (SBRS) in Goleta, California. SeaWiFs is a sensor dedicated to ocean color and biochemistry applications. Most of its spectral bands are included in the MODIS instrument with nearly the same center wavelengths but with much narrower bandwidths. In addition to products similar to the ocean scientific parameters provided by SeaWiFS, the MODIS sensors produce many science products for land and atmospheric applications. The VIRS spectral bands are nearly identical to some of the MODIS bands. All three sensors were designed to include solar diffusers for on-orbit calibration of bands in the VIS, NIR, and SWIR regions.

The following sections provide an overview of luanr observations and data utilization approaches used by MODIS, SeaWiFS, and VIRS with emphasis on the results and their applications. Details are provided on how to use lunar observations to improve the sensors' on-orbit calibration and characterization and sensor calibration inter-comparisons. Specific techniques discussed include the use of lunar observations (MODIS) to monitor the time-dependent change of response vs. scan-angle (RVS), quantitative intercomparison of band responses of sensors in different orbits (Terra MODIS and SeaWiFS) by using the moon as a transfer medium, tracking changes in response (SeaWiFS) via lunar measurements, and the allocation of system level degradation between the sensor and its diffuser (VIRS).

6.2 Lunar Calibration and Characterization of MODIS

6.2.1 MODIS Lunar Calibration Approaches and Applications

The MODIS Level-1B (L1B) product for the reflective solar bands (RSBs) is the top of the atmosphere (TOA) reflectance factor, $\rho_{\mathrm{EV}}\cos(\theta_{\mathrm{EV}})$, which is related to the instrumet's response by

$$\rho_{\mathrm{EV}}\cos(\theta_{\mathrm{EV}}) = m_1(\mathrm{B,D,S,M})dn'_{\mathrm{EV}}(\mathrm{B,D,S,F,N})d_{\mathrm{SE}}^2/\mathrm{RVS}(\mathrm{B,M},\vartheta) \tag{6.1}$$

where ρ_{EV} is the Earth scene's bi-directional reflectance factor (BRF), θ_{EV} is the solar illumination angle of the Earth scene, B, D, S, M, F, and N are indexes for the band, detector, subframe (1 – 4 and 1 – 2 for 250 m and 500 m bands), mirror side, frame (1 – 1354 crosstrack pixels), and scan number, respectively, m_1 (B, D, S, M) is a calibration scaling factor, dn'_{EV} (B, D, S, F, N) is the Earth scene digital response corrected for offset and instrumental temperature effects, d_{SE} is the Earth-sun distance at the time of the measurement in astronomial units (AU), RVS(B, M, ϑ) is the response versus scan angle, and ϑ is the angle of incidence (AOI) on the scan mirror.

By applying Eq. (6.1) to the solar diffuser observations, the scaling factor can be derived by

$$m_1(\mathrm{B, D, S, M}) = \frac{\rho_{\mathrm{SD}}^{\mathrm{prelaunch}}\cos(\theta_{\mathrm{SD}})}{dn'_{SD}(\mathrm{B, D, S, F, N})d_{\mathrm{SE}}^2}\Gamma_{\mathrm{SDS}}\Delta_{\mathrm{SD}} \tag{6.2}$$

with the RVS having been normalized to the SD AOI ($\vartheta_{\mathrm{SD}} = 50.25$), where $\rho_{\mathrm{SD}}^{\mathrm{prelaunch}}$ is the pre-launch measured BRF of the SD, θ_{SD} is the solar illumination angle of the SD, Γ_{SDS} is the solar diffuser screen (SDS) vignetting (or transmision) with the SDS in place, and Δ_{SD} is the SD degradation factor. For SD observations without the SDS, $\Gamma_{\mathrm{SDS}} = 1$. Δ_{SD} is determined using the onboard solar diffuser stability monitor (SDSM).

The moon has a very stable reflectance. It can be used as a light source to calibrate the MODIS RSBs, similar to the SD. The difference between using the moon and the SD is that the moon is not a spatially uniform reflector while the SD (Esposito et al., 2004) has non-uniformities of less than 0.7% (only the SDS causes the SD to be locally non-uniformly illuminated). We use a summation of the instrument's responses from multiple pixels of a lunar view instead of the individual responses so as to minimize the effects lunar non-uniformity. Then we can define the lunar scaling factor as:

$$m_1^{\text{moon}}(\text{B, M}) = \bar{m}_1(\text{B, M}) / \text{RVS}(\text{B, M}, \vartheta_{\text{moon}})$$
$$\propto \Gamma(\varphi, \lambda_{\text{B}})[1 + L(\varphi, \alpha, \beta, \lambda_{\text{B}})] / d_{\text{SM}}^2 d_{\text{MM}}^2 \sum_{D,S,F} dn'_{\text{moon}}(\text{B, D, S, F, N}) \tag{6.3}$$

where $\bar{m}_1(\text{B, M})$ is the scaling factor averaged over detectors and subframes, ϑ_{moon} is the AOI ($= 11.4°$) when viewing the Space View (SV) port, through which MODIS observes the Moon, λ_{B} is the center wavelength of the band, φ is sun-moon-MODIS phase angle and $\Gamma(\varphi, \lambda_{\text{B}})$ describes the averaged effect of the illumination angle of the lunar surface, α and β are lunar longitude and latitude librations, respectively, $L(\varphi, \alpha, \beta, \lambda)$ represents the impact of the lunar librations on the observed response, d_{SM} is the sun-moon distance, and d_{MM} is the moon-MODIS distance. The first line in Eq. (6.3) shows the relationship among the defined lunar scaling factor, the SD scaling factor, and RVS at the AOI of the lunar view while the second line in the equation can be used to trend the lunar scaling factor from lunar observations.

The phase angle correction function, $\Gamma(\varphi, \lambda)$ for the wavelength region of interest has been modeled by Hapke and his coworkers (Hapke, 1963) (Hapke, 1966) and measured by Lane and Irvine (Lane and Irvine, 1973). It can also be derived from the results of Kieffer's lunar model. It is well known that $\Gamma(\varphi, \lambda)$ strongly depends on the phase angle φ. For both Terra and Aqua MODIS lunar calibrations, the phase angle is kept in the region $55° - 56°$ to reduce the impact of possible errors in the phase angle correction. $L(\varphi, \alpha, \beta, \lambda)$ depends on both phase angle and libration. In MODIS lunar calibrations, $L(\varphi, \alpha, \beta, \lambda)$ is derived from the results of Kieffer's model.

The difference between the lunar views and the SD observations is the AOI on the scan mirror. The calibration coefficients derived from the SD calibrations provide information on the entire electronic and optical system response for an $\text{AOI} = 50.25°$ while those derived from lunar views give the response for an $\text{AOI} = 11.4°$. From Eq. (6.3), we can write

$$RVS(\text{B, M}, \vartheta_{\text{moon}}) = RVS(\text{B, M}, \vartheta_{\text{moon}})_{t_0} / RVS_{\text{rel}}(\text{B, M}, \vartheta_{\text{moon}}) \tag{6.4}$$

where t_0 is any selected time, $RVS(\text{B, M}, \vartheta_{\text{moon}})_{t_0}$ is the RVS for the AOI of the lunar views at time t_0, and

$$RVS_{\text{rel}}(\text{B, M}, \vartheta_{\text{moon}}) = \frac{m_1^{\text{moon}}(\text{B, M}) / m_1^{\text{moon}}(\text{B, M})_{t_0}}{\bar{m}_1(\text{B, M}) / \bar{m}_1(\text{B,M})_{t_0}} \tag{6.5}$$

is the relative RVS. $\bar{m}_1(\text{B, M})_{t_0}$ and $m_1^{\text{moon}}(\text{B, M})_{t_0}$ in Eq. (6.5) are the detector and subframe averaged scaling factors and lunar scaling factors, respectively at time

t_0. If we select the time of the first on-orbit lunar observation as t_0, $RVS(\text{B}, \text{M})_{t_0}$ is the prelaunch measured *RVS*. Thus, the time-dependent *RVS* at AOI = 11.4° is determined from lunar views and SD/SDSM calibrations. With a linear assumption, the time-dependent relative *RVS* is expressed as

$$RVS_{\text{rel}}(\text{B}, \text{M}, \vartheta) = 1 + [RVS_{\text{rel}}(\text{B}, \text{M}, \vartheta_{\text{moon}}) - 1](\vartheta_{\text{SD}} - \vartheta)/(\vartheta_{\text{SD}} - \vartheta_{\text{moon}}) \tag{6.6}$$

Similarly, we can also derive a time dependent *RVS* in quadratic form by including the response from the SRCA at an AOI of 38°.

The lunar irradiance observed by MODIS is calculated by

$$I(B) = \sum_{D,F,S} \frac{E_{\text{Sun}}(\text{B}, \text{D}) m_1(\text{B}, \text{D}, \text{S}, \text{M}) dn'(\text{B}, \text{D}, \text{F}, \text{S}, \text{N})}{\pi RVS(\text{B}, \text{M}, \vartheta_{\text{moon}})(705 S_{\text{tot}}^{\text{B}})^2}, \tag{6.7}$$

where $E_{\text{Sun}}(\text{B}, \text{D})$ is the MODIS Relative Spectral Response (RSR) weighted solar irradiance, and w_B, the solid angle of each pixel of band B, is expressed as $1/(705 S_{\text{tot}}^{\text{B}})^2$. $S_{\text{tot}}^{\text{B}}$ is the number of subframes of band B, and 705 km is the nominal orbital altitude. Since the lunar irradiance depends on the lunar viewing geometry and the lunar librations, the observed lunar irradiance varies from time to time and osscilates annually.

Lunar irradiance provides a direct comparison among different instruments if they observe the moon with the same geometry, which depends on time and sensor position. It is impossible for two instruments to view the moon with exactly the same geometry, especially for Terra and Aqua MODIS since they view the moon at different phases. Terra views a waning moon while Aqua views a waxing moon. However, we can use Kieffer's lunar model as a bridge to compare Terra and Aqua performance by assuming the errors of the model are the same for a waning and waxing moon at a phase angle 55°. We can define

$$R(\text{B}) = \frac{\overline{I_{\text{Terra}}^{\text{observed}}(\text{B}) / I_{\text{Terra}}^{\text{model}}(\text{B})}}{\overline{I_{\text{Aqua}}^{\text{observed}}(\text{B}) / I_{\text{Aqua}}^{\text{model}}(\text{B})}} \tag{6.8}$$

where the ratio of the observed lunar irradiance and that calculated using Kieffer's model with the same observation geometry is averaged over all lunar observations for both instruments.

6.2.2 MODIS Lunar Calibration Results

Both Aqua and Terra view the moon approximately monthly using a roll maneuver in order to keep the sun-moon-MODIS phase angle in the range of 55° – 56°. The roll angle is limited to be between – 20° and 0° to avoid Earthshine heating of the radiative cooler. It is also required that the MODIS be on the night side of the Earth when the moon is viewed to avoid direct sunlight. With these requirements, each MODIS can view the moon about 9 times each year. To date,there have been 44 and 20 lunar calibrations for Terra and Aqua MODIS respectively.

Figure 6.1 shows the Terra MODIS lunar scaling coefficients, for MODIS bands B3, B8, B9, and B17. The Terra coefficients are normalized to its first lunar view implemented on Mar. 24, 2000. The discontinuities around day 305 (Oct. 31, 2000) and day 549 (July 2, 2001) in Fig. 6.1 are due to switching of the MODIS electronics from side A to the redundant side B and from side B back to side A. The scaling coefficient for B17 is relatively flat compared to those of the other three bands. Note that the Terra coefficients increase steadily (decreased responsitivity) and that the coefficient of B8 increases the most.

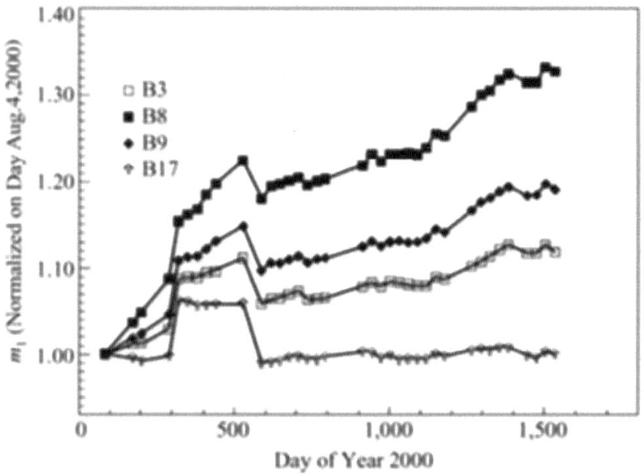

Figure 6.1 Terra band averaged lunar scaling factor for B3, B8, B9, and B17

Figure 6.2 shows Terra MODIS detector averaged SD scaling coefficients for bands B3, B8, B9, and B17. They were normalized to the first SD calibration of Terra MODIS, implemented on Feb. 24, 2000. From the plots in Figs. 6.1 and 6.2, it is seen that the averaged coefficients obtained from the lunar view increase faster than those obtained from the SD view, especially during the first three years on-orbit. For band 8, the SD scaling coefficient (m_1) has increased about 24.5% since Mar. 24, 2000, while that obtained from the lunar view has increased 33%. According to Eqs. (6.4) and (6.5), the difference in the coefficients reflects changes in the RVS, which is always normalized to the AOI of the SD. This indicates that the degradation of the Terra RVS is strongly AOI dependent and that the Terra B8 RVS decreases about 8.5% more at the small frame side (smaller AOIs) than that at the large frame side (larger AOIs).

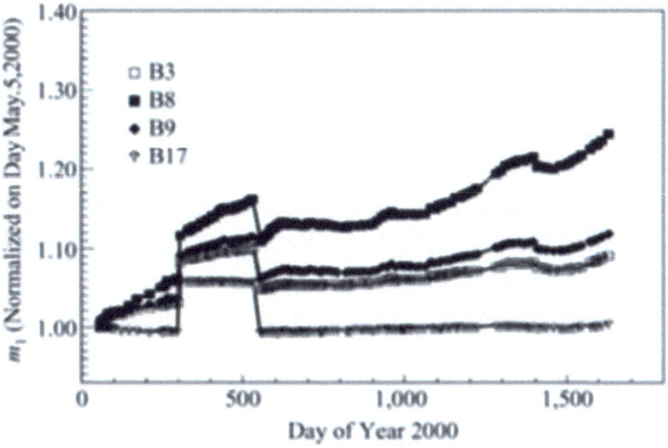

Figure 6.2 Terra band averaged SD scaling factor for B3, B8, B9, and B17

The Aqua lunar and SD observations have shown that the degradation of the Aqua RVS is also AOI dependent and that the pattern of the degradation is slightly different from that of the Terra RVS. At the beginning of the mission, the Aqua MODIS lunar scaling coefficients (m_1) increase slower than the SD scaling coefficients, which is opposite to Terra MODIS, but two years later, the lunar scaling coefficients increase faster than those of the SD, which is the same as Terra MODIS, two years late. Time-dependent RVS Look Up Tables (LUTs) derived from lunar and SD observations are presently being used with the Aqua calibrated, geo-located Level-1B (L1B) code and those derived from lunar, SRCA, and SD observatons are being used in the Terra L1B (the Aqua SRCA data is not stable enough to be used for the time dependent RVS).

Figure 6.3 shows the lunar irradinces observed by Terra B8 over the last four and a half years and those calculated using Kieffer's lunar model with the same view geometries. Both the observed and calculated lunar irradiances depend

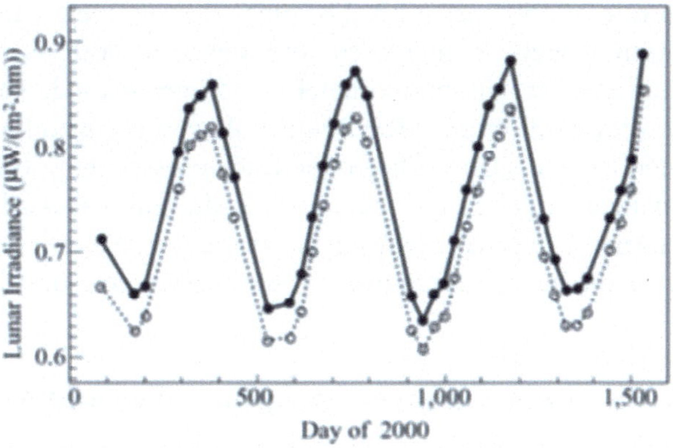

Figure 6.3 Lunar irradiance observed by Terra MODIS B8 (solid line and filled circles) and calculated by Kieffer's model (dashed lines and empty circles)

on the view geometries and osscilate with the seasons. They differ constantly by about 5%. The differences are due to both the errors in both the Terra MODIS measurements and the model.

Figure 6.4 shows the ratio defined in Eq. (6.8) for B1 – B4, B8 – B12, and B17 – B19. Other MODIS RSBs are either saturated when MODIS views the moon (B13 – B16) or have strong eletronic cross talk (B5 – B7 and B26). From Fig. 6.4, it is seen that Terra and Aqua MODIS are consistently within 1% except B2 (1.9%), B3 (2.3%) , and B8 (1.6%).

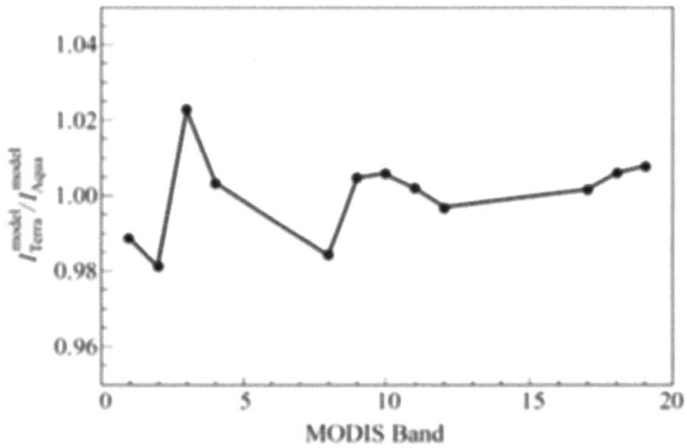

Figure 6.4 Terra and Aqua MODIS lunar irradiance comparison

6.3 Lunar Calibration and Characterization of SeaWiFS

SeaWiFS operates in a sun-synchronous polar orbit, crossing the equator from north to south at local noon. In normal operation the spacecraft is maintained in a nadir orientation, using pitch-axis momentum wheels for attitude control, with a spacecraft pitch rate of 360° per orbit. For lunar measurements, the rotation rate of the momentum wheels is increased and the spacecraft is pitched in the opposite direction at a rate of approximately 0.15° per second. The manuever is started after the spacecraft crosses the South Pole and is timed so that SeaWiFS will view the moon as the spacecraft ground track crosses the sub-lunar point. At the end of the maneuver, when the spacecraft again points toward the Earth, the pitch rate is returned to normal. During the maneuver, SeaWiFS is oriented so that it scans across the lunar surface from west to east in celestial coordinates.

6.3.1 SeaWiFS Lunar Calibration Approaches and Applications

For each lunar calibration, the radiances observed by SeaWiFS are integrated over the lunar images. The time series used to monitor the radiometric stability of the instrument are these integrated radiances for each band, normalized by the integrated radiances of the first calibration. The uncorrected time series are shown in Fig. 6.5. Periodic signals in the time series arise from variations in the geometry of the observations from one lunar calibration to the next. In order to track the radiometric stability of the instrument, the measurements must be normalized to a common viewing geometry. Geometric corrections are computed and applied for the sun – moon and instrument – moon distances, the oversampling of the lunar images in the along-track direction, the phase angles of the observations, and the libration angles of the observations. The current implementation (Eplee et al., 2004) of these corrections for SeaWiFS are improvements on previous geometry correction methodologies (Barnes et al., 1999, 2001).

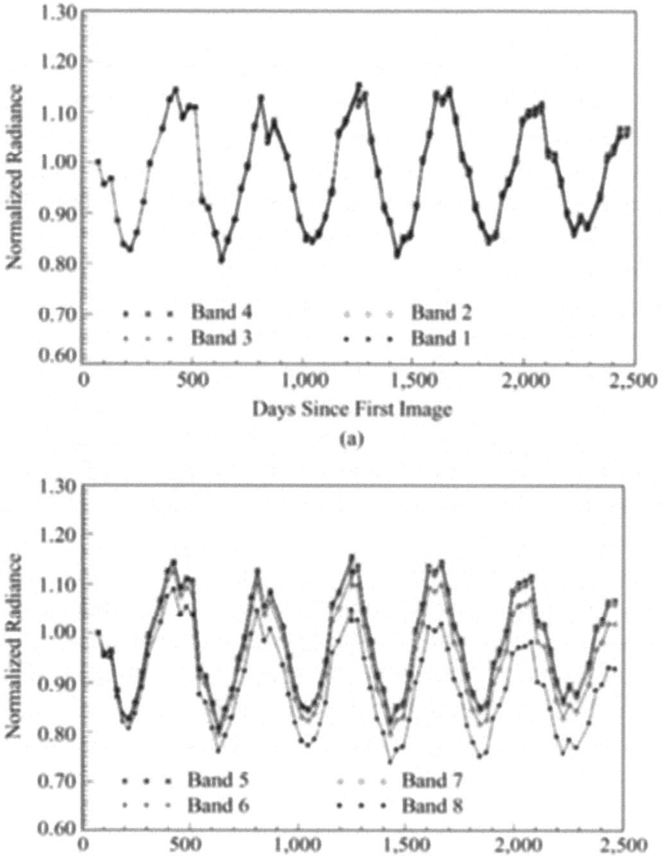

Figure 6.5 SeaWiFS Lunar Calibration Time Series. The periodicities in the uncorrected time series arise from variations in the observing geometry from one calibration to the next

The distance corrections normalize the observations to a sun – moon distance of one Astronomical Unit and to an instrument – moon distance of one mean earth – moon distance (384,401 km). The distance corrections remove the large-scale periodic signal from the time series.

During the lunar calibrations, the spacecraft pitches across the moon (in the along-track direction) at a slower rate than the scan rate of the instrument, resulting in an oversampled image of the moon. The oversampling correction compensates for the pitch rate across the moon. Since the pitch rate is not known during the lunar calibration maneuvers, the correction is computed by dividing the actual size of the moon, as seen from the spacecraft, by the apparent size of the moon in the along-track direction in the lunar image. The mean value of the oversampling correction over the SeaWiFS mission is 0.279 ± 0.007. The oversampling correction reduces some of the high-frequency periodic signal in the time series.

Operational considerations have caused lunar calibration to occur over a phase angle range of 5° – 10°. To date, all but six calibrations have occurred over a range of 6° – 8°, with a median phase of 6.95°. To compute the correction for each band, a quadratic function of the phase angle has been fit to the inverse of the lunar time series over a phase angle range of 4° – 11°. These fits have been normalized to yield values of unity at a phase angle of 7°.

The phase angle corrections become significant as the phase angle of the calibrations depart from 7°. Once the phase angle corrections have been applied, the dominant periodic signal remaining in the time series is due to the libration.

Libration effects are changes in the lunar radiance as seen from the spacecraft due to variation in the side of the moon that faces the Earth during the lunar observations. Commonly, the libration angles are defined to be the selenographic longitude and latitude of the sub-Earth point on the lunar surface. For correcting the lunar calibration time series, the libration angles are defined to be the coordinates of the sub-spacecraft point on the lunar surface. Since the amount of sunlight reflected by the lunar surface is a function of the position of the Sun in the lunar sky, the definition of libration angles is expanded to include the selenographic coordinates of the sub-solar points on the lunar surface. These four sets of coordinates constitute the libration angles for computing the libration effects in the SeaWiFS data. The libration effects are computed from simultaneous linear regressions of the libration angles against the phase-corrected lunar calibration time series for each band. The libration corrections are the inverses of the libration effects.

6.3.2 SeaWiFS Lunar Calibration Results

The geometry-corrected lunar radiances have had the distance, oversampling, phase angle, and libration corrections applied. The resulting time series has the form:

$$\frac{S(t,\alpha,l_{sc},b_{sc},l_{sun},b_{sun},\lambda)}{S(t_i,\alpha,l_{sc},b_{sc},l_{sun},b_{sun},\lambda)} = \frac{S_{obs}(t,\alpha,l_{sc},b_{sc},l_{sun},b_{sun},\lambda)}{S_{obs}(t_i,\alpha,l_{sc},b_{sc},l_{sun},b_{sun},\lambda)} \times f_1(t)\, f_2(t)\, f_3(t,\alpha,\lambda)\, f_4(t,l_{sc},b_{sc},l_{sun},b_{sun}) \tag{6.9}$$

where S—geometry-corrected integrated lunar radiances;
S_{obs}—observed integrated lunar radiances;
f_1—distance corrections;
f_2—oversampling corrections;
f_3—phase angle corrections;
f_4—libration corrections;
t—time of the observation;
t_i—time of the first observation;
λ—SeaWiFS band and center wavelength;
α—phase angle;
l_{sc}—longitude of the sub-spacecraft point;
b_{sc}—latitude of the sub-spacecraft point;
l_{sun}—longitude of the sub-solar point;
b_{sun}—latitude of the sub-solar point.

The corrected lunar calibration time series are shown in Fig. 6.6. The periodic geometry effects have been removed. The corrected time series have been fit using a decaying exponential function for each band. The time series for bands 1,2, and 5 – 8 were fit using two simultaneous exponential functions, while bands 3 and 4 were fit using single exponential functions. These functions have the form:

$$f(t,\lambda) = a_0(\lambda) - a_1(\lambda)[1 - e^{-C_1(\lambda)(t-t_0)}] - a_2(\lambda)[1 - e^{-C_2(\lambda)(t-t_0)}] \tag{6.10}$$

where t—time of the observation;
t_0—reference time for the time series;
λ—SeaWiFS Band;
a_0—initial value of the function;
a_1—amplitude of the first exponential function;
a_2—amplitude of the second exponential function;
C_1—time constant of the first exponential function;
C_2—time constant of the second exponential function.

For the simultaneous exponentials, the time series were fit with a short period time constant of 200 days and with a long period time constant of 1,600 days. For the single exponential, the time series were fit with the long period, 1,600-day time constant. The inverses of these fits are the radiometric corrections that are applied over time to the SeaWiFS top-of-the-atmosphere radiances prior to retrievals of the ocean color data.

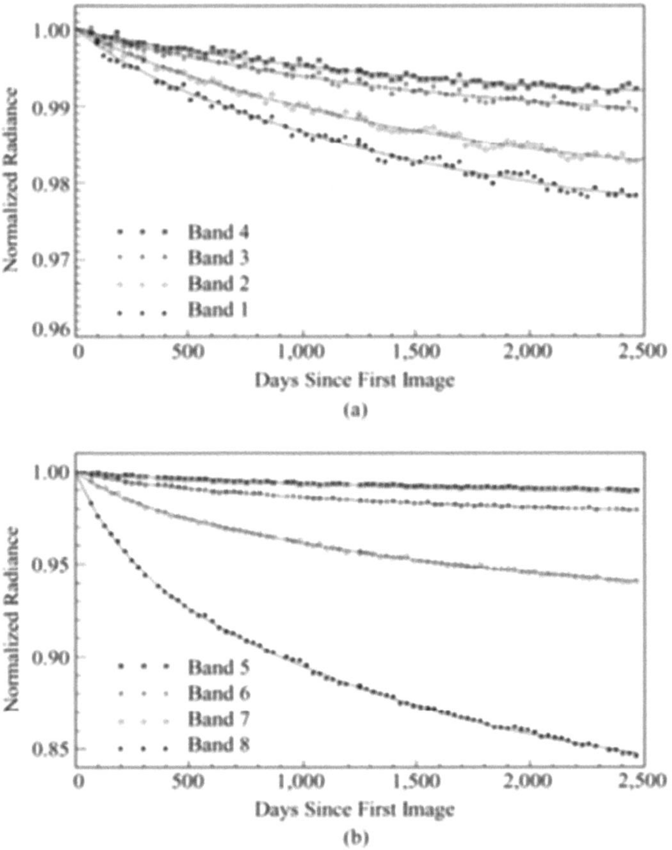

Figure 6.6 SeaWiFS Lunar Calibration Time Series. The periodic effects have been removed from and exponential fits have been applied to the geometry-corrected time series

As an evaluation of the radiometric corrections over time, the uncorrected lunar data (shown in Fig. 6.5) have been calibrated with the exponential fits. These calibrated time series have been processed though the geometric corrections described above. These calibrated radiances have also been processed through the USGS lunar model. The USGS lunar model provides an alternate approach to the geometric corrections used by the SeaWiFS Project. The time series for Band 5 is shown in Fig. 6.7; This time series is typical of the other bands. The SeaWiFS geometric corrections and the USGS lunar model yield results that are equivalent to within the scatter in the observations. The differences arise from the respective approaches to the geometric corrections implemented by the SeaWiFS Project and by the USGS lunar model. These results confirm previous comparisons between calibrated SeaWiFS lunar radiances and the USGS lunar model (Barnes et al., 2004).

The radiances in the calibrated lunar time series are stable to better than 0.07% over the course of the mission, as derived from the SeaWiFS geometric

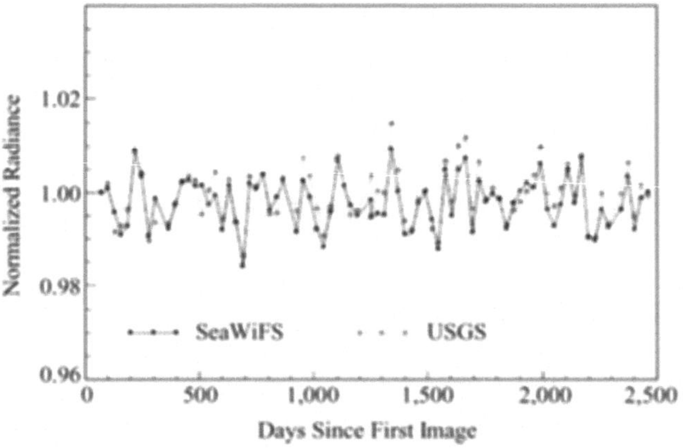

Figure 6.7 Calibrated SeaWiFS Lunar Time Series for Band 5. The plot shows the calibrated time series with libration-corrections applied (SeaWiFS) and the USGS ROLO model output. The two time series are equivalent to within the scatter in the observations

corrections and confirmed from the USGS lunar model output. The residual drifts in the radiometric response of the instrument over time are reduced by factors of 100, and are typically smaller than – 0.004% per thousand days. These results show that the SeaWiFS lunar calibration methodology provides top-of-the-atmosphere radiances for SeaWiFS that are stable to better than 0.07% over the mission. The resulting water-leaving radiances are stable to better than 0.7%, allowing the SeaWiFS Project to implement a vicarious calibration of the water-leaving radiances that is independent of time.

6.4 Lunar Calibration and Characterization of Visible and Infrared Scanner

6.4.1 VIRS Lunar Calibration Approaches and Applications

The Visible and Infrared Scanner (VIRS) is similar to the MODIS and SeaWiFS. The primary approach for on-orbit radiometric calibration of both the VIRS reflected solar bands (RSB), centered at 0.62 and 1.61 μm, and the MODIS RSB is based on linear interpolation between two measured points. For VIRS, the two measurements are the digital counts (dN) from a no-radiance (dark) target and the dN from a view of the solar diffuser. The VIRS RSB radiometric calibration equations have been described by Lyu et al. (2000).

VIRS has, during 6.5 years, displayed a decrease in responsivity of the 0.62 μm band when viewing the solar illuminated SD. Since the Earth-view (EV) data does not include the SD in the optical path, the question arises as to whether the decrease in responsitivity is due to the sensor, to a change in the reflectance of the SD or to both. Meaurements of the moon, as with the EV measurements, do not include the SD in the optical path. VIRS has been acquiring lunar measurements for close to seven years. This data should enable the partioning of the 0.62 μm band's responsitivity decrease between the sensor and the SD.

6.4.2 VIRS Lunar Calibration Results

During the early period of VIRS on-orbit calibration of the reflected solar bands at 0.62 and 1.61 μm (starting in Jan. 1998), calibrations were performed once or twice monthly. After four years, the frequency was decreased to near-monthly. At present, after nearly seven years, there have been 139 solar calibration events. The results of these calibrations are presented in Fig. 6.8. The responsivity in

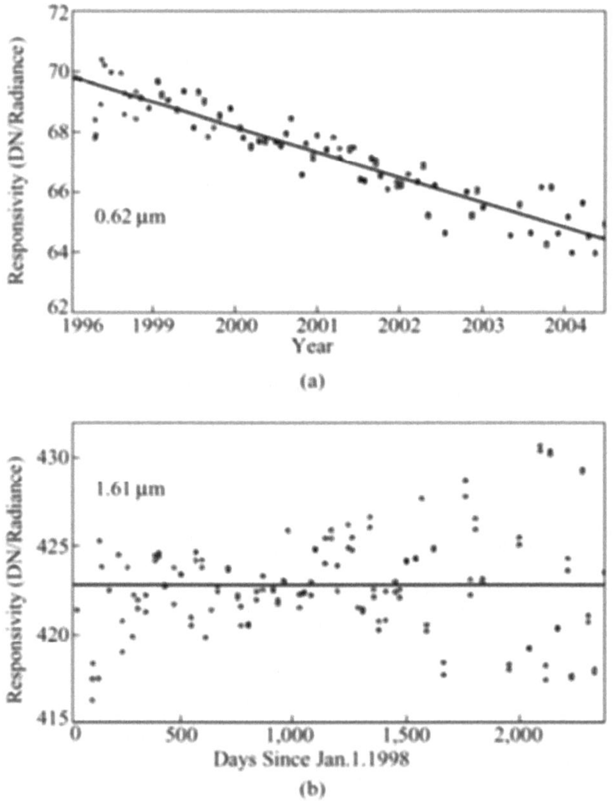

Figure 6.8 (a) Responsivity of VIRS 0.62 μm band determined using 95 on-orbit solar calibrations with a std deviation of 1.07. (b) Responsivity of VIRS 1.61 μm band determined using 95 on-orbit solar calibrations with a std deviation of 2.06

counts (DN) per radiance unit are plotted against days since Jan. 1, 1998. The responsivity used by VIRS to generate spectral radiance is the inverse of the MODIS scaling factor m_1 used to generate its reflectance factor product (see Eq. (6.1)) and therefore decreases with a loss in sensitivity whereas m_1 increases with a loss in system response. As mentioned previously, only the 0.62 μm band exhibits any decrease in response over the VIRS lifetime. This is not too surprising in that both optical components and diffuse surfaces have a tendency to degrade in the shorter wavelengths first. Therefore, the 1.15%/year loss in sensitivity (Lyu and Barnes, 2003), measured while viewing the sun via the diffuser, cannot a priori be divided between the sensor and the diffuser.

The TRMM satellite was launched into a 350-km circular orbit with a 35° inclination angle which is non-sun-synchronous. This has the consequence that the VIRS views of the moon through its space-view port are more frequent than the monthly events viewed by MODIS and SeaWiFS and occur over a wide range of lunar phase angles as shown in Fig. 6.9. The best fit curve of Fig. 6.9 is very similar to the modeled lunar phase curve of Hapke (see Fig. 2(a) of (Lyu and Barnes, 2003)). As discussed in the previous sections of this chapter, both MODIS and SeaWiFS have been careful to perform their lunar measurements at as nearly a constant phase angle as possible. This was done to minimize any errors in normalizing to a single phase angle. This approach was not possible with VIRS. The results of time-ordering the VIRS lunar calibrations and normalizing the integrated values with the best-fit phase curve of Fig. 6.9 are presented in Fig. 6.10. Each point of Fig. 6.10 represents the average of several lunar observations taken during consecutive orbits.

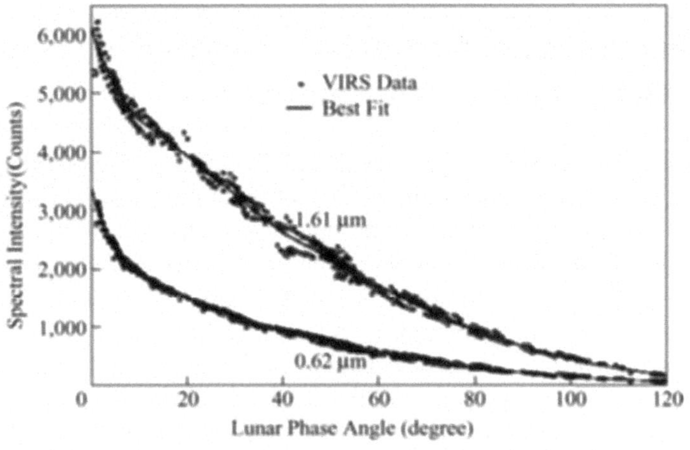

Figure 6.9 Disk-integrated spectral intensity (counts) of VIRS 0.62 and 1.61 μm data vs. lunar phase angle

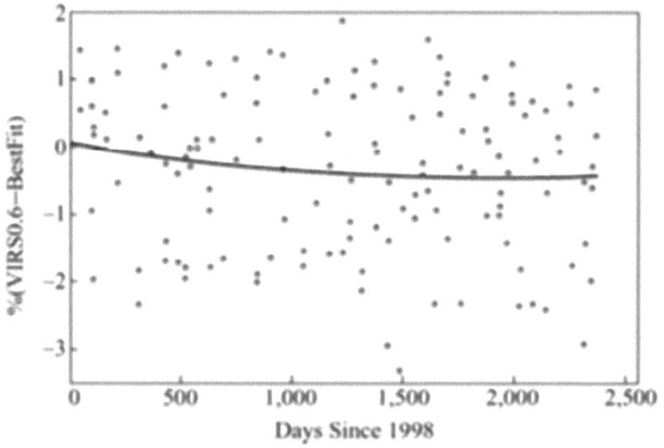

Figure 6.10 Phase-normalized disk-integrated lunar observations of VIRS 0.62 μm band vs. days since Jan. 1, 1998

There have been 490 lunar measurements during 138 observing days spread across 6.5 years. The line through the data (a quadratic best fit) indicates that the VIRS sensor has been degrading approximately 0.15%/year. Since the solar observations of Fig. 6.8(a) indicate a decrease in responsitivity of 1.15%/year, it follows that the diffuser reflectivity is decreasing 1.0%/year. Therefore, the next reprocessing of the VIRS data should include a time-dependent responsitivity that decreases 0.15%/year.

Plans are underway to use the ROLO model (see Section 6.1.1) to furnish the integrated lunar irradiance for each of the VIRS lunar measurements (integrated counts). This will enable tracking changes in sensor response more directly.

6.5 Using the Moon for Inter-Comparison of Sensors' On-Orbit Radiometric Calibrations

Since the lunar surfacr reflectance proporties are extremely stable, the moon can be used for inter-comparing multiple Earth-observaing sensors. The effects due to the lunar phase angles and libration angles are addressed by the ROLO lunar model developed at the US Geologic Survey (USGS) and currently avaliable through NASA's EOS calibration program. Lunar calibration capability is important for sensors with limited or no on-board calibrators.

On Apr. 14, 2003, a deep space maneuver was successfully executed for the Terra spacecraft which allowed all the sensors on-board to perform nearly simutanous lunar calibration. SeaWiFS also participated in this activity with its own maneuver at almost the same time (about 2 min later than the A Terra maneuver). A Terra MODIS luanr calibration is typically performed at 55° phase angle (waxing) and the SeaWiFS lunar calibrations are performed over a phase angle range of 5° – 10°. The quality of a direct comparison of these two sensors is limited by the large differences in their lunar viewing phase angles. The lunar observations from the special maneuvers, with MODIS viewing the moon at 22° phase angle and SeaWiFS viewing the moon at 23° phase angle, have allowed a more direct comparison between these two sensors. Table 6.3 shows the comparison results. The overall agreement is generally within ± 3%, which is within the combined calibration accuracy of both sensors. The MODIS calibration requirement is ± 2% for the RSB reflectance factors and ± 5% for the RSB radiances; SeaWiFS radiance calibration accuracy is ± 4% for all bands.

Table 6.3 Comparison of MODIS viewing the moon at 22° phase and SeaWiFS at 23° phase

SeaWiFS				MODIS				
Band No.	Wavelength (nm)	Measured I (μW/ (m^2 • nm))	Model I (μW/ (m^2 •nm))	Band No.	Wavelength (nm)	Measured I (μW/ (m^2 • nm))	Model I (μW/ (m^2 •nm))	Ratio
1	412	1.790	1.757	8	412	1.805	1.714	0.97
2	443	2.190	2.130	9	442	2.143	2.026	0.97
				3	466	2.465	2.316	
3	490	2.574	2.437	10	487	2.526	2.319	0.97
4	510	2.589	2.458	11	530	2.617	2.463	0.99
5	555	2.776	2.631	12	547	2.704	2.523	0.98
5	555	2.776	2.631	4	554	2.663	2.539	1.01
				1	647	2.596	2.512	
6	670	2.744	2.556	1	647	2.596	2.512	1.04
7	765	2.480	2.266					
8	865	2.009	1.886	2	857	1.974	1.855	1.00
				17	904	1.912	1.705	
				18	935	1.822	1.574	
				19	936	1.815	1.572	

The Visible and Infrared Imaging Radiometer Suite (VIIRS) being developed for inclusion on the National Polar-orbiting Operational Environmental Satellite System (NPOESS) was designed by Raytheon Santa Barbara Remote Sensing based on their experience and lessons learned from the MODIS, SeaWiFS and VIRS.

It has essentially the same on-board calibrators as MODIS with a SeaWiFS-like rotating telescope. The first VIIRS instrument will be on the NPOESS Preparatory Project (NPP), jointly managed by NASA and the NPOESS Integrated Program Office (IPO) as a joint science and risk reduction program. NPP, to be launched in 2009, will provide a data bridge between NASA's EOS missions and the NPEOSS operational missions. There is no doublt that lunar calibration will play a critcal role in ensuring that both the NPOESS and NPP missions produce continuous, long-term climate data records.

6.6 Summary

The moon is a very stable radiometric source. It can be used for space-borne on-orbit radiometric calibrations and calibration inter-comparisons. It has been shown to play a major role in providing a consistent and unique reference among data sets from multiple sensors used in producing long-term climate data records. This chapter reviews issues that need to be addressed in order to correctly apply lunar observations to radiometric calibrations, including the sensors' lunar viewing configuration and geometry. It dicusses the approaches being used to support the on-orbit calibration and characterization of MODIS, SeaWiFS, and VIRS using lunar observations. For MODIS, the moon has been used to monitor the sensor's reflective solar band radiometric calibration stability and to update the sensor's response versus scan-angle when used together with other on-board calibrators. It has also been successfully used to examine the RSB on-orbit calibration consistency between the Terra and Aqua sensors. In SeaWiFs, over six years of lunar observations have been made and the trending results have been used to track the sensor's response change and update on-orbit calibration coefficients. Simialrly, the VIRS has used the moon in an attempt to examine its solar diffuser's on-orbit degradation. All the examples (results) presented have shown that the moon can be effectively used for on-orbit radiometric calibration by using carefully selected lunar observations and proper normalization factors to remove the differences among observation conditions.

References

Barnes RA, Eplee RE Jr., Patt FS, Kieffer HH, Stone TC, Meister G, Butler JJ, and McClain CR (2004) Comparison of SeaWiFS measurements of the moon with the USGS lunar model. Appl. Opt. (in press)

Barnes RA, Eplee RE Jr., Schmidt GM, Patt FS, McClain CR (2001) Calibration of SeaWiFS. I. direct techniques. Appl. Opt. 40: 6,682 – 6,670

Barnes RA, Barnes W, Lyu C, Gales J (2000) An overview of the vsiible and infrared scanner radiometric calibration algorithm. J. Atmos. Oce. Technol. 17(4): 395 – 405

Barnes RA, Eplee RE Jr., Patt FS, McClain CR (1999) Changes in the radiometric stability of SeaWiFS determined from lunar and solar measurements. Appl. Op. 38: 4,649 – 4,664

Eplee RE Jr., Barnes RA, Patt FS, Meister G, McClain CR (2004) SeaWiFS lunar calibration methodology after six years on orbit. In: Barnes WL (ed) Earth Observing Systems IX, SPIE 5,542

Esposito JA, Xiong X, Wu A, Sun J, Barnes WL (2004) MODIS reflective solar band calibration uncertainty analysis. In: Barnes WL (ed) Proceedings Earth Observing Systems Ⅸ, SPIE 5,542

Hapke BW (1966) An improved theoretical lunar photometric function. Astron. J. 71: 333 – 339

Hapke BW (1963) A theoretical photometric function for the lunar surface. J. Geophys. Res. 68: 4,571 – 4,576

Lane AP, Irvine WM (1973) Monochromatic phase curves and albedos for thel unar disk. Astron. J. 78:267 – 277

Lyu C, Barnes W (2003) Four years of TRMM/VIRS on-orbit calibration and performance using lunar models and data from Terra/MODIS. J. Atmos. Oce. Technol. 20(3): 333 – 347

Lyu C, Barnes W, Barnes R (2000): First results from the on-orbit calibration of the tropical rainfall measuring mission. J. Atmos. Oce. Technol. 17(4): 385 – 394

Kieffer HH, Stone TC, Barnes RA, Bender S, Eplee RE Jr., Mendenhall J, Ong L (2003) On-orbit radiometric calibration over time and between spacecraft using the Moon. In: Fujisada H, Lurie JB, Aten ML, Weber K (eds) Proceedings Sensors, Systems, and Next-Generation Satellites Ⅵ. SPIE 4,881: 287 – 298

Kummerow C, Barnes W, Kozu T, Shiue J, Simpson J (1998) The Tropical Rainfall Measuring Mission (TRMM) sensor package. J. Atmos. Oce. Technol. 15(3): 809 – 817

Stone TC, Kieffer HH, (2002): An absolute irradiance of the Moon for on- orbit calibration. In: Barnes WL (ed) Proceedings Earth Observing Systems Ⅶ. SPIE 4,814: 211 – 221

Sun J, Xiong X, Guenther B, Barnes W (2003) Radiometric stability monitoring of the MODIS reflective solar bands using the moon. Metrologia 40: 85 – 88

Xiong X, Sun J, Xiong S, Barnes W (2003) Using the moon for MODIS on-orbit spatial characterization. In: Roland Meynart, Steven P. Neeck, Haruhisa Shimoda, Joan B. Lurie, Michelle L. Aten (eds) Proceedings Sensors, Systems, and Next Generation Satellites Ⅶ, SPIE 5,234: 480-487, Barcelona, Spain

Xiong X, Sun J, Barnes W (2002) MODIS on-orbit characterization using the moon. In: Fujisada H (ed) Proceedings Sensors, Systems, and Next Generation Satellites Ⅷ, SPIE 4,881: 299 – 307

7 A Review of Remote Sensing Data Formats for Earth System Observations

Wenli Yang

7.1 Introduction

Earth observing remote sensing involves a variety of diverse instruments and platforms, ranging from hand-held close-range spectrometers to imagers and sounders on board satellites tens of thousands kilometers above the earth. The data format for a remote sensing mission is usually chosen based on a number of considerations, including requirements of the sensing system, mission objective, the design and technology of data processing, archiving, and distribution systems, as well as community data standard. Currently, earth observing data, or geospatial data, are stored in a wide variety of formats. For most end users, the inconsistencies in the formats of data obtained from different sources can cause problems, sometimes significant difficulties, in prompt and effective use of the data. It would be desirable if all data can be generated in one best format that is universally applicable for all systems, missions, and product levels. Such a "best format" is, however, not available and seems not likely to emerge due to the diverse nature of the earth system and the cultural differences between disciplines, not to mention the ever evolving computing and archiving technologies. Fortunately, there are some commonly used data formats which are considered as somewhat standardized formats in several communities. Knowledge in these common formats will help the use of data products stored in such formats and the analysis/processing software developed for such formats, and the development of tools to perform conversions among such formats. This chapter will review some of these data formats.

7.1.1 Vector and Raster (or Feature and Grid) Data

Geospatial data can be classified into different categories based on different schemas, such as land, ocean and atmospheric, and topographical, climatological, and meteorological. From data type point of view, they are traditionally grouped into two classes: vector data and raster data. Vector data refers to data that represents the spatial configuration of geospatial features as a set of directed line segments. It is usually used to represent such features as rivers, contours, profiles, transportation lines, and boundaries of natural and political regions. This data holds both spatial and attribute properties of the feature it represents, such as the geographic coordinates and the surface type of a road. It often provides a topological relationship between features. For example, the identification codes of the left and right sides of a polygon boundary when the boundary is traversed at a given direction. Raster data refers to data that are stored as multi-dimensional arrays. It contains a matrix of values representing the attribute property or properties (e.g. temperature, pressure, ozone concentration) at the locations where the matrix elements locate. The spatial property (i.e. the location of each matrix element) of raster data is implicitly included in the spatial arrangement (i.e. relative positions) of the matrix elements.

While terms vector and raster have been traditionally used, several different (and/or new) terms are used for these two types of data, especially in geospatial data and information technology communities. Vector data is now more frequently referred to as feature data because it is mainly used to represent or delineate geographic features. The term raster is now more frequently replaced by term "grid", which is more generally defined as "a network composed of two or more sets of curves in which the members of each set intersects the members of the other set(s) in a systematic way that divides a space" (ISO, 2003). By using the word "curve", this definition effectively includes grid data whose cells are non-rectangular in shape and may also be irregular (varying) in both sizes and shapes. The term "imagery" is referred to as a kind of grid data whose attribute values are a numerical representation of the physical parameter measured by an instrument from which the data are transmitted (ISO, 2004b). This definition excludes thematic grid data such as land cover classification and soil erosion potential data which are often derived from the imagery data. Because this chapter will cite different documentation and format specifications where different terms are often used, terms vector and feature are used interchangeably and terms raster, grid, imagery and image are used interchangeably in the context of this chapter. Similarly, terms cell and pixel will also be used interchangeably to mean an atomic spatial element (i.e. not further dividable spatially) of a grid, although the latter is more frequently used together with image or imagery. Most earth observing remote sensing data are stored in grid format. This chapter will focus on formats designed for grid data.

7.1.2 Georectified Data and Georeferenced Data

All geospatial grid data is associated, implicitly or explicitly, with some sort of information that makes the data to be spatially identifiable in a particular geospatial coordinate reference system, such as the geodetic coordinate or projected coordinate systems. Based on the characteristics of grid cell's spatial arrangement, geospatial grid data can be classified into two subclasses: Georectified data and Georeferenced data. In Georectified grid data, the cells are regularly spaced in reference to certain earth or projected map coordinate system. Any cell can be uniquely geolocated, given the location of any one cell in the grid (usually one of the grid corner cells), the cell spacing, and the grid orientation. In Georeferenced grid data, the cells are irregularly spaced in reference to any earth or projected map coordinate system. There is no a priori association between one cell's location and that of another. Thus, the geolocation of a cell in a Georeferenced grid data cannot be determined based on another cell's location. Cell locations of a Georeferenced grid can be uniquely determined through certain Georectification algorithms, such as polynomial fitting, using information provided with the grid data. Most raw remote sensing data are in this form.

7.1.3 Metadata

Another important kind of data is Metadata, which is simply defined as data about data. Geospatial data is multidisciplinary in nature and is often used by individuals or organizations who are not the providers of such data. Proper description of the data is essential for data providers to better manage the data and for perspective data users to better understand and adequately use the data. Metadata serves the purpose of providing necessary information about the data so that the data can be better stored, updated, searched, accessed, used, and reused. There are many different ways to define metadata for a data set and to present the metadata information to data users. The simplest form of metadata is an ASCII text file accompanied with a data file. For example, a data producer may provide users a grid data set with 180 rows and 360 columns. The provider then writes an ASCII text metadata file saying that it is a 16-bit unsigned integer type data of simulated land surface temperature in the unit of 0.1 K, with a filling value of 0 K for ocean surface, and that the spatial resolution of the data is 1- by 1-degree, with upper left corner located at 0-degree longitude and 90-degree latitude. While metadata like this can provide accurate description of a data set in many cases, they are not appropriate for large volumes of diversified geospatial data because ASCII metadata written by different people are usually not consistent in many aspects such as terminology, definition, format, and language style. They are not appropriate for automated processing. Significant efforts have been made to produce standardized terminologies and definitions for geospatial metadata. Several standards are currently available and some are under development (FGDC, 1998, 2002; ISO, 2003, 2004a; NASA, 1994). Methods of storing metadata with science data are important design considerations of geospatial data formats and data systems. All the data formats introduced in this chapter include schema of packing metadata with data sets, which make the formats more or less self-descriptive and interoperable.

7.2 Hierarchical Data Format

Hierarchical Data Format (HDF) is an object-based format developed by the National Center for Supercomputing Applications (NCSA) at the University of Illinois. It is designed to facilitate data access in heterogeneous computing environments. Currently there are two major versions of HDF, version 4 (HDF4) and version 5 (HDF5). This section describes HDF4. HDF5 will be introduced in Section 7.4. For simplicity, acronym HDF is used specifically for HDF4 in this chapter.

HDF consists of a file format specification that describes the physical layout of an HDF file and a software package that is used to manipulate an HDF file and its data objects. The HDF software package consists of a set of libraries that allow HDF users to access and manage HDF data without knowing the physical location of the data and how it is stored in the file.

7.2.1 The Physical Layout of HDF

An HDF file contains a file header, at least one data descriptor block, and zero or more data elements as shown in Fig. 7.1 (from Fig. 2(a) in (NCSA, 2003a)).

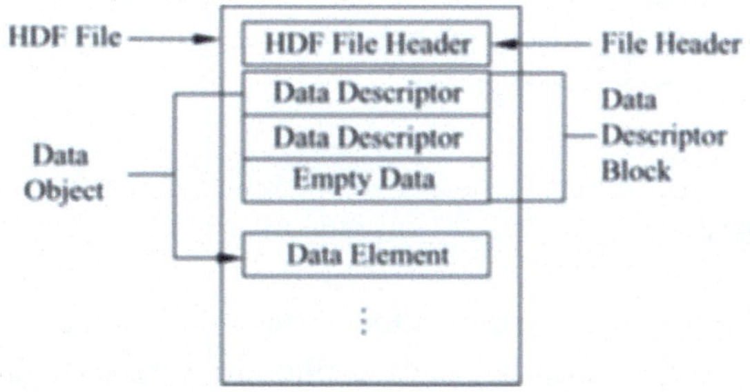

Figure 7.1 Physical layout of an HDF file (from Fig. 2(a) in (NCSA, 2003a))

7.2.1.1 File Header

The file header is a four-byte field located at the beginning of the file used to identify that the file is an HDF file. The four bytes are ASCII representations of control characters: ^N, ^C, ^S, and ^A (or $0 \times 0e$, 0×03, 0×13, and 0×01 in hexadecimal).

7.2.1.2 Data Object

A data object is comprised of a data descriptor and a data element. The data element contains raw data. The data element's type, location, and size are specified in the data descriptor. A data descriptor consists of 96 bits. The first

16-bit stores a tag which specifies the type of data element. The following 16-bit is a reference number used to distinguish objects with the same tag. The next 32-bit following the reference number specifies the location of the data element in the file and the last 32-bit specifies the size of the data element. The largest possible size of a data element is two gigabytes.

7.2.1.3 Data Descriptor Block

Data descriptors are stored as blocks called data descriptor blocks, each of which is preceded with a 16-bit "block size" field and a 32-bit "next block" field. The "block size" tells the number of descriptors contained in the current block and the next block field specifies the location of next data descriptor block. The next block field of the last block contains a value of zero.

7.2.1.4 Grouping of Data Objects

Although each data object contains all information necessary to access the raw data stored in an HDF file, data objects are usually grouped together to provide "meaningful" data rather than just numbers. For example, the following objects are necessary in order to provide complete information about a raster image: an image object containing the image data itself, a dimension object specifying the number of rows and columns of the image, and a group object identifying the members of the group. Grouping in HDF is conceptual but not physical. An object belongs to one group can also belong to another group. Groups can be further grouped to form a larger group.

7.2.2 Attribute

Attributes are used to provide information about a data object, a group, and a file, such as whether a scale factor has been applied to a data set. In most cases they can be considered as metadata. Attributes take "name = value" format, where "name" is a character string serves as the name of an attribute and "value" contains one or more entries of values. For example, the following attribute of a data set indicates that a scale factor, 0.01, needs to be applied to the data set to get its true values: "scale_factors = 0.01". Some attribute names are reserved for specific meaning and attributes with those reserved name are called predefined attributes. For example, valid_range is a reserved attribute name used to specify the maximum and minimum values of a data set. Other attributes are called user-defined attributes. Attributes attached to a file are called global attributes. The information provided by a global attribute usually applies to all components in the file, such as the production date and time of the file, the name of processing center, and the name of the instrument used in data collection. Group and object level attributes usually provide information for the related group and object, such as the fill value used for a specific data array.

7.2.3 HDF Data Models

HDF data models or data structures are designed for different types of data. There are six data models: (1) a scientific data set (SDS) model used to store multidimensional data; (2) a raster image model used to store image data; (3) a palette data model used to store color lookup tables to be associated with a palette image data; (4) a Vdata model that used to store tabular or "vertex data" similar to the aforementioned vector data; (5) an annotation data model used to provide metadata information about data objects, groups, and files; and (6) a Vgroup model used to associate related data objects and data models (e.g. linking three related SDSs, such as temperature, latitude, and longitude, in one Vgroup). For each model, the HDF library provides an Application Programming Interface (API) to manipulate the data. Among the six data models, the SDS model is most relevant to the earth observing remote sensing data because earth observing remote sensing data are predominantly multidimensional array (i.e. grid) data. Thus, only the SDS model will be introduced.

7.2.4 The HDF SDS Data Model

The SDS data model is designed to store multidimensional arrays (i.e. grid type) data sets. Information needed to describe a multidimensional array includes the number of dimensions, the size of each dimension, and the data type of the array. The SDS data model in HDF is thus designed to include four required components (see Fig. 7.2): (1) a multidimensional data array object containing the grid data itself; (2) a dimension object specifying the number of data dimensions and the size of each dimension; (3) a data type object specifying the type of the data; and

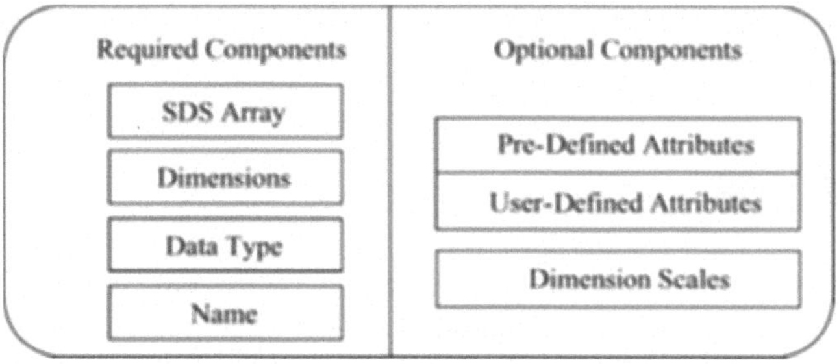

Figure 7.2 The contents of an SDS (from Fig. 3(a) in (NCSA, 2003a))

(4) a name identifying the SDS. In addition to these required components, a SDS can also include three types of optional components: predefined attributes, user defined attributes, and dimension scale.

The attribute components are used to provide metadata information about a SDS, such as descriptions of the nature of the data, unit of data values, and calibration coefficients. The dimension scale is used to assign values to a SDS data array's dimension and it is useful when a dimension represents an irregularly spaced coordinate values, such as a "depth" dimension representing an irregular depth intervals (e.g. 1, 4, 10, and 18 meters).

7.2.5 The HDF SD API and Programming Model

The SD API in the HDF library provides functions that create, write, read, and inquiry SDSs and their components in a HDF file. The API functions can be grouped into different categories based on their functionalities: (1) access functions that initiate/terminate SD interface and create a new or select an existing SD; (2) basic I/O functions that read/write SDS data arrays; (3) general inquiry functions that obtain general information of a file or a SDS such as SDS names and data types; (4) dimension functions that get or set dimension related information; and (5) attribute operation functions that read/write attributes (user-defined or predefined) and obtain information about existing attributes. In addition, the SD API also includes other functions such as data compressing, data chunking/tiling, and setting data block size. The complete SD API functions are listed in Table 3A of the HDF User's Guide (NCSA, 2003a).

The basic procedures involving SDS programming model include:

(1) Initialize the SD interface and open an existing file or create a new file.

(2) Access to an existing SDS using the SDS name and/or index, or create a new SDS.

(3) Perform desired operations on the SDS such as inquiring/reading/writing data, attributes, and dimension information.

(4) Terminate access to the data set.

(5) Terminate access to the SD interface and close the file.

The C code in Appendix B.1 shows how to use the SD API to manipulate SDS in an HDF file.

7.3 HDF-EOS

While HDF is an effective format for general scientific data, it does not provide a standardized structure to store spatial location (i.e. geolocation) information, which is vital to geospatial data. In order to effectively link the earth observing data and their derived scientific products to locations where the original observations are taken, the NASA's Earth Observing System (EOS) program extended the native HDF to HDF-EOS (Hierarchical Data Format-Earth Observing System). The primary difference between HDF and HDF-EOS is that HDF-EOS defines specific structures of storing geolocation information in an HDF file. There are standardized ways in HDF-EOS to store and retrieve geolocation information and therefore any software developed based on HDF-EOS structure can easily read and write such information. The EOS program also provides APIs that create, access, and manipulate HDF-EOS files. Three data models, point, swath, and grid, are defined in HDF-EOS. Among the three models, the point model is primarily designed for discrete data points and the swath and grid models are for multidimensional array type data. In the following paragraphs, the point model will only be briefly introduced. Readers interested in this data model can refer to the HDF-EOS Library User's Guide (NASA, 2004a) for more detailed information. The swath and grid models will be introduced in a greater detail.

7.3.1 The Point Data Model

The Point data model is used for those data sets in which different data points are not spatially or temporally associated in a systematic way (e.g. regularly spaced or time-ordered in reference to certain coordinate, either geospatial or non-geospatial coordinate). The spatial or temporal location of each data point in this model needs to be explicitly specified. The point data model can be used in simple discrete observation situations such as temperatures and pressures measured at different locations being organized in a table with a series of (x, y, T, P) records, where x and y specify a location and T and P represent temperature and pressure values at that location. It can also be used in more complex data organizations such as a number of measurements taken at different times at certain location. For example, a point data set in the form of (x, y, $t1$, $M1$, $t2$, $M2$,…), where one location (x, y) is considered one *point* and ($M1$, $M2$, …) are measurement values at different times ($t1$, $t2$, …) at this location. More complexly, a point data set can contain measurement obtained from different times and different locations by certain platform, where a platform, e.g. a moving ship, is considered one Point and all data values associated with the platform belong to this point. HDF-EOS point model uses different index levels to deal with different complexity levels of the point data. Graphical presentations of the above examples are shown in the HDF-EOS Library Function Reference Guide (NASA, 2004a).

7.3.2 The Swath Data Model

The swath data model is designed to store multidimensional data array where the data points are spatially (or temporally) associated in certain systematic way but the sizes and/or shapes (e.g. the ground instantaneous fields of view or ground sampling distance) of different data points in a swath are different in reference to any geospatial coordinate reference system. Thus, the model is used for georeferenced but not georectified data. The spatial location of one data point in a swath cannot be directly inferred from that of another. External geolocation information must be presented in order to determine the geolocations of all swath data points. A swath structure contains two primary components: data fields and geolocation fields (NASA, 2004a). Data fields are the scientific data stored as multidimensional arrays. Geolocation fields contain geospatial position information of the data fields and are usually composed of two fields representing two geospatial coordinates such as latitude and longitude. The geolocation fields and data fields are linked through associated dimensions. For example, Fig. 7.3 shows a swath containing a two dimensional data field, myDataField, with the two dimensions named as dataRow and dataColumn. The geolocation fields in the swath, named Latitude and Longitude, respectively, are two-dimensional arrays with their two dimensions named as geoRow and geoColumn. The dataRow dimension in the data field corresponds to the geoRow dimension in the geolocation field and the dataColumn dimension in the data field corresponds to the geoColumn dimension in the geolocation field. The associations between dimension pairs of geolocation and data fields (i.e. geoRow to dataRow and geoColumn to dataColumn) and correspondence between positions in a dimension pair are defined by the dimension mapping relationship, which includes names of related dimensions and the Increment and Offset values, in a swath's structure metadata (see Fig. 7.3). With dimension mapping information, the latitude and longitude coordinate values of a pixel in the myDatafield field can be obtained by finding the values of the corresponding pixels in the geolocation fields.

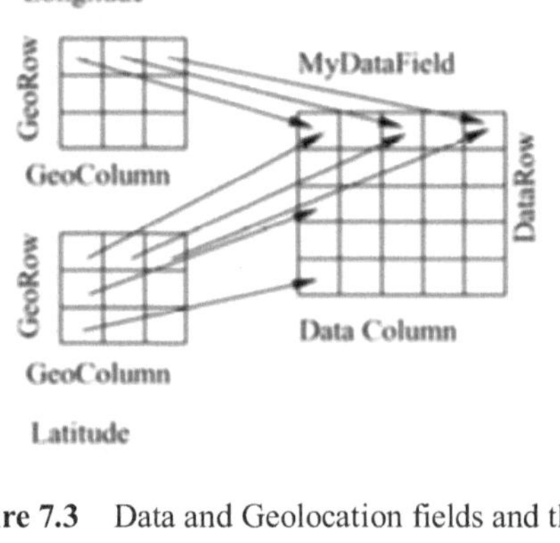


```
Group=DimensionMap
Object =DimensionMap_1
GeoDimension= "GeoRow"
DataDimension= "DataRow"
Offset=0
Increment=2
End_Object =DimensionMap_1
Object =DimensionMap_2
GeoDimension= "GeoColumn"
DataDimension= "DataColumn"
Offset=0
Increment=2
End_Object =DimensionMap_2
Group=DimensionMap
```

Figure 7.3 Data and Geolocation fields and their dimension mappings in a swath

The Offset value in a dimension mapping is the offset position along a spatial dimension (e.g. dataColumn) of a data field pointed by the first geolocation element at the corresponding geolocation dimension (e.g. geoColumn). Increment represents the interval in data-dimension for every element in geo-dimension. In addition to the regular dimension mapping in which the Offset and Increment apply to the entire range of a dimension, HDF-EOS also allows irregular mapping between a geo-dimension and a data-dimension by using an index array that specifies which geolocation element points to which data point. For example, an index array (0, 3, 8) indicates that there are three elements in the geo-dimension data and that the values of these three elements are geolocation values of the first, the fourth, and the ninth elements, respectively, of the corresponding data-dimension.

7.3.3 The Grid Data Model

The grid data model is also designed to store multidimensional data array where the data points are spatially associated in certain systematic way. The key difference between grid and swath models is that the data in grid model is georectified while that in swath model is georeferenced. In the grid model, spatial location of a grid point can be derived from that of any reference point (usually one of the four corner points) in the same grid given the distance between adjacent points in the grid. Thus, no external geolocation information is needed for data sets in the grid structure. For example, in Fig. 7.3, if all pixels in the data field, myDataField, are of the same size in reference to the latitude and longitude coordinates and the coordinate values of the center of the upper left corner pixel are known, then the location of the center of any pixel in the data field can be calculated without the two geolocation fields. The main components of a grid structure include one or more data fields, information of the spatial coordinate reference coordinate system (e.g. geodetic and projected map coordinate systems, ellipsoid, datum), the cell size of the grid point, and the dimension information telling which dimension of the grid corresponding to which spatial coordinate. It should be noted that the term "grid" in HDF-EOS is different from that used in the "grid type data" as described in Section 7.1.1. The latter is in contrast to feature (traditionally called vector) type data and means a more broad group of data that includes both georeferenced and georectified grid data (i.e. raster, matrix or multidimensional array type data) while the former only applies to georectified gird data sets.

7.3.4 The HDF-EOS APIs and Programming Models

The HDF-EOS library provides API functions to create, write, read, and inquiry point, swath and grid data. The naming convention of the HDF-EOS functions is similar to that of the native HDF functions. All function names start with a 2-character prefix (upper case for C functions) that identifies the data model to which they belong, i.e. PT for point, SW for swath, and GD for grid models. The programming models for the three APIs are also similar and functions in each API can be grouped into a number of categories based on the functionalities: (1) access functions that initiate/terminate an interface and create/open/close a file or a Point, Swath, or Grid; (2) definition functions that define various components such as data/geolocation fields and dimension mappings in a Swath, origin and projection type in a Grid, and levels and linked fields between two levels in a Point; (3) basic I/O functions that read/write data and attributes; (4) inquiry functions that obtain information about a data model such as number of data/ geolocation fields, number of dimensions, dimension mapping relationship in a Swath and project type and parameters in a Grid; and (5) subset functions that define and access a subset region in a data set. A complete list of HDF-EOS API functions are listed in Tables 1-1, 1-2, and 1-3 of the HDF-EOS Function Reference Guide (NASA, 2004b).

The basic procedures involving HDF-EOS programming models include:

(1) Initialize an interface and open/create a file.

(2) Access to or create a SW/GD/PT data structure.

(3) Perform desired operations on the data such as defining/inquiring/ reading/writing data, attributes, and dimension information.

(4) Terminate access to the data structure.

(5) Terminate access to the interface and close the file.

The C code in Appendix B.2 shows how to use the SW API to create a swath and write geo- and data-fields into the swath. The API for grid, GD, is very similar to the SW API. The most significant difference between the two is that the GD API includes functions for projection information while the SW includes functions for geolocation and dimension mapping information. The C code in Appendix B.3 shows how to use GD API to inquire and read information from an existing grid.

7.3.5 HDF-EOS Versus Native HDF

Although HDF-EOS is selected by the NASA EOS program as the standard data format, not all data processing software generates EOS data and its derived products in HDF-EOS format. Particularly, quite a few EOS data products are produced in native HDF format by data centers outside NASA. Many EOS data users ask such questions as "I am using NASA EOS data but the HDF-EOS tool I obtained does not work on my data. How do I know if my data is in HDF-EOS format"? The easiest way to tell an HDF-EOS file from a native HDF file is to look for the HDF-EOS structure metadata named as "StructMetadata.0" and stored as a file level global attribute in the form of Object Description Language (ODL). Whenever a HDF-EOS API is initiated to create an HDF-EOS file (i.e. PTopen, SWopen, or GDopen), this global attribute will be created and the value of this attribute will include three empty groups named SwathStructure, GridStructure, and PointStructure, respectively. Later when a data structure is actually defined, the corresponding group will be filled with actual parameters. For example, if a swath is defined using SWcreate (fileID, "mySwath"), a subgroup for the new swath, mySwath, will be created within the SwathStructure group. Since an HDF-EOS file is a valid HDF file, any HDF tool that can display HDF global attribute, such as the ncdump utility and the HDFView package developed by NCSA, can be used to view this HDF-EOS structure metadata. Theoretically, as long as such a metadata exists, the file is in HDF-EOS format. But the real HDF-EOS data products, in addition to the three data structures, include more ECS (EOS Core System) specific metadata such as inventory and archive metadata. Interested readers are encouraged to read the ECS metadata standard (NASA, 1994) for more information about ECS specific metadata.

7.4 HDF5

After HDF Version 4.1 was released, NCSA started to develop HDF Version 5. HDF5 addresses some of the limitations and deficiencies in old versions of HDF to meet with the requirements of the current and anticipated computing systems and applications. The improvements in HDF5 include larger file size, more objects, multi-thread and parallel I/O, unified and flexible data models and interfaces, etc. Although inheriting the old version numbering, HDF5 is a new data format and is not back compatible with old versions of HDF. HDF5 consists of a software package for manipulating an HDF5 file, a file format specification describing low-level objects in a physical disk file, and a user's guide describing high-level objects as exposed by HDF5 APIs.

7.4.1 The Physical Layout of HDF5

At lowest level, an HDF5 file consists of the following components: a super block, B-tree nodes, object headers, a global heap, local heaps, and free space (see Fig. 7.4). The HDF5 file physical format is specified with three levels of information. Level-0 is file identification and definition information. Level-1 provides information about the infrastructure of an HDF5 file. Level-2 contains the actual data objects and the metadata about the data objects (NCSA, 2003c).

7.4.1.1 Level-0: File Metadata

Level-0 provides information about a file, i.e. file level metadata. It is composed of a super block and an optional file driver information block. The super block contains a file signature followed by file version information, information about the byte numbers of address offsets and object sizes, and information about the

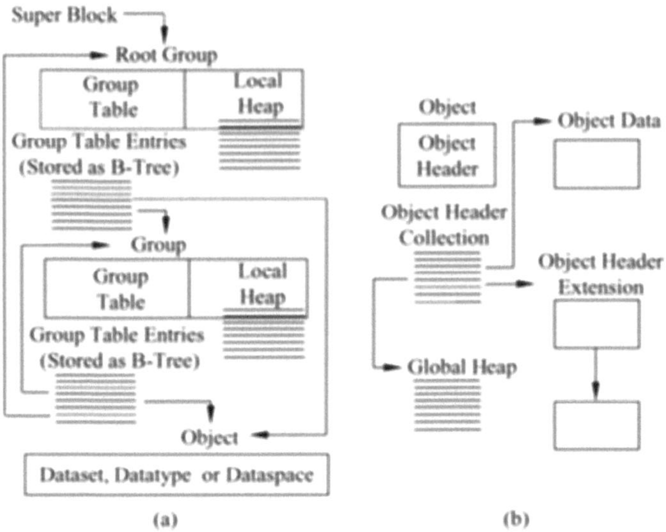

Figure 7.4 Relationships among the HDF5 Root group, other groups, and objects (a) and the HDF5 Objects (b) (from Figs. 8.1 and 8.2 in (NCSA, 2003c))

group B-trees. Among these, the file signature field is used to identify the file as an HDF5 file and also used to detect certain types of data corruption. This field consists of eight bytes which are, in hexadecimal, 0×89, 0×48, 0×44, 0×46, $0 \times$ 0d, $0 \times$ 0a, $0 \times$ 1a, and $0 \times$ 0a. Unlike HDF4 where the file identification field is the first four bytes, the file signature in an HDF5 file may begin at certain predefined offsets within the file. The byte offset is 0, or 512, or a multiple of 2 to the previous location, i.e. 1,024, 2,048, 4,096, and so on. The optional file driver information block contains information needed by the file driver in order to reopen a file.

7.4.1.2 Level-1: File Infrastructure Information

Level-1 contains information about B-link trees, groups, local and global heaps, and a free space index. A B-link tree is used to flexibly insert objects into a file and to quickly search the objects stored in a file. The B-link tree information includes an ASCII string node signature, "TREE", and node's type, level, number of children, and addresses information. A group is an object internal to the file that allows arbitrary nesting of objects within the file, including other groups. Group nodes are pointed to by a B-link tree. Information about a group node includes an ASCII string node signature, "SNOD", version number, number of symbols, and group entries. A heap is a collection of small heap objects. The local heap consists of an ASCII string heap signature, "HEAP", offset and address of the heap data segment, and offset of the first free block within the data segment. The global heap is used to store information shared between different datasets. It consists of a 4-byte magic number "GCOL", a version number, the size of the global heap, various stored objects in the heap and, at the end of the heap, a free space if it exists. The free space index is a collection of unused block data dispersed throughout the file.

7.4.1.3 Level- 2: Data Objects

A data object is comprised of object header and data. The data object header contains information needed to interpret and describe the data, such as data storage, number type, byte order, number of dimensions and dimension sizes, and other object level metadata or metadata pointers. The data component contains the scientific or other types of data itself that is considered by end users as data rather than certain kind of description or metadata information. The data in a data object is stored separately from the object header and may not necessarily be in the HDF5 file itself (see Fig. 8.4b).

7.4.2 HDF5 Data Models

HDF5 comprises of two primary data models: "group" and "dataset" (NCSA, 2003d). They are highest level HDF5 objects that are exposed to HDF5 API.

7.4.2.1 The HDF5 Group and Dataset

HDF5 group structure includes a group header, which contains a group name and group attributes, and a group symbol table which includes a list of HDF5 objects belong to the group. HDF5 dataset structure includes a dataset header and a data array. Groups can be grouped together to belong to a larger group and can be linked to form a directed graph. One particular group, the root group, is pointed by the super block. An HDF5 file can be viewed as an UNIX-like tree structure with each object being linked to the root path "/ ". In this tree structure, groups are the internal nodes while dataset are at the leaf level. Figure 7.5 shows an example of such a tree structure. In this figure, "/fool" signifies a member of root group and "/foo2/dat3" signifies a member of the group "foo2", which in

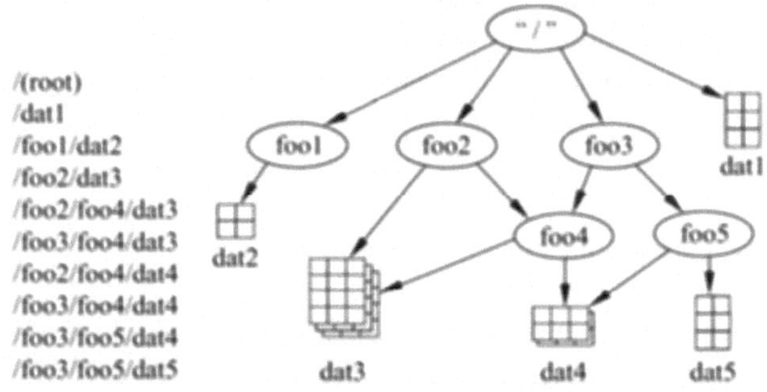

Figure 7.5 HDF5 objects viewed in a directed graph

turn is a member of root group. The difference between this graph and an UNIX directory structure is that an object in this graph does not have a unique upper directory.

7.4.2.2 The HDF5 Dataset Header

An HDF5 dataset header comprises of four classes of information: dataset name, datatype, dataspace, and storage layout. They provide enough information to identify and interpret the data included in a dataset.

The dataset name is an ASCII character string used as the name for a dataset.

The datatype specifies the number type of the data in a dataset. There are two primary data types: an atomic datatype that can not be decomposed at datatype interface level (e.g. integers, floats, and strings) and a compound datatype that is made up of atomic datatypes (i.e. similar to a struct in C). In addition, there are also a NATIVE datatype that is a system-specific instance of an atomic datatype and a Named datatype that is to be shared among several datasets.

The dataspace provides information about the dimensionality of the dataset array, which includes number of dimensions, the actual dimension sizes, and the maximum dimension sizes. A dimension can either be fixed, whose actual and maximum size are the same, or unlimited which indicates that it is extendible. A dataspace can either be a simple dataspace which is regular N-dimensional array of data points or a *complex* dataspace which is a collection of data points organized in another manner. A scalar dataspace is a special case of the simple dataspace and is defined to be a 0-dimensional single data point.

The storage layout describes how a dataset array is stored. It includes continuous, compact, and chunked formats. The continuous format indicates that data is stored in the same linear way as it is organized in memory. The compact format is designed for small data array so that it can be stored with the dataset header. The chunked format means that the dataset array is divided into equal-sized chunks that are stored separately.

7.4.2.3 The HDF5 Attribute

HDF5 attributes are named datasets. But they are special kinds of datasets that are designed to be attached to primary objects (i.e. datasets, groups, or named datatypes) as metadata and are always contained in the header of the object they are attached to. An attribute is comprised of a name and a value part. The latter contains one or more data entries of the same data type.

7.4.3 HDF5 API and Programming Model

The HDF5 API provides functions to open/create and close HDF5 files and create, write/read groups, datasets, and attributes to/from HDF5 files. The current release (release 1.4) is implemented in C. All functions begin with a 3-character prefix H5* where "*" is a single letter that identifies the category to which the function belongs.

The following categories of functions are included in the HDF5 library:

(1) H5F: File-level access functions that create, open, close, and set various (e.g. creation and access) properties of an HDF5 file.

(2) H5G: Group functions that create, open, and close groups and perform other operations on group objects such as link/unlink datasets and set/query comments.

(3) H5T: DataType functions that provide a mechanism for describing the storage format of individual data points. These include creating, opening, closing, and operating on a datatype, such as committing a datatype to a file (making it a named datatype), and inquiring the size and byte order of a datatype and properties of a member datatype in a compound datatype.

(4) H5S: DataSpace functions that create, close and manipulate a dataspace. These include defining or extending the extent of a dataspace and selecting subsets of a dataspace for other operations.

(5) H5D: Dataset functions that manipulate a dataset and the raw data in a dataset, such as dataset creation/opening/closing, raw data I/O, setting storage layout, memory and transfer properties, and querying of datatype, dataspace and creating property list of a dataset.

(6) H5P: Property list functions that manipulate object creation and access properties.

(7) H5A: Attribute functions that create/close/delete attributes and to perform attribute inquiries such as the number of attributes in an object and their name(s) and type(s).

(8) H5Z: Compression registration functions that define and query compression filters for chunked datasets.

(9) H5E: Error handling functions.

(10) H5R: Reference functions that perform operations on referenced objects and dataset regions.

(11) H5I: Identifier interface that currently includes a function of retrieving the type of object referenced by an identifier.

The basic procedures involving creating an HDF5 file containing a multidimensional array data set include:

(1) Create a new file.

(2) Create a dataset.

(3) Set the dimensionality and data type for the data set (i.e. dataspace, datatype).

(4) Write data into the data set.

(5) Close the data set and other associated objects (e.g. dataspace, datatype)

(6) Close the file.

A C code example is given in Appendix B.4 that demonstrates how to create an HDF5 file and datasets, writing data and attributes and read a subset of data from the file.

7.5 HDF5-Based HDF-EOS

Corresponding to HDF5, the ECS Project has released a new version of HDF-EOS library, designated as version 5.x, which provides APIs to work with HDF5-based HDF-EOS data (referred to as HDF-EOS5 hereafter). Similar to the old HDF-EOS (referred to as HDF-EOS2 hereafter) format, the main difference between HDF5 and HDF-EOS5 is that the three EOS data models, i.e. Point, Swath, and Grid, are incorporated into the HDF5 files to provide a standardized mechanism for geolocating different types of earth observation data. In addition, HDF-EOS5 includes a zonal average (ZA) structure intended to be used for data constructed based on time sequence or along a sensor's ground tract direction. Similar to the swath structure, the relationship between time/track with data field(s) is described by the dimension mapping relationship in the structure metadata. But the two regular geolocation fields, e.g. latitude and longitude, are not present in the ZA structure. There is no difference between the HDF5 and HDF-EOS5 in terms of file's physical layout, i.e. an HDF-EOS5 file is a valid HDF5 file.

Although the new version of HDF-EOS5 library essentially works with a completely new file structure, its interface is kept very similar to that of the HDF-EOS2 so that users familiar with the HDF-EOS2 APIs can work with the new APIs relatively easily without having to have in-depth knowledge of the HDF5 format and its APIs. Nevertheless, there are still several important differences between HDF-EOS2 and HDF-EOS5 APIs, especially those functions that involve the use of new HDF5 feature such as new data type, external file, and data chunking and tiling. Those differences are summarized in Section 3.5 and Table 3.1 in the HDF-EOS5 Interface Overview (NASA, 2004c). Complete lists of API functions for the PT, SW, GD, and ZA are listed in Tables 4.1, 5.1, 6.1, and 9.1 of the same reference (NASA, 2004c) and also in Tables 1.1, 1.2, 1.3, and 1.5 of the accompanying Function Reference Guide (NASA, 2004d).

7.5.1 HDF-EOS5 Data Structure

When a new HDF-EOS5 file is created using one of the four data structure's file open functions (i.e. HE5_SWopen, HE5_GDopen, HE5_PTopen, HE5_ZAopen) the structure as shown in Fig. 7.6 is established, except for the "mySwath" group which is created using the swath creation function HE5_SWcreate. There are two groups under the root group, an "HDFEOS" group and an "HDFEOS INFORMATION" group. The "HDFEOS" group contains four groups for the four data structures and an "ADDITIONAL" group, with which file level attributes can be attached. The "HDFEOS INFORMATION" group contains an attribute named "HDFEOSVersion" which provides version information and a dataset named "StructMetadata.0", which, similar to the same metadata in HDF-EOS2 file, provides information about the SW, GR, PT and ZA structures in the file. This metadata initially contains four empty groups for these four data structures and is updated when a data structure is established. Theoretically, one HDF5_EOS file can contains all four data structures but usually only one is present.

7.5.2 HDF-EOS5 Programming Model

The basic procedures involving HDF-EOS5 programming models include:

(1) Initialize an interface and open/create a file.

(2) Access to an existing or create a new SW/GD/PT/ZA object.

(3) Perform desired operations on the object such as defining/inquiring/reading/writing data, attributes, and dimension information.

(4) Terminate access to the object.

(5) Terminate access to the interface and close the file.

Appendix B.5 gives a C code example of using HDF-EOS5 API to create an HDF-EOS5 file and define/write data and geolocation fields, dimension mappings, and attributes. The swath structure generated using this code is shown in Fig. 7.6 starting from the mySwath object, except that the two index mapping objects are not drawn.

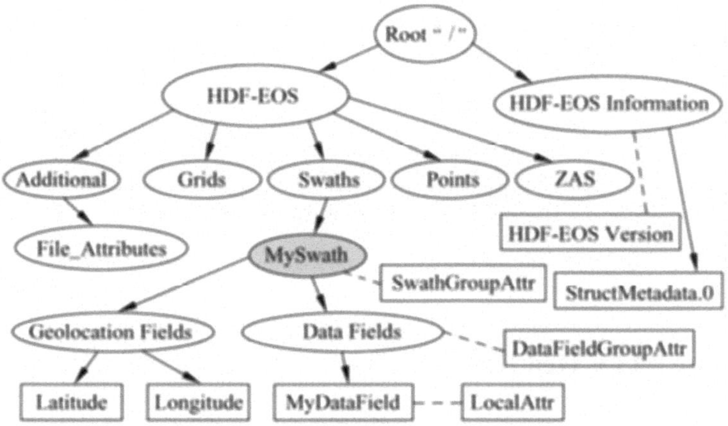

Figure 7.6 HDF-EOS5 file data structure (dashed lines indicate attached attribute)

7.6 NITF

The National Imagery Transmission Format (NITF) is designed primarily by the National Geospatial-Intelligence Agency (NGA), formerly named National Imagery and Mapping Agency (NIMA), and is a component of the National Imagery Transmission Format Standard (NITFS). It is adopted by ISO as an international standard known as Basic Image Interchange Format (BIIF) (ISO/

IEC 12087-5). NITF is aimed primarily to be a comprehensive format that shares various kinds of imagery and associated data, including images, graphics, texts, geo- and non-geo-coordinate systems, and metadata, among diverse computing systems and user groups. The format is comprehensive in contents, implementable among different computer systems, extensible for additional data types, simple for pre- and post-processing, and minimal in terms of formatting overhead (NIMA, 1998).

7.6.1 The Physical Layout of NITF

The top level NITF file structure includes a file header and one or more data segments which can be image, graphics, text, data extension, and reserved extension (see Fig. 7.7).

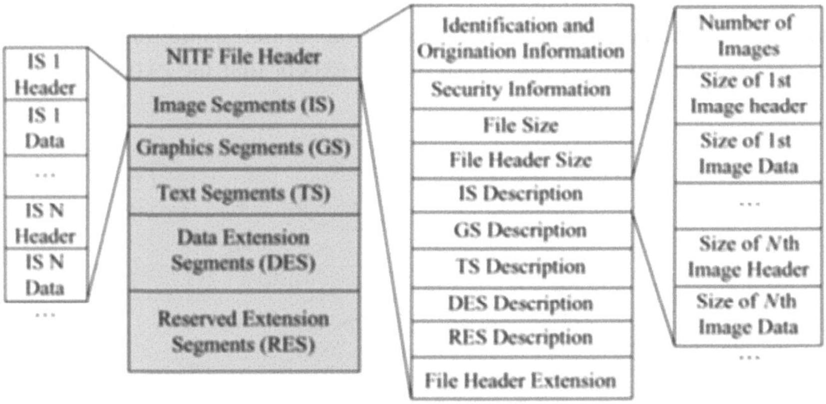

Figure 7.7 The NITF file structure (based on Fig. 8.3, 8.5, and 8.7 in NIMA, (1998))

7.6.2 The NITF Header

The NITF file level header consists of the following components: (1) identification and origination information such as version number, originating station identifier and date and time; (2) security information including security classification and classification system, security authority, control number, etc.; (3) total file size and file header size; (4) descriptions of sub-segments including number of sub-segment components (e.g. number of images) and the sizes of component header and component data; and (5) file header extension containing information about tagged recorded extensions, which are data components whose identification are recorded with or controlled by NGA. A complete field level description of the header is presented in Table A-1 of the NITF specification (NIMA, 1998).

7.6.3 The NITF Image Data Segment

The first three components following the NITF header are designed for image, graphics, and text types of data, respectively. The last two are intended for adding future data storing capabilities unspecified by the current NITFS. Each segment includes one or more sub-segments that consist of a header and a data component. The header of a sub-segment provides detailed information of that sub-component, such as data type, bit per pixel, dimensionality, organization, and various metadata. Among the three specified data segments, the image segment is most relevant to the earth observation data and it will be introduced in the following paragraphs.

7.6.3.1 The NITF Image Header

The NITF image header contains more than fifty required fields and a number of conditional fields that describe various aspects of the image data. These include: (1) general metadata such as image identification, category, creation date and time, origination, and title and comments; (2) security information fields similar to those in the file level header; (3) data type, dimensionality and organization including number of rows/ columns/bands, dimension order (e.g. band sequential and band interleaved by rows); (4) image coordinate system and geographic location information; (5) compression information; (6) image color representation, display and lookup table information; (7) image block information; and (8) user defined data pertaining specifically to the image. A complete field level description of the image header is presented in Table A-3 of the NITF specification (NIMA, 1998).

7.6.3.2 The NITF Image Data

The image data contains the data composing the image and is stored in the format as specified by the image header. It should be noted that in the current version (V2.1) the maximum number of bits allowed is 16 for integer and 32 for floating point data, which are not appropriate for certain earth observation and geolocation data.

7.6.4 The NITF Related Software

Three publicly available software tools, CIVA, Mr. BIFF, and SNIPER (http://164.214.2.51/ntb/baseline/software/jitctest.html), can be used to display and validate the content of NITF files, including header, subheader, and data contents. These tools are developed and used by the Joint Interoperability Test Command's NITFS Certification Test Facility & Lab in the certification process of NITF imagery systems. A number of commercial image processing software packages, such as ERDAS/IMAGING and ENVI, also include modules to read, write and display NITF data.

7.7 TIFF and GeoTIFF

The Tagged-Image File Format (TIFF) is designed for raster image data (Adobe Developers Association, 2003). It is primarily used to describe the unsigned integer type bi-level, gray scale, palette pseudo color, and three-band full color image data but can also be used to store other types of raster data. Although TIFF is not considered as a geospatial data format, its extension, GeoTIFF, which includes standardized definition of geolocation information, is one of the most popular formats for earth observing remote sensing data.

7.7.1 The Physical Layout of TIFF

The TIFF physical layout includes four components: (1) an 8-byte TIFF header containing byte order, TIFF file identifier, and the offset address (in byte) of the first Image File Directory (IFD) in the file; (2) one or more IFDs, each containing the number of directory entries, a sequence of 12-byte directory entries, and the address of the next IFD; (3) directory entries each having a tag number indicating the meaning of the tag, a data type identifier, a data value count containing number of values included in this tag, and an offset containing the file address of the value or value array; and (4) the actual data of a tag. Because the offset is of 4-byte size, the actual value of a tag is directly put in the offset field if and only if there is only one value and the value fits into 4 bytes (see Fig. 7.8).

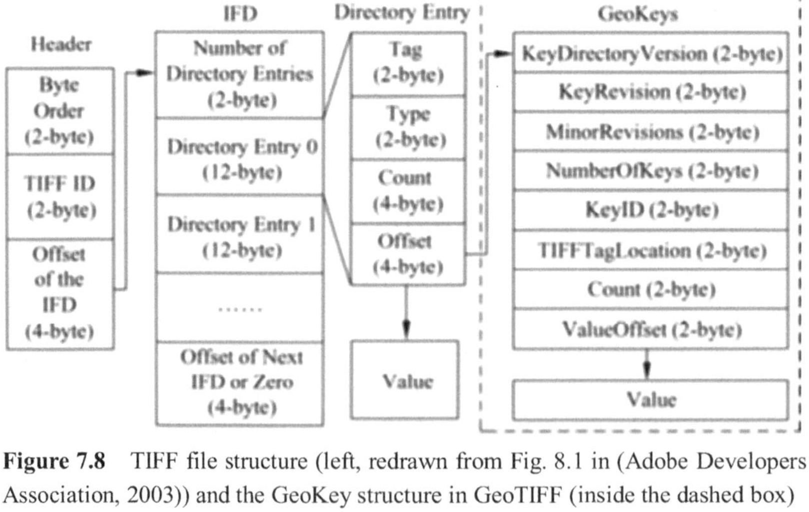

Figure 7.8 TIFF file structure (left, redrawn from Fig. 8.1 in (Adobe Developers Association, 2003)) and the GeoKey structure in GeoTIFF (inside the dashed box)

7.7.2 The TIFF Data Model

TIFF uses tagged structure to define an image data set and its storage. An image is specified by a collection of tags. For example, to describe an image's storage in a file, the following parameters are needed: column and row numbers, data type, and file address of the image data. These parameters are presented in TIFF as tags using ImageWidth, ImageLength, BitsPerSample, and StripOffset. These are similar to the dimension (containing numbers of rows and columns), the data type, and the SDS data array objects in HDF4 or the dataspace, datatype, and dataset objects in HDF5.

TIFF defines four types of images: bi-level, gray scale, palette color, and full color images. Most required tags are the same for different images but additional tags may be needed for different types of images. The following tags are required tags for all image types: ImageWidth, ImageLength, Compression, PhotometricInterpretation, StripOffset, RowsPerStrip, StripByteCounts, XResolution, YResolution, and ResolutionUnit. Among these tags, RowsPerStrip, StripOffset and StripByteCounts specify how many strips an image is divided, number of bytes in each strip, and the address of a strip data in the file (Note: in Version 6.0, strips can be replaced with tiles so that an image can be divided not only at the row direction but also at the column direction. This is, however, an extension and is not supported by Baseline TIFF readers); Compression specifies whether an image data is compressed or the compression type if it is; ImageWidth and ImageLength tell the actual dimensions of an image; and the rest of the parameters define the image displaying specifications in a display device. For images other than the bi-level type, a BitsPerSample tag is necessary to define the bit-depth of unsigned integer type data, which is default TIFF image data type. A SampleFormat can be used to indicate other data types.

A single IFD does not support general multidimensional scientific grid data with more than three dimensions. In fact, TIFF is aimed primarily at either single band (i.e. two dimensional, 2D) or three band (i.e. three dimensional with a fixed size of 3 for the third dimension) images. A SamplePerPixel tag is required for a three dimensional image and this value is usually equal to 3. For images with more than three 2D components, an ExtraSamples tag must be used to describe the meaning of the extra components (samples). For example, if a full color image has SamplesPerPixel value of 5, then it must contain an ExtraSamples tag with a value of 2. In order to use TIFF to store high (>3) dimensional grid data, multiple IFDs need to be used to break the higher dimension(s) (i.e. those >3) into sub-files. This capability is supported by TIFF Version 6.0 but not required by a Baseline TIFF reader, which may not read any IFD beyond the first one. Another issue of using TIFF as a general grid data format is that the popular Band Sequential (BSQ) organization, which can be defined by specifying the PlanarConfiguration tag value to be 1, in science data is also treated as an extension. Thus, baseline TIFF readers may not support the BSQ data organization, which is undesirable for earth observing remote sensing data.

A number of TIFF tags are used to store metadata information like the HDF attributes, such as DocumentName, dateTime, ImageDescription, Copyright, MaxComponentValue, and MinComponentValue.

TIFF tag numbers range from 0 to 65,535. Tags numbered 32,768 or higher are usually referred to as private tags and can be reserved and registered, through the TIFF administrator, for an organization to store particular information. Such registration can avoid compatibility issue since otherwise different people may use the same tag for different meanings.

7.7.3 GeoTIFF

GeoTIFF is an extension to the TIFF 6.0 specification aiming to provide a standard mechanism of describing geolocation information which is not specified in TIFF. Similar to the relationship between HDF and HDF-EOS, a GeoTIFF file is a valid TIFF file and inherits the file structure and data model of TIFF, but it provides a framework for specifying geographical spatial reference systems in TIFF. An easiest and most straightforward approach to defining geographic information could be to use the private TIFF tags. However, TIFF tag is a limited resource and should be used sensibly. GeoTIFF designers adopted a wise way to use just six tags to store a broad range of geospatial referencing information. GeoTIFF uses GeoKeys, whose structure is essentially the same as that of standard TIFF tags, to store such geospatial information as projection, coordinate system, datum, and ellipsoids (Ritter and Ruth, 2000).

7.7.3.1 The TIFF Tags Reserved for GeoTIFF

Six TIFF tags are reserved for GeoTIFF:

(1) ModelPixelScaleTag (33,550);
(2) ModelTransformationTag (34,264);
(3) ModelTiepointTag (33,922);
(4) GeoKeyDirectoryTag (34,735);
(5) GeoDoubleParamsTag (34,736);
(6) GeoAsciiParamsTag (34,737).

Among these reserved tags, the GeoKeyDirectoryTag is an entry point for all GeoKeys, i.e. all GeoKeys are accessed through this tag.

7.7.3.2 The GeoKeyDirectoryTag and the Structure of GeoKey

The four fields of the GeoKeyDirectoryTag (see the directory entry in Fig. 7.8) contain following information: (1) tag value = 34,735, (2) data type = 3, indicating unsigned short integer, (3) *N*, an integer larger or equal to 4 indicating the number of values contained in this tag, and (4) the file address of its values. Here the values pointed to by this tag are the GeoKey values. The GeoKey values include an array of four or more unsigned short integer values, as indicated by the second and third fields of the GeoKeyDirectoryTag, that are grouped into blocks of 4 as the following: {{KeyDirectoryVersion, KeyRevision, MinorRevision, NumberOfKeys}, {KeyID, TIFFTagLocation, Count, ValueOffset} (see Fig. 7.8),..., {KeyID, TIFFTagLocation, Count, ValueOffset}}. The first group contains the key's header information, including version number and number of GeoKeys. Following the header are blocks of keys each containing four fields. The number of blocks is indicated by the NumberOfKeys value in the header. The four fields in each block (i.e. each GeoKey) are designed very similar to the fields in a TIFF tag. The first field, KeyID, is the GeoTIFF key identification number, which functions identically as the TIFF tag number but is completely independent of TIFF tag space. The KeyID tells the meaning of the value(s) contained or pointed to by this key, just as a TIFF tag does. The difference is that TIFF tag provides such parameters as ImageWidth while GeoKey gives geospatial related information such as the GeogPrimeMeridianGeoKey that specifies the location of the prime meridian for a geographic coordinate system. The second field, TIFFTagLocation, indicates from which TIFF tag the GeoKey value(s) can be obtained (note that in TIFF all values must be retrieved through tags). Similar to the TIFF tag, if this field contains a value zero, then it tells that there is only one unsigned short integer value in the key and the value is stored directly in the fourth field of the key, the ValueOffset field. The third field, Count, provides the number of values contained in the key. The last field, ValueOffset, points to the position into the value array of the TIFF tag referred to by the TIFFTagLocation field (note that, unlike the TIFF tag's offset field, it does not directly indicate file byte offset). The following example shows how the GeoKey is used:

The value of the GeoKeyDirectoryTag = (1, 1, 2, 1, 2057, 34736, 1, 0)
The value of the GeoDoubleParametersTag (34,736) = (6,378,206.4)

The first block of four values of the GeoKeyDirectoryTag indicates that this is the Version 1.1.2 GeoTIFF GeoKey directory and that there is only one GeoKey in this tag. The second four values show that the GeoKey (2,057) is a GeoSemi-MajorAxisGeoKey, meaning that the value of the key represents an ellipsoid semi-major axis value. Field value 34,736 tells that the value of this GeoKey is obtained from TIFF tag 34,736, which is one of the six GeoTIFF reserved tags, i.e. the GeoDoubleParametersTag, used to store double precision floating point values. The last field value, 0, indicates that the GeoKey value is stored in the first (0th) position of the data value array pointed to by the offset field in the GeoDoubleParametersTag, i.e. 6,378,206.4.

Similar to TIFF tags, the range of GeoKey ID number is from 0 to 65,535, with numbers 32,768 and higher being private keys. The other key numbers include: (1) reserved keys from 0 to 1,023 and from 5,120 to 32,767; (2) GeoTIFF configuration keys between 1,024 and 2,047; (3) Geographic/geocentric coordinate system (CS) parameter keys between 2,048 and 3,071; (4) projected CS parameter keys between 3,072 and 4,095; and (5) vertical CS parameter keys between 4,096 and 5,119.

7.7.3.3 The Geolocation Information in GeoTIFF

GeoTIFF can be used for both georectified and georeferenced data. Geolocation information in georectified data are represented using geographic CS GeoKeys, projected CS parameter GeoKeys, and vertical CS GeoKeys. For georeferenced but not georectified data, three GeoTIFF reserved TIFF tags, ModelPixelScaleTag, ModelTiepointTag, and ModelTransformationTag, are used to give information linking raster (i.e. grid, image) coordinate space with model (i.e. geographic/ projected) coordinate space. The ModelTransformationTag provides transformation matrix to transform raster coordinate values to model coordinate values while the ModelTiepointTag provides coordinate values in both rater and model space for a number of tie-points., which can be used to derive transformation relationships. The ModelPixelScaleTag is used when there is a linear relationship between raster and models space. A single tie-point is necessary in linear transformation case.

7.8 Summary

This chapter provides a review of a number of data formats commonly used for grid type geospatial data, with an emphasis on the HDF and HDF-EOS formats. The latter is selected by the NASA EOS program as its standard data product format. There are a variety of other data formats that are suitable for storing raster image data such as the newly emerging JP2 format based on the JPEG2000 Image Coding System, an ISO standard, and proprietary image formats used in commercial image processing software packages. However, it is difficult to cover most commonly used public data formats in a brief chapter as this. It is hoped that the material presented in this chapter is helpful to those who do not have much experience in geospatial grid data formats, especially the HDF4- and HDF5-based HDF-EOS formats and their programming models.

Acknowledgements

The materials presented in this chapter are mainly summarized and/or directly cited from the listed references. I would like to thank the individuals and groups who authored/prepared these documents and made them available to the public. I also wish to thank Mr. Larry Klein of the EOSDIS Maintenance and Development Project and two anonymous reviewers for their constructive comments and suggestions.

References

Adobe Developers Association (2003) TIFF Revision 6.0, http://www.adobe.com/Support/TechNotes.html

FGDC (2002) Content Standard for Digital Geospatial Metadata: Extensions for Remote Sensing Metadata. FGDC-STD-012-2002, Reston, Virginia

FGDC (1998) Content Standard for Digital Geospatial Metadata. FGDC-STD-001-1998, Reston, Virginia

ISO (2004a) Geographic Information—Metadata – Part2. ISO 19115-2, Geneva, Switzerland

ISO (2004b) Geographic Information—Sensor and Data Models for Imagery and Gridded Data. ISO 19130 Committee Draft, Geneva, Switzerland

ISO (2003) Geographic Information—Metadata. ISO 19115: 2003(E), Geneva, Switzerland

NASA (2004a) HDF-EOS Library User's Guide for the EMD Project, Volume Ⅰ: Overview and Examples. Technical Paper 170-EMD-001, Revision 2, Greenbelt, Maryland

NASA (2004b) HDF-EOS Library User's Guide for the EMD Project, Volume Ⅱ: Function Reference Guide. Technical Paper 170-EMD-002, Revision 2, Greenbelt, Maryland

NASA (2004c) HDF-EOS Interface Based on HDF5, Volume Ⅰ: Overview and Examples. Technical Paper 175-EMD-001, Revision 02, Greenbelt, Maryland

NASA (2004d) HDF-EOS Interface Based on HDF5, Volume Ⅱ: Function Reference Guide. Technical Paper 175-EMD-002, Revision 2, Greenbelt, Maryland

NASA (1994) Proposed ECS Core Metadata Standard Release 2.0. NASA Technical Paper 420-TP-001-005, Greenbelt, Maryland

NCSA (2003a) HDF User's Guide, HDF4 Release 2.0. NCSA, Champaign, Illinois

NCSA (2003b) HDF User Reference Manual, HDF4 Release 2.0. NCSA, Champaign, Illinois

NCSA (2003c) HDF5 File Format Specification, http://hdf.ncsa.uiuc.edu/HDF5/doc/H5.format.html

NCSA (2003d) Introduction to HDF5 Release 1.4, http://hdf.ncsa.uiuc.edu/HDF5/doc/H5.intro.html

NIMA (1998) National Imagery Transmission Format Version 2.1. NIMA Document MIL-STD-2500B, p 110

Ritter N, Ruth M (2000) GeoTIFF Format Specification. Revision 1.0, http://www.remotesensing.org/geotiff/spec/geotiffhome.html

8 A Simple, Scalable, Script-Based Science Processor

Christopher Lynnes

8.1 Genesis of the Simple, Scalable, Script-Based Science Processor

In 1999, the impending launch of the Terra satellite, combined with concerns about the ability of commercial software to process its high data volumes, led the Goddard Earth Sciences Distributed Active Archive Center (GES DAAC) to develop a contingency science processing system. Severe time and money constraints forced the GES DAAC towards a minimalist architecture that eventually becomes the Simple, Scalable, Script-based Science Processor (S4P). It was named Simple for its architecture and small size (a few thousand lines of code), Scalable for its ability to scale up to heavy processing loads or down to modest automation tasks, Script-based for its reliance on the Perl scripting language for its infrastructure, and Science Processor for its genesis in running scientific algorithms.

Nevertheless, the commercial software option was used when data first began flowing to the GES DAAC from the Moderate Resolution Imaging Spectroradiometer (MODIS) instrument aboard Terra. In the meantime, however, S4P was found to be useful for a number of science-processing-related tasks: a data mining system for processing data from the Tropical Rainfall Measuring Mission (Lynnes and Mack, 2001); an on-demand subsetting system for distributing MODIS data subsets; and a system for processing MODIS data received by Direct Broadcast stations.

In 2001, it became apparent that the commercial software would be too expensive to scale up for MODIS reprocessing job rates. As a result, NASA funded the GES DAAC to develop a full-scale processing system to handle the reprocessing campaigns. This system, the S4P for Missions (S4PM), ultimately took on processing for the forward MODIS data stream and the processing for the Atmospheric Infrared Sounder (AIRS), Humidity Sounder for Brazil (HSB), and Advanced Microwave Sounding Unit (AMSU). Thus, the commercial software was completely replaced by S4PM at the GES DAAC for processing science data resulting in substantial cost savings. Added benefits have accrued as well: S4PM has proven easier for staff to operate, and local software engineers are able to provide 24×7 support, supplying fixes for bugs within hours of their manifestation.

8.2 Architecture and Design

8.2.1 The S4P Kernel

The S4P system is a data-driven processing system. That is, the system is (almost) completely idle in the absence of input data. Conversely, the arrival of data triggers the system to begin processing. Thus, there is little notion of planning within the system; the system is managed by regulating the resources available to it and the input data that feeds it. Once the data arrive, the actual production that takes place is controlled by predefined production rules. The data-driven paradigm is particularly appropriate for 24×7 processing with high rates of job execution, both of which are not conducive to "hand-crafted" production plans. The GES DAAC's processing environment operates continuously, with as many as 16,000 executions of various science algorithms occurring in a 24-hour period.

The S4P system is based on the metaphor of a factory assembly line. An implementation is made up of a number of "stations", each of which performs a particular type of processing on its inputs, sending the outputs to one or more downstream stations (see Fig. 8.1). Of course, such a virtual assembly line has two advantages over a physical assembly line: multiple tasks can be performed simultaneously at a given station, and the output products can be easily duplicated and sent to multiple downstream stations.

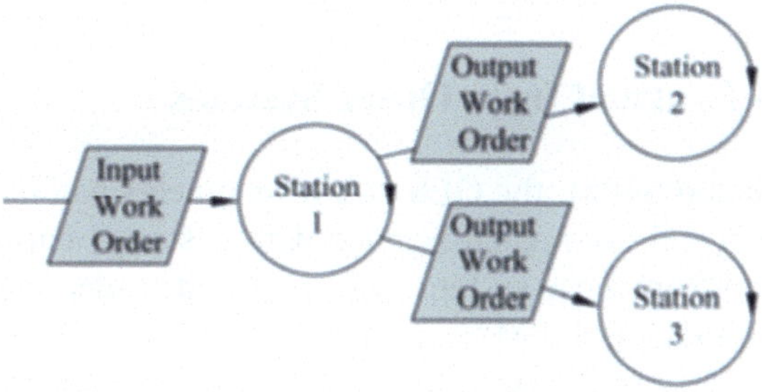

Figure 8.1 S4P stations process ASCII work orders as input, forking one job for each work order. Output work orders are shipped downstream to other stations. A given work orders can be sent to multiple stations

The assembly line metaphor is implemented by mapping stations to file system directories. Within each station, one or more scripts are assigned to operate when input files arrive. A continuously running daemon process, named the "stationmaster", monitors the station for input files and executes the appropriate scripts when they arrive. The output files are then sent to the appropriate downstream stations.

This design was inspired by the Advanced Very High Resolution Radiometer (AVHRR) Data Acquisition and Processing System (ADAPS), developed at the EROS Data Center (http://edcdaac.usgs.gov/1KM/ops.html). The ADAPS, developed in the late 1980's and early 1990's, also uses a number of daemons scanning directories for input data. One key modification in S4P is the movement of small ASCII "work order" files instead of the much larger data files. The work orders contain not only information about input files, but can incorporate additional useful information such as metadata and production rules.

An S4P system generally consists of two parts, the kernel and the data-specific implementation. The S4P kernel is generic, consisting of the stationmaster daemon, a graphical user interface for monitoring and controlling stations, and a number of utility programs and subroutines. The data-specific implementation consists of the particular station configuration and the various scripts and executable images that run within those stations. This allows S4P to support a range of scientific applications, from complex science data production systems to very simple processing consisting of just a few steps. Thus, S4P is scalable down as well as up. An example is a small ad hoc S4P based system developed in only a few days in 2001 to provide MODIS data rapidly to field experiments as part of the Asian Atmospheric Particulate Environment Change Studies. On the other end, the most robust S4P implementation fielded to date is the S4P-Missions, which handles all MODIS and AIRS science data production that is performed by the GES DAAC.

In developing both the S4P kernel and the S4PM system, we followed several design principles that were informed by the lessons learned from other systems.

8.2.2 Lessons Learned from Other Systems

In developing S4P and S4PM, the GES DAAC engineers had the luxury of direct experience with several automated science data systems, dating back to 1991. The lessons learned from these systems informed and guided the architecture and implementation of S4P and S4PM:

(1) Lesson 1: Most of the operations staff's time is spent troubleshooting. A reasonably automated system should require little or no effort from the operators if all conditions are nominal. Therefore, most of the operational staff's time is likely to be engaged in troubleshooting problems within the system. This principle led S4P developers to make a concerted effort to make troubleshooting and failure recovery as easy as possible for the operators.

(2) Lesson 2: More code leads to more code errors. Many of the most commonly used software reliability models scale the predicted failures with some

measure of the code size (Musa, 1975). (The increase can be more than linear, as more code also typically implies more code modules and thus errors within modules are supplemented by errors between modules.) As a result, the S4P and S4PM developers made a conscious attempt to trade off potential functionality against the probability of introducing additional sources of failure.

(3) Lesson 3: Errors in the science data processing system are often precipitated by errors in the data archive. That is to say, when the science processing system is tightly integrated with the data archive system, errors in the latter will often propagate into the former. Of course, one way to address this problem is through rigorous error checking. Alternatively, one can decouple the processing system from the archive as much as possible.

(4) Lesson 4: Reliance on Commercial Off-the-Shelf (COTS) software tends to limit the hardware options available. This is a simple function of the fact that the set of hardware options is the intersection of the hardware supported by the various COTS software being employed. The more COTS software used, the smaller the intersection is likely to be. This can be alleviated by employing client-server architecture, but at the expense of introducing multiple single-points-of-failures into the system. Alternatively, open-source software and locally developed software can expand the hardware options available. In the first case, popular open-source products often spawn porting efforts to additional platforms. In the second case, the local development team controls the platforms to be supported, rather than a software vendor. Back in 1999, the impending rise of inexpensive, powerful Intel/Linux-based systems made platform availability an important consideration.

8.3 Design Principles

8.3.1 Design for Trouble

One of the most fundamental design principles used was "Design for Trouble". With such a large number of processing jobs running, the S4P system needed to be sufficiently automated so that virtually no user input was required for nominal processing operations. The corollary to this is that most operator actions are in response to failures in the system. Therefore, it made sense to optimize the system for troubleshooting. In practice, this led to the following design guidelines.

1. Keep Everything Visible

Many complex systems rely on shared memory, remote procedure calls or inter-process communication to pass information or commands between programs. However, these communications are very difficult for operators to see. Worse yet, they are often ephemeral, so that if something goes wrong in the communication, the evidence is lost, or confined to that reported by one side. In contrast, S4P was designed to pass all information via text files. Thus the inputs to any software failure are always available to the person troubleshooting the problem.

2. Keep Related Things Together

One of the most time-consuming aspects of troubleshooting complex systems is tracing errors through a multitude of log files, configuration files and program code. In S4P, this is mitigated by using the station directories as an organizing principle for the various files. As a result, all files relating to particular job execution can be found in the subdirectory formed for that particular job or its parent station directory.

3. Enable Quick Drill-Down, Diagnosis, and Recovery

The graphical user interface (GUI) for S4P (named tkstat) is the focal point for drilling down into problems. Its top level (see Fig. 8.2) is designed to provide the operator with a view of an entire processing stream. Each row corresponds to a station, and each grid box to a job execution, either failed (red), overdue (yellow), running (green) or queued (blue).

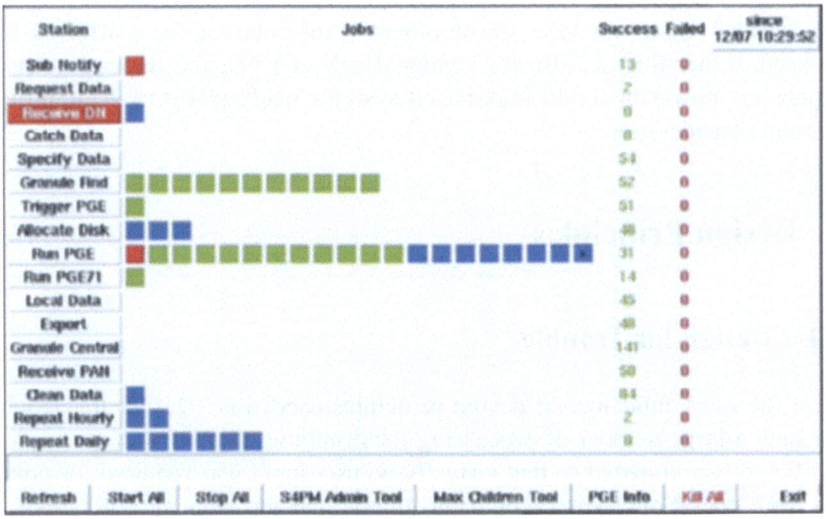

Figure 8.2 A graphical user interface provides a view of nearly all jobs in an S4P system. Each row represents an S4P station, with the left-hand button indicating the station name. Stopped stations are shown as red buttons. Failed jobs are shown as red boxes, and always appear on the left hand side of a row. Green boxes indicate currently running jobs and blue boxes indicate queued jobs. Clicking on a station button or job box invokes the drill-down interface

In almost all cases, drill-down to the lowest level of job-related information can be accomplished with only two clicks of the mouse: the first click brings up a window to view the contents and status of the job directory in question, and the second click displays the contents of a log or configuration file (Fig. 8.3). For log files, the information most germane to the failure is most often at or near the end of the log file. Therefore, the tkjob file viewer was designed to automatically scroll to the bottom of any log file selected by the user.

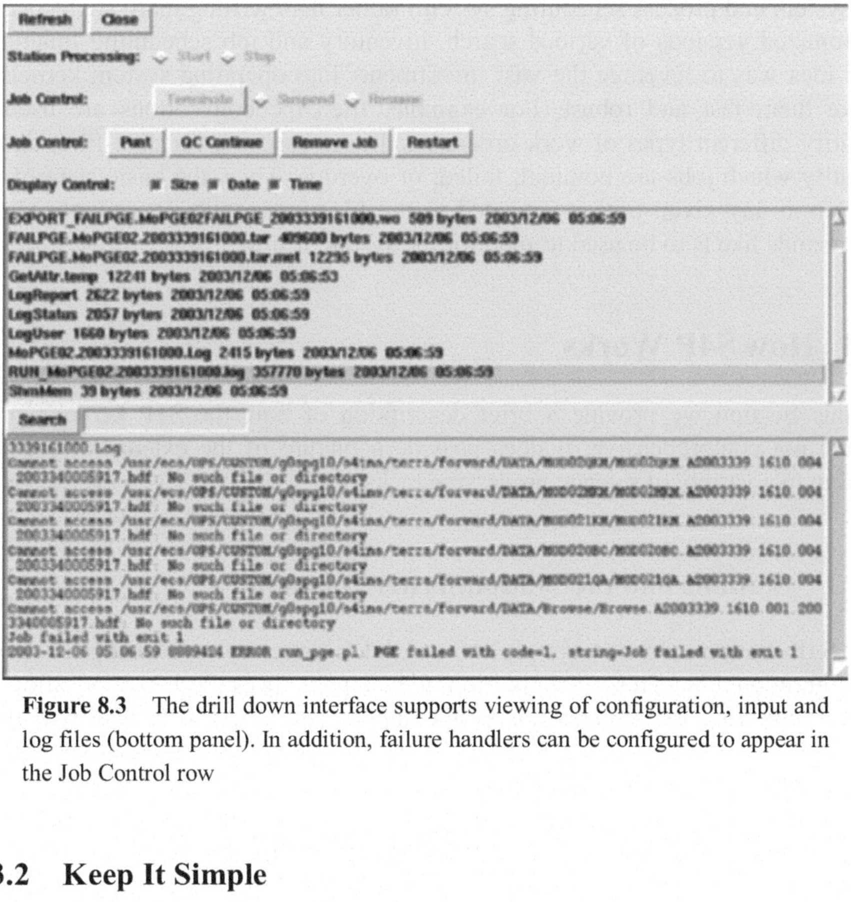

Figure 8.3 The drill down interface supports viewing of configuration, input and log files (bottom panel). In addition, failure handlers can be configured to appear in the Job Control row

8.3.2 Keep It Simple

The "Keep It Simple (Stupid)", or KISS, principle has been promoted in a variety of engineering disciplines. However, its vagueness often makes it difficult to translate into action. Accordingly, it was broken down into several specific guidelines for S4P implementation.

1. More Code = More Bugs

The most obvious KISS guideline is simply to keep the system small, not through overly elegant (or obscure) code styles, but rather by implementing only the bare minimum functionality required. On a day-to-day basis, this was typically translated via the Extreme Programming rule "You aren't going to need it" or YAGNI (Beck, 2000), or the 80/20 rule (Juran and Gryna, 1951) that 80% of functionality can be achieved in the first 20% of code.

2. Use the Operating System Wherever Possible

Coupled with the 80/20 rule, this involved using the operating system (i.e. the filesystem and process scheduling system) rather than writing more sophisticated, customized versions of various search, inventory and job scheduling functions. The idea was to leverage the vast investments into operating system kernels to make them fast and robust. For example, filename conventions are used to identify different types of work orders, facilitate searching for data files, and to identify which jobs are nominal, failed, or overdue. Thus, the basic state of the system at any given time is encoded in the filesystem, allowing simple UNIX commands like ls to be used to monitor the system health.

8.4 How S4P Works

In this Section we provide a brief description of how the S4P kernel works. While not comprehensive, it does provide a picture of the extensibility of the system to a variety of science processing tasks.

8.4.1 Stations and the Stationmaster Daemon

An S4P station consists simply of a UNIX directory containing a station configuration file. This configuration file actually uses Perl syntax, allowing syntax checking by running the Perl compiler (i.e. "perl -c"). The two most important variables in the configuration files are two associative arrays: cfg_commands, which maps input work order "types" to the programs to be executed with the work orders, and cfg_downstream, which maps output work order types to the downstream stations to which they are to be routed. (The associative arrays of Perl provide a particularly compact way of representing these maps in the station configuration file.) Work order types are encoded in the filename conventions of S4P work orders, which follow the pattern DO.< work_order_type >.< job_id >.wo. Input work orders supplied by non-S4P systems usually do not follow this convention, so an additional method of selecting the proper executable is provided using pattern matching constructs.

Each station directory is monitored by its own stationmaster daemon. This is a Perl program that polls the directory on a fixed interval (typically five seconds) for the arrival of files that are recognizable as work orders. On detection of a work order, stationmaster creates a unique subdirectory for the job. This uses the prefix "RUNNING" to allow easy monitoring either from the command line or the tkstat interface. It also serves to prevent interference by multiple jobs in a subdirectory and keeps the various files for that job together in one place.

After changing into the job subdirectory, the stationmaster then forks off a child process, which runs the appropriate program based on the cfg_commands associative array. The program specification can be any combination of an executable and static arguments; the input work order filename is then appended as the last argument. The executable is responsible for parsing and interpreting the contents of the input work order. Thus, the work order can be in any format, and may even represent the data itself. In practice, however, most S4P implementations use structured ASCII formats, as they are easier to construct for testing and easier to read for troubleshooting.

If the job runs to completion and exits with zero, the stationmaster child looks in the subdirectory for files recognizable as output work orders. Using the cfg_downstream associative array, it moves each output work order, along with a copy of the job log, to one or more other stations. If the job fails, either by not running, terminating abnormally, or exiting with a non-zero exit code, the stationmaster child replaces the "RUNNING" prefix with "FAILED".

8.4.2 Monitoring Stations and Jobs

The use of naming conventions allows basic monitoring of an S4P system from the command line using the "ls" command, or using a graphical filesystem browser. However, it is useful to be able to see the status of a whole S4P system in a single glance, particularly those jobs that have failed. Accordingly, tkstat, a simple graphical interface program written in Perl/Tk was developed as the main monitoring tool (see Fig. 8.2). This shows the system as a grid, with each row generally representing a station and each grid box representing a running, failed or queued job. For stations that run multiple different types of jobs, depending on the work order type, each job type can be given its own row.

In the default configuration, failed jobs are indicated by red grid boxes, overdue jobs by yellow boxes, currently running jobs by green and queued (pending) jobs by blue boxes. (Grayscale or symbol-based schemes can also be used for color-blind users.) Since the emphasis in S4P is on detecting problems ("Design for Trouble"), the jobs are sorted so that the failed jobs are placed to the left, followed by overdue jobs.

Clicking on the box for a failed job brings up an interface for drilling down into problem jobs (Fig. 8.3). At its root, the drilldown interface is a file browser, but with a couple of key features. The first is that all log files are displayed starting at the bottom of the file, rather than the top. This is because the log messages most pertinent to a failed job tend to be the last ones written to the log file. A second key feature is the ability to specify error handlers for each station. An associative array in the station configuration file (cfg_failure_handlers) maps button names to the scripts to be executed when the button is pushed by the operator. The most common actions are to remove the whole job or to restart it by resubmitting the work order to the parent station. However, station-specific failure handlers can be specified as well, such as specific cleanup procedures or file repair. Similarly, manual override procedures can be associated with jobs that are still running. Finally, graphical user interface programs can be associated with a station, again by an associative array (cfg_interfaces) in the station configuration file.

8.4.3 Station Configurability

In addition to the basic configuration variables referred to above, the station configuration file supports a number of additional configuration parameters:

(1) Maximum children: the maximum number of child processes running in the station at any one time. This is useful for throttling stations that run particularly CPU-intensive jobs (such as the science processing algorithms themselves.) Setting this value to 1 has the effect of making the station "single-threaded", which is useful for processes that update existing files.

(2) Job priority: the order in which queued jobs will be forked. Priority may be based on the type of job or the arrival time. The default order is alphabetical, which is useful for data processing when the data start (or end) time is included in the filename.

(3) Reservations: reserves a certain number of job slots for a given job type (preventing starvation by job priority).

(4) Maximum time: the elapsed time beyond which a job is deemed "overdue". This can be configured to be specific to different job types within a station.

(5) Polling interval: overrides the default 5-second interval between stationmaster checks for new work orders.

8.5 S4P Reuse

Even in its early development as a contingency system, S4P was designed with extensive reuse in mind, so that the effort would not be wasted if it were not deployed on behalf of Terra. As a result, it was developed with an open architecture: that is, it is easy for someone other than the original developers to reuse and, more importantly, extend S4P. A key contributor to this aspect is the Perl scripting language in which S4P is written. Perl is not only available for a remarkable array of platforms, but it runs remarkably similarly in all of them. However, it is important to note that no language restriction is placed on the scripts or executables that run within S4P stations. These can be any program that runs from a command line (i.e. non-interactively). S4P is also not dependent on any COTS software whatsoever. Aside from the standard Perl distribution, the only system requirements are a few open-source Perl modules that are readily available from the Comprehensive Perl Archive Network.

These open architecture features lend themselves to an Open Source model of ongoing development. The S4P system has been approved for public release by NASA, and several efforts both within NASA and elsewhere have been reusing and extending S4P, feeding modifications back to the S4P baseline at the GES DAAC. Because S4P is not restricted to a particular science processing framework, it has been put to use in a number of different contexts to satisfy a variety of data management tasks. These are described below to illustrate the range of processing systems that can be accommodated.

8.5.1 On-Demand Subsetting

One of the earliest S4P-based processing systems was designed to support orders for data subsets. The sizes of many data files in the GES DAAC archive are quite large, some exceeding 800 MB. In many cases, users want just a few geophysical parameters, or a limited spatial extent, and sending the whole file is both wasteful of bandwidth and difficult for the users to handle. In response, the DAAC developed an automated system to redirect certain data requests to a subsetting system, whereupon the user's desired subsets are extracted and sent off to the user instead of the original data files (Fig. 8.4).

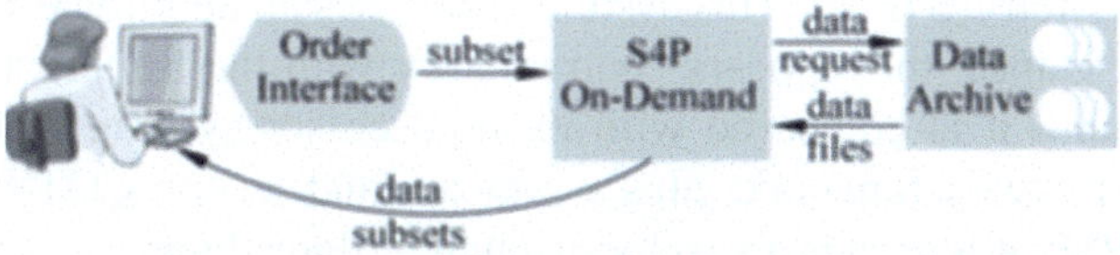

Figure 8.4 The S4P-based On-Demand Subsetting system intercepts orders for data subsets, obtains the data from the data archive, extracts the requested subset and ships the subsetted data back to the user. The result is substantial savings in data volume, making transmission and handling easier

8.5.2 Near-Archive Data Mining

Another early implementation was also targeted at reducing the distribution volume requirements. The Near-Archive Data Mining system was first implemented to support data from the Tropical Rainfall Measuring Mission (TRMM). The system allows a user to submit a data mining algorithm to be run at the GES DAAC on all files of a given data set (Lynnes and Mack, 2001). The results are then made available to the user for FTP download. This can reduce the volume that is distributed to the user by orders of magnitude. The initial implementation required the user to submit the algorithm to the GES DAAC for manual integration into the data mining system. A subsequent implementation to support data from the Earth Observing System provides a web-based interface that allows the user to upload, build, integrate and test the algorithm interactively.

8.5.3 Direct Broadcast Processing at IMaRS

The open architecture of S4P in particular lends itself to reuse by external organization. One such example is the system developed to process MODIS and SeaWiFS data obtained via direct broadcast by the Institute for Marine Remote Sensing (IMaRS). This system automatically processes raw downlink data to geophysical parameters such as sea surface temperature and chlorophyll concentration. One of the unique aspects of the IMaRS implementation is the ability to dynamically distribute the load among a cluster of computers by using a "router" station that sends input work orders to the node that will produce the result most quickly (Taylor, 2002).

8.5.4 S4P for Missions

Perhaps the most elaborate S4P implementation is the one fielded at the GES DAAC for standard processing of data from the Terra and Aqua missions, known as S4P for Missions (S4PM). This system comprises an additional 10,000 lines of code. At first glance, this appears contradictory with the Simplicity principle of S4P design, until it is compared with its predecessor system, which comprises several times more custom code, plus a database and several COTS products.

This S4PM system runs a variety of science algorithms:

(1) MODIS Level-1 for Direct Broadcast;

(2) MODIS Level-1 and Cloudmask standard products;

(3) Several subsetters and subsamplers;

(4) Level-1 and Level-2 standard products for AIRS, HSB and AMSU;

(5) Attitude and ephemeris pre-processors.

The S4PM system is specifically designed to run 24 hours a day, seven days a week, with a minimum of operator intervention. In addition to the standard tkstat monitor, it includes a number of station-specific graphical user interfaces to make the system easier to operate. Furthermore, as various failure modes have been identified, customized failure handlers have been introduced to handle them.

8.6 S4P for Missions Implementation

The ability of S4P to scale up is demonstrated in the S4PM implementation. This system runs continuously, 24 hours a day and seven days a week. Up to 16,000 algorithm executions each day are managed by S4PM, covering a fairly wide range of science processing, and spread over three Silicon Graphics servers and two Linux workstations. Though originally designed to process only MODIS data, S4PM was later enhanced (with minimal modification) to process AIRS, HSB and AMSU data, in addition to algorithms for preprocessing attitude and ephemeris data. The implementation is described below as an illustration of a full-featured S4P science processing system.

The overall design of the S4PM system can be represented as a data flow diagram (see Fig. 8.5), where each station implements a transformation, and input/output work orders implement the data flows among stations. The diagram has been arranged into functional areas for the purpose of discussion. The overall data flow through the system proceeds generally in a clockwise fashion beginning from the lower left. The primary data source and sink is the EOSDIS Core System data archive.

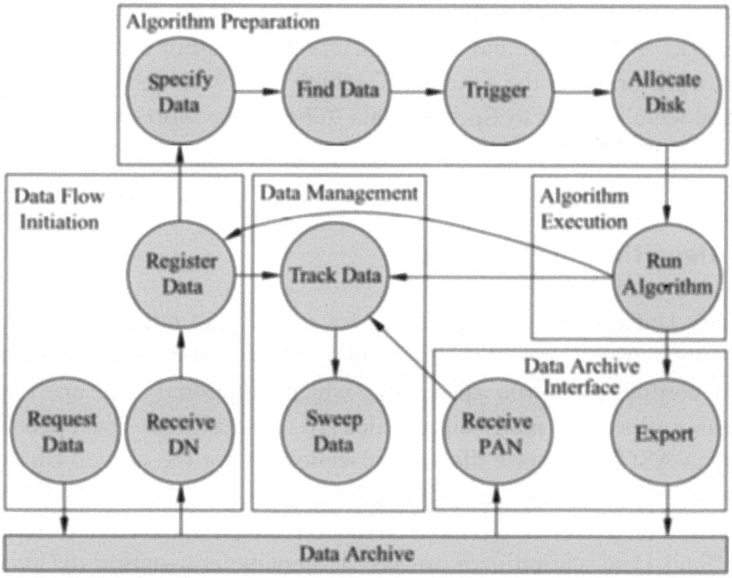

Figure 8.5 Station schematic of the S4P for Missions system. The overall flow of data through the system is clockwise, beginning at the lower left

8.6.1 Data Flow Initiation

Data flow into the system may be initiated in one of three ways:

(1) A standing order (subscription) in the ECS data server for FTP Push.

(2) An automated request mechanism that is triggered by a subscription notification that new data has been ingested into the ECS data server.

(3) An operator-initiated request using a Perl/Tk interfaces (tkcompreq.pl).

The second of these is the normal case for forward processing, while the last is used for reprocessing or filling in holes in the data (such as those caused by failed subscriptions). The first method, FTP Push subscription is used only for Direct Broadcast processing to eliminate a possible source of latency. The third case, the operator uses a GUI to select a time range and the input data types. These are submitted as work orders to the Request Data station, which monitors the input data area, metering out the requests as space becomes available. This allows the operator to create long-term "plans" without flooding the input disk area. The Request Data station also filters out data that have recently been requested by itself to prevent inadvertent duplicate orders.

Once the data are pushed to the S4P system, the ECS data server sends a Distribution Notice via email. The Distribution Notice (DN) is intercepted by a mail filter, and placed in the Receive DN station directory. Its arrival is detected by the Receive DN stationmaster, which forks a child to parse the notice and check the size of the files within it. The Register Data station applies quality control checks to the data, and then registers them within the S4PM system via an output work order to the Granule Tracking station. It then creates an output work order, which is sent to Specify Data to begin the work-order-driven part of S4P.

8.6.2 Algorithm Preparation

The Specify Data station is responsible for identifying other input data needed for processing a given algorithm. Each algorithm has production rules encoded in an ASCII configuration file specifying which additional data are required or optional, as well as how many output files are to be created. That specification is written to an ASCII template file with placeholders for the needed files, and sent as a work order to the Find Data station to be filled in with the actual locations of the files.

The Find Data station uses an S4PM filename convention to "predict" the filename (or rather the filename pattern) and attempts to locate the file in a specific directory within the filesystem. If found, it fills in the template with the actual directory and filename information. The station can account for required vs. optional data, with timers if necessary. Manual override routines are provided through the tkstat program to enable the release of jobs with optional data on timers.

If and when all the necessary data are found, the filled-out template with the full pathnames is passed to the Trigger station. The Trigger station converts the template to a Process Control File (PCF) following an EOSDIS toolkit standard (Arrington, 2000). The directories for the output data are left with recognizable placeholders. The Process Control File is sent as a work order to the Allocate Disk station, which allocates the output disk area and fills in the placeholder directory paths. This fully qualified PCF is then sent to the Run Algorithm station to execute the algorithm.

8.6.3 Algorithm Execution

The Run Algorithm station executes the algorithm and handles a number of bookkeeping tasks relating to the output, such as maintaining production history and linking browse files to data files. A work order describing the new data is sent to a second instance of the Register Data station, beginning the whole cycle again for downstream algorithms. The Run Algorithm station sends the same work order to the Export station to begin the archive process.

8.6.4 Data Archive Interface

The Export and Receive Product Acceptance Notice (PAN) stations implement an EOSDIS standard interface with the archive, known as the Science-Investigator-led Processing System (SIPS) interface (Earth Science Data and Information Systems Project, 2002). This consists of depositing a Product Delivery Record (PDR) file in an area to be polled by the archive. The archive responds by depositing a Product Acceptance Notice in a predefined area, either noting successful archiving, or describing which files failed and why. The Receive PAN station parses these notices, generating PDR files for resubmission of failed files when necessary.

8.6.5 Data Management

A key task of automated processing systems is deleting the input and output data when they are no longer needed, so that the disk space can be freed up. This can be complex when data are used by more than one algorithm. S4PM uses a central Track Data station to manage the data files while they are in the system. The initial registration from the Register Data station includes the number of times the data are expected to be used during nominal processing. As each algorithm runs, the Run Algorithm station sends a work order to the Track Data station so that the uses may be decremented. When the uses reach zero, Track Data sends a work order to the Sweep Data station to delete the data and deallocate the disk space.

8.7 Future Development

8.7.1 Case-Based Reasoning

Although S4P-based systems are relatively easy to monitor and troubleshoot, our ideal goal would be to run the systems in a "lights-out" mode, i.e. with virtually no manual intervention. On the surface, this may seem difficult given the variety of data, hardware, configuration and software errors that can cause failures. However, years of responding to operator calls for troubleshooting assistance have indicated that while the variety of errors is large, the resultant symptoms are fewer and the different possible recovery actions are fewer still. This kind of situation lends itself to a Case-Based Reasoning approach to automated failure recovery (Howe, 1995). Currently, S4P has the ability to log all job failures together with the associated manual failure recovery action. As we learn more about the failure mode ontogeny, we will be able to add software to enable an S4P system to learn automatically what kind of recovery action to take for a given fault. After a certain number of failures of a given type have been followed by the same manual recovery action, the S4P system will begin to execute the recovery action automatically.

8.7.2 Open-Source S4PM

The adaptation of the original MODIS-based system to handle AIRS algorithms indicates that it is fairly easy to adapt to new missions or data processing streams. Thus, the S4PM system could be a useful resource as NASA's Earth Science Enterprise evolves toward a more distributed, heterogeneous model for data systems. As a reusable system, S4PM holds a number of key advantages:

(1) free of proprietary / commercial code;

(2) small, simple and easy to understand;

(3) scalable both up to very complex processing systems and down to very simple ones;

(4) easily modified; and

(5) platform-independent.

As a result, work is underway to prepare S4PM for release as open source code. The informal open source arrangement with S4P kernel users has already proven valuable for feeding bug fixes and enhancements into the code, and such an arrangement is likely to be similarly fruitful for S4PM.

8.8 Conclusions

The development and deployment of both S4P and S4PM demonstrate that a robustly operational system need not require large amounts of complex code.

Indeed, the compact nature of the system makes the code more robust, as there are fewer potential sources of error. Of course, such a small system was possible only by making maximum use of existing capabilities, in this case within the operating system itself, a higher level language (Perl) and a number of contributed Perl modules. The other key component was an occasionally ruthless approach to screening possible requirements for inclusion. However, the result has been a highly flexible system that has been reused in a number of instances. Work is currently underway to better document the S4P system from a programmer's standpoint in order to enable further reuse.

References

Arrington D (2000) Release 6A SDP Toolkit Users Guide for the ECS Project. EOSDIS Core System Project Document 333-CD-600-001 http://edhs1.gsfc.nasa.gov/waisdata/sdp/html/cd33360001.html

Beck K (2000) Extreme Programming Explained. Addison-Wesley, Boston, p. 224

Earth Science Data and Information Systems Project (2002) Interface Control Document between the EOSDIS Core System (ECS) and the Science Investigator-Led Processing Systems (SIPS) Volume 0: Interface Mechanisms, NASA/GSFC

Howe AE (1995) Improving the reliability of artificial intelligence planning systems by analyzing their failure recovery. IEEE Transactions on Knowledge and Data Engineering 7(1): 14 – 25

Juran JM, Gryna FM (1951) Juran's Quality Control Handbook. McGraw-Hill, New York

Lynnes CS, Mack RM (2001) KDD Services at the Goddard Earth Sciences Distributed Active Archive Center. In: Grossman RL, Kamath C, Kegelmeyer P, Kumar V, Namburu RR (eds) Data Mining for Scientific and Engineering Applications. Kluwer, Dordrecht, 165 – 182

Musa J (1975) A theory of software reliability and its application. IEEE Trans. Software Eng. 1(3): 312 – 327

Taylor J (2002) Parallelization and Load Balancing of the Simple, Scalable, Script-Based Science Processor (S4P) with a Dynamic Job Router. University of South Florida Honor's College Thesis

9 The MODIS Reprojection Tool

John Dwyer and Gail Schmidt

9.1 Introduction

The MODIS Reprojection Tool (MRT) is designed to help individuals work with MODIS Level-2G, Level-3, and Level-4 land data products. These products are referenced to a global tiling scheme in which each tile is approximately 10° latitude by 10° longitude and non-overlapping (Fig. 9.1). If desired, the user may reproject only selected portions of the product (spatial or parameter subsetting). The software may also be used to convert MODIS products to file formats (generic binary and GeoTIFF) that are more readily compatible with existing software packages. The MODIS land products distributed by the Land Processes Distributed Active Archive Center (LP DAAC) are in the Hierarchical Data

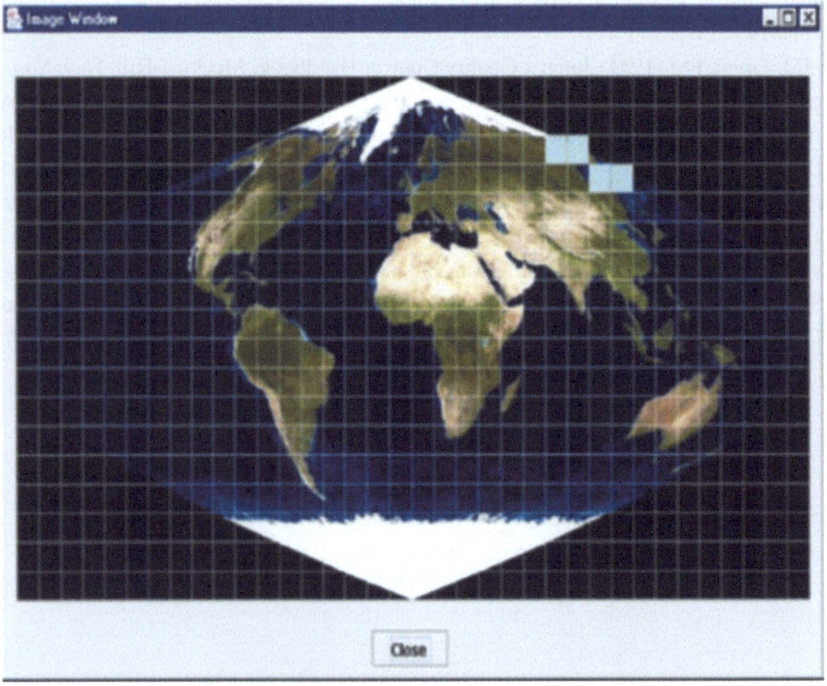

Figure 9.1 The view selected tiles popup from the ModisTool

Format - Earth Observing System (HDF-EOS), developed by the National Center for Supercomputing Applications at the University of Illinois at Urbana Champaign for the NASA EOS Program. Each HDF-EOS file is comprised of one or more science data sets (SDSs*)* corresponding to geophysical or biophysical parameters. Metadata are embedded in the HDF file as well as contained in a .met file that is associated with each HDF-EOS file. The MRT supports 8-bit, 16-bit, and 32-bit integer data (both signed and unsigned), as well as 32-bit float data. The data type of the output is the same as the data type of each corresponding input SDS.

The MRT software is highly portable and currently runs under four different operating systems: Windows 95/98/2000/NT/ME/XP, Linux (Redhat), Solaris, and Irix. The look and feel of the software is consistent across the different platforms. The MRT may be invoked from a user friendly graphical user interface (GUI) or from a powerful command-line interface. The GUI allows novices and users with light processing requirements to reproject, subset, or reformat MODIS products. The scriptable command-line interface, with its variety of options, is likely to be the method of choice for reprojecting large numbers of files.

9.2 MRT Functional Capabilities

9.2.1 The MRT GUI

The ModisTool refers explicitly to the MRT GUI (see Fig. 9.2), although MRT functions can be invoked using a command-line interface or in a background mode through the use of command-files. The GUI allows convenient construction of a parameter file to define the desired processing to be applied to the input data. The parameter file is an ASCII text file that can be easily modified or constructed for use in command files. The MRT User Guide (available at http://edcdaac.usgs.gov/landdaac/tools/modis/index.asp) provides detailed information on the MRT functional capabilities, methods of execution, and associated file contents and formats (Schmidt, 2003). This section reviews the MRT options by reference to the ModisTool primary interface shown in Fig. 9.1. Note that the contents on the left side relate to the Source (input) data and the fields on the right side correspond to the Destination (output) data.

9.2.1.1 Source

Default directories may be specified for the Input File, Output File, and Parameter File by using the *Settings* option in the ModisTool toolbar. Once a file is defined and highlighted under Input File, the file may be opened, the metadata can be viewed, or the geographic location of the selected tile may be viewed by selecting View Selected Tile. A global map of tiles will pop up showing the

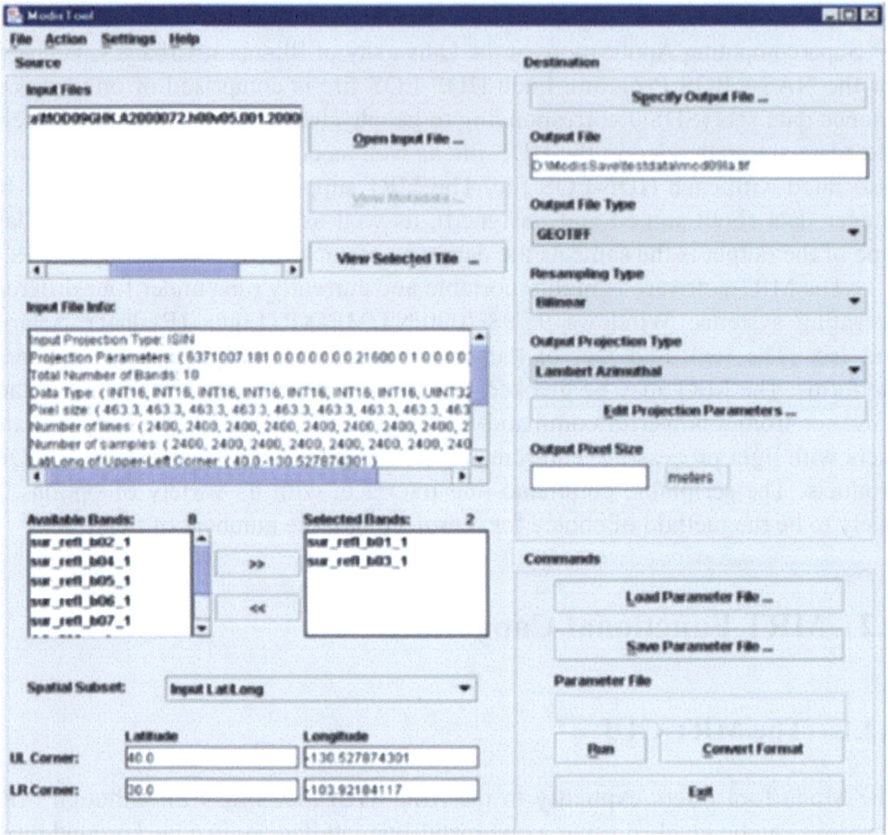

Figure 9.2 The main screen for the ModisTool, which is the graphical user interface for the MODIS Reprojection Tool

selected tiles in light blue (see Fig. 9.1). Basic input file information may be examined by scrolling the Input File Info box, in which the input projection type, projection parameters, corner points (in latitude/longitude), number of SDSs, image dimensions, pixel size, input data type, corner coordinates and datum information are displayed. Below the Input File Info box are two windows that show what SDSs are *Available* and which have been *Selected* for processing. Next is a drop-down menu that allows the option to specify the type of Spatial Subset to be applied to the input file. Below the spatial subset drop-down menu are four boxes in which to enter the upper left (UL) and lower right (LR) corner coordinates to be used to define the subset area. Note that the headings for these boxes will change with the type of subset option that is chosen.

9.2.1.2 Destination

Clicking on Select Output File opens a dialog box in which to specify the name of an output file. The appropriate extension must be used for the type of output file to be written: HDF-EOS (*.hdf*), GeoTiff (*.tif*) or generic binary (*.hdr*). In the generic (raw) binary and GeoTIFF formats, unlike the HDF-EOS format, only one image band is stored per output file. However, the user does not need to specify a new output file name for each selected SDS; the SDS name is automatically appended to the base name output file.

The Output File Type, Resampling Type, and Output Projection Type are selected from the pull-down menus. If necessary, the desired output projection parameters may be modified by activating the Edit Projection Parameters dialog box, which includes the datum specification (see Fig. 9.3). The Output Pixel Size may also be specified, and the units will depend on the output projection. The Load Parameter File and Save Parameter File enable the use or editing of an existing parameter file or the creation of a new one. The name of this file will then appear in the Parameter File window. To reproject and resample the input file click on the *Run* button, otherwise *Convert Format* is used to write the data to a different output format without any resampling. Spatial and/or band (SDS) subsets may be specified for either processing option.

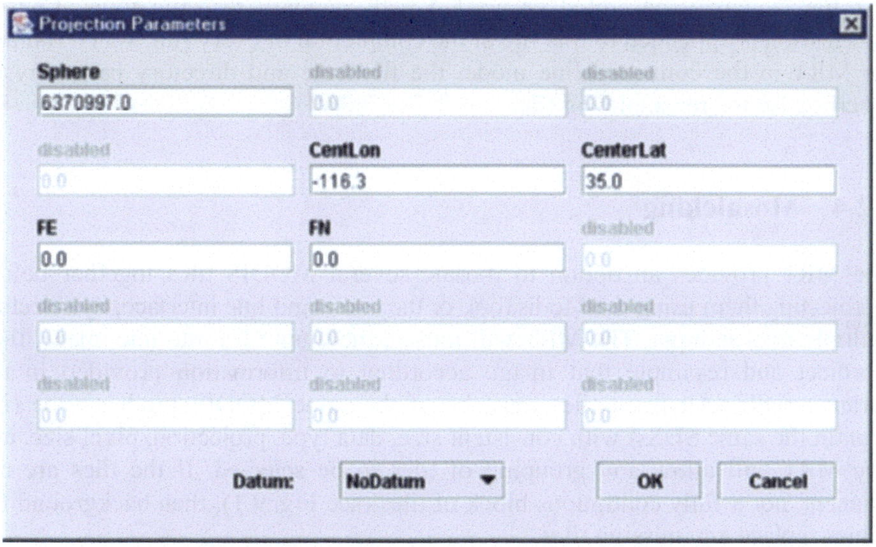

Figure 9.3 ModisTool dialog box for editing projection parameters

The Convert Format button is inactive at startup of the MRT. Once the input and output file information is provided, the Convert Format button will be available for use. Note: The Run button is not available until all the necessary processing parameters are specified, but the Convert Format button is available once the output filename has been provided. A Convert Format command is also an option under the main ModisTool "Action" menu.

9.2.2 Parameter Files

The parameter file contains all the information necessary to read gridded MODIS products in the Integerized Sinusoidal (ISIN) or Sinusoidal (SIN) projections from an input data file, transform the data into a specified output projection, and write the results to an output file. The parameter file contains the file names and file types of the input and output data files, bSDS and/or spatial subsetting information, output projection type, output projection parameters, output UTM zone, output resampling type, and output pixel size. Parameter files are ASCII text, and may be created, printed, and edited using the ModisTool or any text editor. The parameter file format is described fully in the User's Guide for the MODIS Reprojection Tool.

9.2.3 Log File

The MRT writes status information to a separate window when running the ModisTool (see Fig. 9.4) and also to a log file (resample.log). This is an ASCII text file, and may be edited or printed with standard text file tools. Logged information is appended to this file at the completion of every run. When running the MRT in the command-line mode, the filename and directory path may be specified for the resample.log file.

9.2.4 Mosaicking

The MRT provides an option to mosaic several MODIS tiles together before reprojecting them using the ModisTool, or the command-line interface, by selecting multiple tiles as input. The MRT will mosaic the input tiles into one image, then reproject and resample that image according to information provided in the parameter file. All input tiles must be of the same MODIS product type (i.e. contain the same SDSs) with consistent size, data type, projection, pixel size, etc. The MRT will allow any grouping of tiles to be selected. If the tiles are not adjacent, nor a fully contiguous block of tiles (see Fig. 9.1), then background fill values replace any missing tiles.

In the above example, the first row will have two fill tiles (right two tiles) and the second row will also have two fill tiles (left two tiles). The output image will be based on a rectangle consisting of two rows and four columns of tiles. When processing several tiles, the Input File Info, Available Bands (SDSs), and Spatial Subset boxes will be populated with coordinate information corresponding to the geographic extent of the mosaic. The input files can be reprojected and resampled (via the Run button) or just mosaicked and converted to a different format. Spatial and/or SDS subsets along with output file information may be specified, just as when processing one single input file.

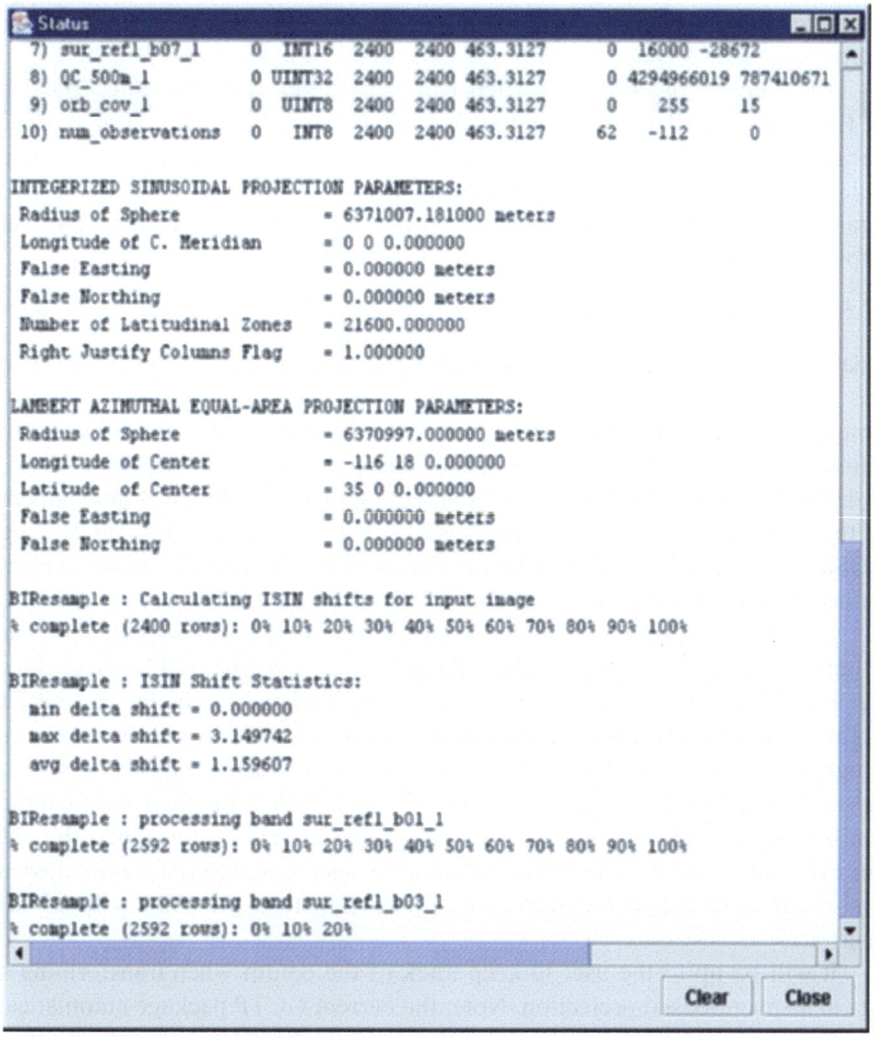

Figure 9.4 The Status window displays the resampling progress when running the ModisTool

9.2.5 Map Projections

The MRT makes use of the Generalized Cartographic Transformation Package (GCTP) that had been modified to incorporate the Integerized Sinusoidal projection used for Terra MODIS Version 1 and 3 land products. The MRT allows transformation to and from the following map projections: Albers Equal Area, Equirectangular, Geographic, Hammer, Integerized Sinusoidal, Interrupted Goode Homolosine, Lambert Azimuthal, Lambert Conformal Conic, Mercator, Mollweide,

Polar Stereographic, Sinusoidal, Transverse Mercator, and Universal Transverse Mercator.

The output projection parameters can be modified with the ModisTool using the Edit Projection Parameters button. A dialog box (see Fig. 9.3) allows for editing the 15 projection parameters as well as specifying a datum for the output projection. Integer values will automatically be converted to floating point. Parameter fields will be grayed out when they are not necessary for a particular output projection.

9.2.5.1 Datum Conversions

A datum defines the position of the spheroid relative to the center of the earth. The datum may be defined explicitly or by defining the semi-major and semi-minor axes of an ellipsoid (Snyder, 1987). A spheroid-based map projection, defined by a single value for the radius of the sphere, does not require a datum to be defined. A limited number of input and output datum references are supported by the MRT for datum conversions: NAD27, NAD83, WGS66, WGS72, and WGS84 (Snyder, 1987). The datum for the MODIS land products is not explicitly defined in any metadata, but WGS84 has been implied.

If the input files are to be transformed to a sphere-based projection, then datum conversions are not allowed and the *NoDatum* option will be invoked (see Fig. 9.3). The radius of the sphere must be specified as a value in the first field of the projection parameters. Except for Universal Transverse Mercator and Geographic projections, the first two projection parameters specify the semi-major and semi-minor axes. If the No Datum option is selected and the semi-major and semi-minor axes are not specified in the projection parameters, then the MRT will exit with an error. Similarly, if a datum is specified and the semi-major and semi-minor axes are specified as projection parameters, the MRT will exit with an error.

It will be up to the user to keep track of the datum when transforming the data to a sphere-based projection. Note: the current GCTP package automatically uses the radius of Sphere 19 (6370997.0) for sphere-based projections, except for Sinusoidal and Integerized Sinusoidal; for these projections, the radius is contained in the product metadata and read by the MRT. The user is allowed to specify a sphere radius value, but the No Datum option must be used.

Once the MRT has verified that the input and output projection parameters and datum combinations are valid, then the projection transformation and datum conversion will be performed as follows: input data are transformed to the Geographic projection; the input datum is converted to the specified output datum; and then data are transformed from the Geographic projection to the specified output projection. The datum value will be written to HDF-EOS, GeoTiff, and generic binary output data files. For the HDF-EOS files, the datum will be written to the embedded HDF metadata.

9.2.6 Resampling Process

The resampling process determines how the transformed (destination) pixel values are derived from the input (source) data. An output pixel size may be specified and this will apply to all of the SDSs selected for processing. The default uses the input pixel sizes. For the Geographic projection the pixel size must be specified in decimal degrees; for all other projections the pixel size must be specified in meters.

Resampling is a two-step process. First is the spatial transformation (see Fig. 9.5), which is the mapping of an (x, y) in output space back to an (x', y') in input space. This can result in (x', y') being a fractional coordinate. The second step is resampling the input space pixels in the neighborhood of the fractional coordinate to derive a value for the output pixel. Three methods of resampling are available: nearest neighbor, cubic convolution, and bilinear resampling. If the majority of values that lie under the resampling kernel are background fill values, then a background fill value is output. Otherwise, resampling is performed only from non-background fill values, and the weights of the resampling kernel are adjusted accordingly.

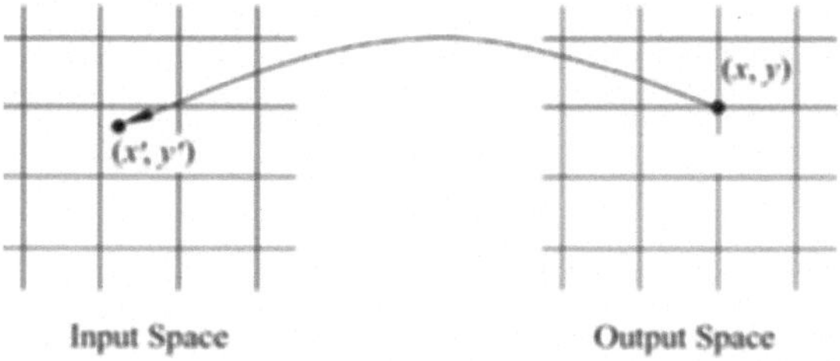

Figure 9.5 Mapping the output pixel location to the input space

Nearest neighbor resampling uses the pixel value of the closest integer coordinate (upper-left corner of pixel) neighbor (see Fig. 9.6).

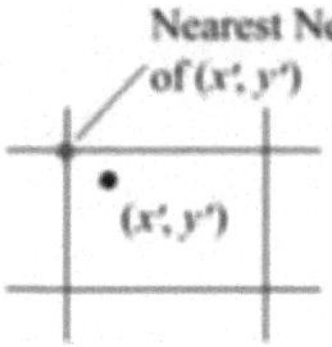

Figure 9.6 Nearest neighbor resampling. Note: Nearest neighbor resampling should be specified for all bands (SDS) that are of nominal and ordinal data types, which includes the bit-mapped quality control (QC) data

Bilinear resampling uses the pixel values of the four nearest neighbors. A set of linear weights are used so that the closest neighbors are weighted more heavily in the final pixel value (see Fig. 9.7). The following equations specify how (x', y') is calculated using bilinear interpolation:

$$x1 = int\,(x') \qquad x2 = x1 + 1$$
$$y1 = int\,(y') \qquad y2 = y1 + 1$$
$$\text{pixel value of } (x',y') = w1 * \text{pixel value at } (x1,y1)$$
$$+ w2 * \text{pixel value at } (x2,y1)$$
$$+ w3 * \text{pixel value at } (x1,y2)$$
$$+ w4 * \text{pixel value at } (x2,y2).$$

The weights $(w1, w2, w3, w4)$ are fractions that all add up to one (1.0). The closer the point (xn, yn) is to (x', y'), the higher the value for wn.

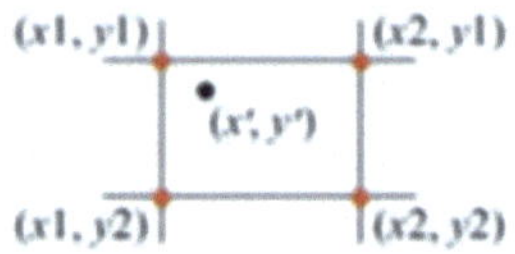

Figure 9.7 Bilinear interpolation resampling

Cubic convolution resampling uses the pixel values of the 16 nearest neighbors. A set of weights is also used so that the closest neighbors have more weight in the overall pixel value (see Fig. 9.8).

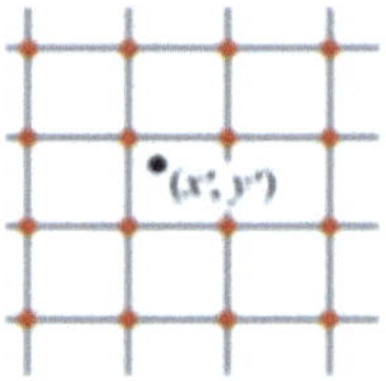

Figure 9.8 Cubic convolution resampling

Now consider a case in which a full product (entire tile) is transformed to a given projection, and then a spatial subset extracted from the same tile is transformed to the same projection using the same parameters. The key to ensuring that the transformed full tile and subset overlay exactly, is in how the spatial subset is defined. By specifying beginning/ending line and sample, the grid sampling remains intact. If the subset is defined by latitude and longitude or output projection coordinates, these may translate to sub-pixel points and a shift in the center of the resampling kernels, resulting in a different value for a given pixel in the reprojected subset relative to the "same" pixel in the full tile that was transformed. If the latitude and longitude specified for the subset match the input grid spacing on an integer rather than fractional increment, only then would the two transformed data sets align.

Differences around the edges will exist in the resampled products because zero-fill values will be included in the resampling kernel for edge pixels in the subset. If more than 50% of the pixel values to be used in the interpolation are background fill values, then (x', y') is assigned the value of background fill. In the case that fewer than 50% of the pixel values to be used in the interpolation are background fill, then fill values are ignored and only pixel values within the valid range are used. In this latter case, the weights for the valid pixels are also readjusted to account for the fact that fewer pixel values are being used in the interpolation.

9.2.6.1 ISIN Shift

The early versions (V001, V003) of the Terra MODIS land products were in the ISIN projection, which included a linear shift between lines on a per pixel basis. Although this shift is linear for each line, the amount of shift will vary from line to line. The amount of pixel shift for a given line can be determined by calculating the difference in the amount of shift between the first and last pixels, and then dividing this difference by the number of pixels in the line. The amount of shift for SDSs of the same resolution is the same, so this only has to be calculated once for each resolution. The shifts are recalculated when a new resolution is encountered in the image.

9.2.7 SDS Subsets

The SDSs contained within the HDF-EOS files may represent geophysical or biophysical parameters, quality control flags, or sensor observation information, any of which may be selected for mosaicking, reprojection, or format conversion. The default is to reproject all input SDSs. Immediately below the *Input File Info* box in Fig. 9.2, SDSs may be selected by clicking on the desired SDS name and using the " " to move it to *Available* or likewise use " " to select an SDS from the *Available* list.

9.2.8 Spatial Subsets

A spatial subset is defined by the upper-left and lower-right corner coordinates for the rectangular region to process. Corners may be specified as input projection latitude-longitude, line-sample pairs, or as output projection coordinate pairs (see Fig. 9.2, Spatial Subset pull-down menu). The other two corners' points (upper right and lower left) are computed automatically. The default is to process the entire data set using the input latitude/longitude corner points. Bounding tiles represent a special case (see Section 9.3.1).

When defining a subset using line-sample pairs (starting line/sample, ending line/sample), a zero-based coordinate system is used where the upper left corner is (0, 0). The four corner points are reprojected to output space, using the specified map projection and datum. A minimum-bounding rectangle is then computed and contains the four reprojected points referenced to the output projection space. All points inside this rectangle, referenced to the output space, are mapped back into input space for resampling (see Fig. 9.5). The default is to reproject the entire input image.

To define a spatial subset using output projection coordinates, specify the upper-left and lower-right coordinates in the output projection units of decimal degrees for geographic and meters for all other projections. The upper right and lower left corners are computed from these coordinates to define a rectangle in the output projection space, and these are mapped back to input space for resampling. The output image corner coordinates will agree with the user-specified output spatial subset corners.

9.2.9 Format Conversion

The MRT supports input of MODIS Level-2G, Level-3, and Level-4 land data products in HDF-EOS format or the generic binary file format created by the MRT. MODIS Level-1 and Level-2 data and MODIS Atmosphere and Ocean products are not supported. The MRT will write processing results in HDF-EOS, GeoTIFF, and generic binary file formats.

No reprojection or resampling of the input data will be performed if files are only selected for reformatting. With format conversion, the projection remains unchanged and the output projection parameters, pixel size, and data type remain the same as the output file. Format conversion will, by default, convert all SDSs. Similarly, the Resampling *Type* (see Fig. 9.2) need not be specified. If the output projection parameters, pixel size, data type or resampling method have been specified, they will be ignored if the Convert Format option is executed. If the QC SDSs are to be unpacked using the MODIS Land Data Operational Product Evaluation (LDOPE) software tools, then the QC SDSs must be retained in HDF-EOS format.

9.2.9.1 Three- and Four-Dimensional Data Products

The MODIS land data products may be 2-, 3-, or 4-dimensional, and the MRT will process these to 2-dimensional data products for generic binary, GeoTiff, and HDF-EOS output formats.

The naming convention for the 3rd and 4th dimension slices are as follows:

< SDS name > . < 3rd dimension name > #. < 4th dimension name > _#

where # is a two-digit value representing the data slice for the associated dimension. If the product only has three dimensions, the < 4th dimension name > _# will not appear. This naming convention is used for the output filenames, the output HDF-EOS SDS names, and the SDS names displayed as Selected or Available in the ModisTool.

For example, the Terra MODIS Albedo 16-Day L3 Global 1km ISIN Grid (MOD43B3) product has two SDSs: albedo, which is 4-dimensional; and BRDF albedo quality, which is 3-dimensional. The four dimensions for the albedo SDS are sample, line, spectral bands (7 solar reflective bands and 3 broad bands), and albedo (directional hemispherical reflectance or "black-sky albedo" and bihemispherical reflectance or "white-sky albedo"). When the albedo SDS is written to the output file as a series of 2-dimensional arrays, the resulting filename for the "black-sky albedo" for the 3rd broad band would be "Albedo.Num_ Albedo_Bands_ 10.Num_Albedos_01".

The 3-D and 4-D naming convention produces long file names when the SDS name, 3rd dimension name, and 4th dimension name are all of substantial length themselves. The HDF-EOS library for the Windows platform currently supports only 57 characters for SDS names. To get around this constraint, the naming convention is then changed as follows to shorten the output file names: < SDS name > .3_#.4_#.

9.2.9.2 Generic Binary Data Output

Generic binary data and metadata are stored in separate files. The header file contains information required by the MRT for processing MODIS data in the generic binary file format. Header files consist of ASCII text in a format described in a Section 9.2.10.1 "Header File Content and Format". When MODIS data are written in generic binary format, each SDS is stored as individual data files, i.e. one file per SDS. Data are stored in row-major order, starting at the upper-left (0, 0) corner of the image. The data type may be 8-bit integer, 16-bit integer, 32-bit integer, or 32-bit float. Integer values may be signed or unsigned. Two-byte and four-byte data types are stored in big-endian order (high byte followed by low byte).

By convention, all related files in raw binary file format are given the same base filename of the input file. Header files are given the*.hdr* extension. Data filenames are generated from the basename and the SDS name, and are given the .dat extension: basename.SDS_name.dat. The implicit association of the various files by the base filename enables tracking and easier use of output files by other applications software or for further processing by the MRT, For example:

MOD09GHK.A2000072.hdr —header file
MOD09GHK.A2000072.sur_refl_b01_1.dat —data file (Band 1)
MOD09GHK.A2000072.sur_refl_b02_1.dat —data file (Band 2)

9.2.10 Metadata

Metadata contained in the .met file that accompanies the HDF-EOS file can be viewed or edited with an ASCII text editor. The metadata embedded in the .hdf file can be viewed using the View Metadata button on the ModisTool GUI. Output file metadata support is currently limited to HDF-EOS files only, in which the output metadata will be followed by the original input file metadata. The input structure, core, and archive metadata information are respectively stored under the HDF attributes OldStructMetadata, OldCoreMetadata, and OldArchiveMetadata.

After selecting an input file with the ModisTool, the metadata embedded in the .hdf file may be examined by clicking on the *View Metadata* button (see Fig. 9.2). In addition to scrolling up and down through the metadata, you may also search for a character string (see Fig. 9.9). Type the desired search string into the *Find* box and click the *Find* button to perform the search. Here is an example of searching for "Latitude" in the metadata (note that the search is case sensitive).

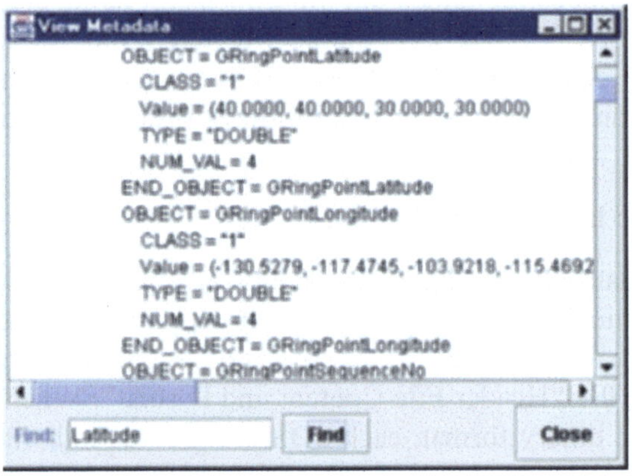

Figure 9.9 The View Metadata popup window in the ModisTool allows searching for case-sensitive character strings in the metadata embedded in the HDF file

When mosaicking tiles, the MRT will output the structure, core, and archive metadata for each of the input tiles. The metadata are renamed to add the relative tile location in the mosaic. Thus the tile which resides in row 0 and column 3 of the output image will specify that information in the output structure, core, and archive metadata names.

9.2.10.1 Header File Content and Format

The header file contains information required by the MRT for processing data as generic binary. Header files may be created and modified by directly editing the header file with an ASCII text editor. When using the ModisTool GUI to process generic (raw) binary as output, header files are created automatically. An Object Description Language-like format is used, as illustrated by the following:

```
#HeaderfileforMOD09GHK.A2000072.h08v05.001.2000084105003
PROJECTION_TYPE = ISIN
PROJECTION_PARAMETERS = ( p1 p2 p3 p4 p5 p6 p7 p8 p9 p10
p11 p12 p13 p14 p15 )
```

```
# COORDINATE_ORIGIN = UL
# lat/long in decimal degrees
UL_CORNER_LATLON = ( lat lon )
UR_CORNER_LATLON = ( lat lon )
LL_CORNER_LATLON = ( lat lon )
LR_CORNER_LATLON = ( lat lon )
# projection coordinates
# UL_CORNER_XY = ( x y )
# UR_CORNER_XY = ( x y )
# LL_CORNER_XY = ( x y )
# LR_CORNER_XY = ( x y )
NBANDS = n
BANDNAMES = ( band1 band2 ... bandn )
DATA_TYPE = ( t1 t2 ... tn )
NLINES = ( r1 r2 ... rn )
NSAMPLES = ( c1 c2 ... cn )
PIXEL_SIZE = ( s1 s2 ... sn )
MIN_VALUE = ( v1 v2 ... vn )
MAX_VALUE = ( v1 v2 ... vn )
BACKGROUND_FILL = ( f1 f2 ... fn )
```

** All text following a "#" are comments for information purposes only, some of which are generated automatically by the MRT when writing output as a generic binary file.*

** The BANDNAMES field is optional; by default, bands are named according to each corresponding input SDS name. Lower-case items represent numeric and string values for the various fields, many of which require arrays of n values for products containing multiple SDSs. Different SDSs may have different data types, dimensions, resolutions, etc.*

** The MIN_VALUE, MAX_VALUE, and BACKGROUND_FILL fields are also optional; by default, no background fill values are used during resampling.*

Only one input projection type is permitted and all (15) projection parameters must be specified. More information on the projection parameters is available in the User's Guide for the MODIS Reprojection Tool. If the data set is in the Universal Transverse Mercator (UTM) projection, the UTM_ZONE or first two projection parameters may be used to specify the zone. Valid UTM_ZONE values range between –60 and 60. If the UTM_ZONE is not defined, then the first two projection parameters must be specified. If both the UTM_ZONE and first two projection parameters are specified, then UTM_ZONE takes precedence. A DATUM may be specified for the input data (NAD27, NAD83, WGS66, WGS72, WGS84) or not (NODATUM). If not specified, then WGS84 will be used and the first two values in the projection parameters must specify the semi-major and semi-minor axes of the spheroid.

The COORDINATE_ORIGIN is an optional comment field that is not used by the MRT on input, but is written on output for information purposes only. It specifies the location of the coordinate origin as one of the four corners (UL, UR, LL, LR). The UL is assumed to be the origin for generic binary data. The

CORNER_LATLON coordinates are required (latitude/longitude in decimal degree units). The CORNER_XY coordinates (in projection units) likewise are optional comment fields.

Valid DATA_TYPE values are INT8, UINT8, INT16, UINT16, INT32, UINT32, and FLOAT32.

9.3 Special Considerations

9.3.1 Bounding Tiles

The MODIS Level-2G, Level-3, and Level-4 land data products are referenced to a global tiling scheme. Each tile is approximately 10° latitude by 10° longitude and tiles are non-overlapping (see Fig. 9.3). The tiles are indexed from left to right (horizontal) and top to bottom (vertical) starting at 0.0. The horizontal tiles range from h00 to h35, and the vertical tiles range from v00 to v17. Oftentimes a MODIS HDF-EOS filename will contain .h##v## to specify the horizontal and vertical location of the tile. Tiles that lie along the periphery of the projection are referred to as bounding tiles.

The bounding tiles are unique in that they contain corner points with projection coordinates that do not map to a valid latitude/longitude. In this case, the MRT will issue a warning to the user that a certain corner point does not have a valid latitude/longitude. To compensate, the MRT uses the bounding rectangle coordinates from the metadata as the corner points for the tile. Different products put these bounding rectangle coordinates in different metadata structures, so the StructMetadata, CoreMetadata, and ArchiveMetadata (embedded in the HDF file) are read and searched for the first occurrence of bounding rectangle coordinates. For example, the MOD09A1 products store their bounding rectangular coordinates in the HDF ArchiveMetadata structure, but the MOD12Q1 data products store their bounding rectangular coordinates in the CoreMetadata structure.[①]

9.3.2 Crossing the International Dateline

Bounding tiles, by definition, also cross the international dateline. For example, the bounding rectangular coordinates reported for the file MOD09A1.A2001001.

① Using the bounding rectangle coordinates often times adds fill data to the output image. The bounding rectangle coordinates are used to create a rectangle in input latitude/longitude space, which is then transformed to the output projection space, rather than using the latitude/longitude coordinates of the corners of the input file to define the corners of the rectangle in input space. The coordinates of the bounding rectangle will define an area much larger than the rectangle based on the exact corner points. This will also be the case when implementing a format conversion on a bounding tile.

h30v13.001.2001088044433 are:

NORTHBOUNDINGCOORDINATE = – 40.000000
SOUTHBOUNDINGCOORDINATE = – 48.189583
EASTBOUNDINGCOORDINATE = – 180.000000
WESTBOUNDINGCOORDINATE = 156.651809

In such cases the MRT will find the longitude value that is closest to the dateline and change its origin. In this case, the "EASTBOUNDINGCOORDINATE" would be changed from – 180.0 degrees to + 180.0 degrees and this information will also be printed as a warning. Similarly, when several tiles are mosaicked together and cross the international dateline, the MRT will change the sign on the longitude value closest to the dateline.

9.4 Summary

The MODIS Reprojection Tool was developed in anticipation of needs by the user community to be able to transform the MODIS Level-2G, Level-3, and Level-4 land data products to different projections, convert the data to alternative formats compatible with commonly used applications software, and to be able to extract spatial subsets or particular SDSs for processing. The source code and binary executables are available at no cost and without any restrictions on reuse. More than 1,600 individuals from around the world have registered for access to the MRT software. Thus several enhancements have been implemented in response to user requests. The software may be compiled to run under most common operating systems without sacrificing user interfaces. The MRT can be used to generate global mosaics from individual MODIS product tiles, although this may require more detailed attention due to the two gigabyte file size constraints imposed by the HDF-EOS format.

Acknowledgements

This work was made possible in part by Science Applications International Corporation under U.S. Geological Survey contract 1434-CR-97-CN-40274.

References

Schmidt G (2003) The MODIS Reprojection Tool User's Manual. http://edcdaac.usgs.gov/landdaac/tools/modis/index.asp

Snyder JP (1987) Map Projections—A Working Manual, US Geological Survey Professional Paper 1395. US Government Printing Office. Washington, D.C. p. 383

10 A Tool for Conversion of Earth Observing System Data Products to GIS Compatible Formats and for the Provision of Post-Processing Functionality

Larry Klein, Ray Milburn, Cid Praderas and Abe Taaheri

10.1 Introduction

EOS Terra and Aqua, launched in 1999 and 2001, respectively, carry instruments that provide observations of the earth at many wavelengths and spatial resolutions. The data are used to provide scientists, policy makers and others with a continuous record of parameters used to characterize physical processes and effects of human activities such as, land use, pollutant transport and climate change.

The standard format for EOS instrument data is HDF-EOS, which is based on the National Center for Supercomputing Applications (NCSA) Hierarchical Data Format (HDF) (HDF 4.1R3 Users Guide, 2003). HDF is a self-describing, portable format for scientific data. HDF-EOS applies standards for attaching geolocation and temporal information to scientific data. The format was developed to provide a convention for geo-locating data from disparate instruments. Most EOS data products are stored and distributed in this format. HDF-EOS files also contain core and product-specific metadata. Core metadata are also stored in databases for user search and access.

HDF-EOS formatted data products are not generally accessible to common Geographical Information Systems (GIS) applications, such as ArcInfo, ENVI and ERDAS. Although the latter two tools have HDF-EOS interfaces, many variations and inconsistencies in storage, unanticipated during development of HDF-EOS, cause many products to be inaccessible. In order that the data are useful to GIS users, the most commonly used data must be accessible to application tools commonly used by that community. One such format is GeoTIFF.

We note that HDF-EOS has worked well as a storage and interchange format. It does not, however, meet all the requirements of specialized end users.

Land products from the Advanced Spaceborne Thermal Emission and Reflection Radiometer (ASTER) and Moderate Resolution Imaging Spectroradiometer (MODIS) on the EOS Terra platform, come in HDF-EOS Swath and Grid formats. Swath products are generally multibanded (multi-dimensional) and contain geolocation information in geographic coordinates. Swath data are commonly organized by a temporal index. Grid data are projected onto a regular grid in one of many coordinate systems.

MODIS products are in Swath and Grid formats, but in early versions the Level-3 (Grid) Land products were in the Integerized Sinusoidal Projection. This format projection was developed by the MODIS team and was not compatible with most GIS software. For the lower level MODIS products, complete geolocation information, is in a separate physical file, although coarser geolocation information is contained in the products themselves. Geolocation data for MODIS Ocean products is also contained in separate files.

ASTER products are generally uniform in format; however they use both geocentric and geodetic coordinates. An additional challenge for tool developers comes from the fact that some products use a variety of storage methods, often mixing HDF-EOS objects and generic HDF objects in the same file. Multi-angle Imaging Spectro-Radiometer (MISR) data are stored in "blocks", of up to 180 per dayside path. In order to make MISR data more amenable to comparison with other data, blocks must be extracted, stitched together and subsetted.

There are also detailed differences in data storage. For example, MISR data can use 14 of 16 bits to store real numbers and the remaining two bits store special information. Quality bits can be stored in varying number, e.g. MODIS land products. If a user does not know this beforehand, information could be lost on translation or conversion.

Issues such as those outlined above have caused many difficulties for GIS users and for application tool developers. EOS Distributed Active Archive Centers (DAACs), instrument science teams and other data users are developing specialized software to convert certain product to GIS-compatible formats. These software tools address products accessed by their respective data users or team members. A more general purpose tool is required that can process a wide variety of products from numerous EOS instruments. This tool had to provide the same functionality across many EOS products and without compelling users to understand internal storage formats and inconsistencies across products.

The HDF-EOS to GIS Format Conversion Tool (HEG) has been developed in response to these needs. It allows a user working on his/her workstation to reformat, re-project and perform operations such as subsetting and stitching on selectable HDF-EOS objects stored in EOS products. More than 70 EOS products from ASTER, MISR, MODIS and radar instruments were initially tested with HEG functionality. The output GeoTIFF files produced by the tool is readable by commonly used GIS applications. The Tool also allows the user to select other output formats, such as binary and HDF-EOS Grid. The output data can be transformed to other projections selected from the US Geographical Survey (USGS) General Cartographic Transformation Package (GCTP) (Taaheri, 2003). Access to and processing of HDF-EOS input files is controllable from a GUI or from a command line interface. The GUI or command line interface allows selection of file and objects within the file to be processed, the projection for output, and a latitude/longitude box to define a spatial subset, the output file format, and the output file location. In addition, many files in HDF-EOS Swath or Grid format can be mosaiced, (stitched together) to form composite output products. For a detailed explanation of a Swath or Grid object, please refer to the HDF-EOS Library User's Guide, Vol. 1 and Vol. 2 (Klein, 1996a,b).

It is expected that a user will have access to a number of files in his/her local directory. The HEG tool can access and process files in this local directory. Processed files will contain original and derived metadata, which describe the output product. Assuming that minimal metadata required by EOS is contained in the new file, it can be re-ingested into EOS archives (see metadata chapter in (Taaheri, 2003)).

The tool described in this section is "stand-alone" and down-loadable to a variety of workstations. A version of HEG is also available at EOS archive centers. This version can access a subset of EOS archived data directly. The direct access storage is called the EOS Data Pool. The Data Pool contains many terabytes of selected data (Marley et al., 2003). See Section 10.4 for access instructions to both standalone and Data Pool versions of the tool.

There are no Commercial-off-the-Shelf (COTS) dependencies in the HEG tool. It incorporates some of the software and functionality of the MODIS Reprojection Tool, developed by the Land Processes Distributed Active Archive Center (LP DAAC), at the US Geological Survey Data Center.

HEG is built in C and Java. Graphical User Interface (GUI) and command line interfaces are provided.

10.2 Functionality

The primary function of HEG is to provide data users with the capacity of converting EOS standard product formats into GeoTIFF output ingestible into GIS tools. Users will also perform pre-processing functions on data and may or may not upload to a GIS tool. For example, users often require spatial coverage available in adjacent files, or contained within single files. Other pre-processing functions, such as re-projection of original data, subsampling, and interpolation are required. The tool must work on a variety of common platforms. The tool must work in stand-alone and client-server modes. A summary of functionality provided in HEG is given:

(1) Reprojection (to UTM, Polar Stereographic, Geographic, Transverse Mercator, Mercator, Lambert Azimuthal, Lambert Conformal Conic, State Plane, Sinusoidal);

(2) Conversion of MODIS data from Sinusoidal projection (uses MODIS Reprojection Tool code);

(3) Output written in HDF-EOS Grid, native binary and/or GeoTIFF formats;

(4) Stitching (Mosaicing);

(5) Subsetting Grid and Swath data, by temporal or band parameter and geolocation;

(6) Resampling;

(7) Production of single- or multiple-band GeoTIFF files;

(8) Extraction of MISR "blocks";

(9) Metadata preservation and creation;

(10) Operates on 4-D data sets (e.g. MOD043);

(11) Performs conversion to a single-band or multi-band GeoTIFF's;

(12) Conversion of HDF-EOS Swath to HDF-EOS Grid;

(13) Conversion of HDF-EOS Swath or Grid to GeoTIFF;

(14) Conversion of HDF-EOS Swath or Grid to generic binary (with metadata file);

(15) Reprojection (USGS, MODIS Integerized Sinusoidal, EASEGrid);

(16) Subsampling (Subsample stacks);

(17) Boundary smoothing using bi-linear or cubic convolution interpolation;

(18) Support for MODIS, MISR, ASTER and Digital Elevation Model products;

(19) Allows file selection from users local storage;

(20) Details of internal file structures transparent to users;

(21) Operable on Sun, SGI, Windows and Linux (MAC OS X support is planned);

(22) Integrated into EOS Data Pools, on-line storage at EOS archive centers;

(23) Java-based GUI; and

(24) Command line interface.

10.3 GUI Overview

The HEG has a graphical user interface (GUI) that uses Java Swing technology. The Java Swing widget set is comparable to the widget set of X-Windows and MS-Windows. It is designed so that a HEG user should be able to become a proficient user of the tool after a short period of time. The interface consists of two main windows. The first window is for the selection of HDF-EOS files for conversion into the GeoTIFF format. The second window is to be used to stitch and subset common HDF-EOS files.

The purpose of the HEG tool interface is to assist users in creating files that are more useful and informative. The HEG tool works specifically on HDF-EOS Swath and Grid objects. An HDF-EOS object is a structure, built on HDF arrays, tables and attributes. The object is self-described by machine- and human-readable metadata. This is "structural metadata" in HDF-EOS parlance. The objects are assessable via a software interface, the HDF-EOS library (Klein, 1996a,b).

The HEG GUI has primarily two functionally separate windows which can be selected by the *Tool* button. The first window is used when working with one

HDF-EOS data set at a time. The second window is used to work with more than one HDF-EOS data set by Stitching/Mosaicing. The number of data sets able to be stitched is limited by the memory in the local computer.

The data in each window may be converted to GeoTIFF, Non-standard binary, or HDF-EOS Grid formats. When a dataset is displayed in the GUI the user does not need to know what type of dataset it is (i.e. MISR, ASTER, etc.) or the format it is in (i.e. Grid or Swath). The user can pick the selected bands, subset geographically, select a projection and an output format, and have the conversion performed. The user is not compelled to perform the conversion immediately. The desired selections may be saved in a "parameter" file and run from the command-line at a later time.

Figure 10.1 is a snapshot of an actual MISR Terrain HDF-EOS data set loaded and ready to be *Run* which will produce a GeoTIFF image. The resulting image would correspond to Fig. 10.3. To get to this point, the user first loads the data set through *File→Open* button. Information on the file can be seen in detail by clicking on the *Object Info* button. The Object and the Field of interest are then selected. Subsetting values are entered (the full image can be used as default). In the middle of the window, various options were selected such as output format type, output projection, pixel resolution, resampling type (i.e. nearest neighbor, bilinear interpolation, etc.). The output filename is also specified. Once all the options are determined, then the user clicks on the *Accept* button, and the "job" is

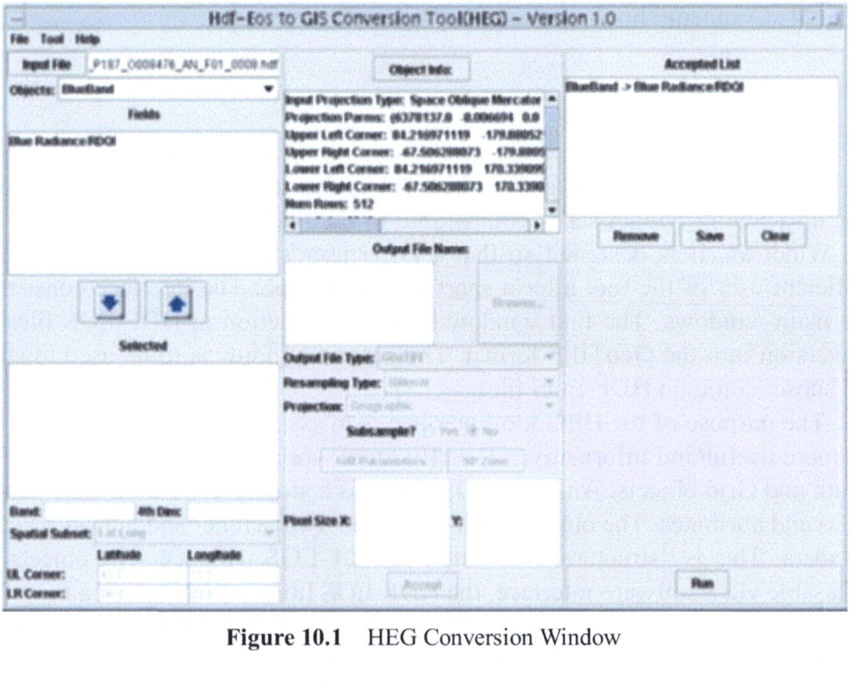

Figure 10.1 HEG Conversion Window

stacked in the *Accepted List*. A user can stack multiple jobs of all types, from different input data sets if desired. Once the user is satisfied with the content of the "Accepted List", the "Run" button is clicked.

Figure 10.2 is a snapshot of the HEG GUI *Stitch/Subset* window. It shows two ASTER L1B data sets to be stitched, spatially subsetted, and output to GeoTIFF format. The output corresponds to the image in Fig. 10.6. The input data sets are stacked in the *Input Files* window. Similarly to the example in Fig. 10.1, the user makes various selections starting with the Object and Field of interest, subset values if desired, output projection type, output format type, output filename, etc. When the user is satisfied, the *Accept* button is clicked and job is loaded into the *Accepted List*. In Fig. 10.2, the user can select *Run* to run the job, or select *Save* to save a parameter file, which can be used to run on the command line.

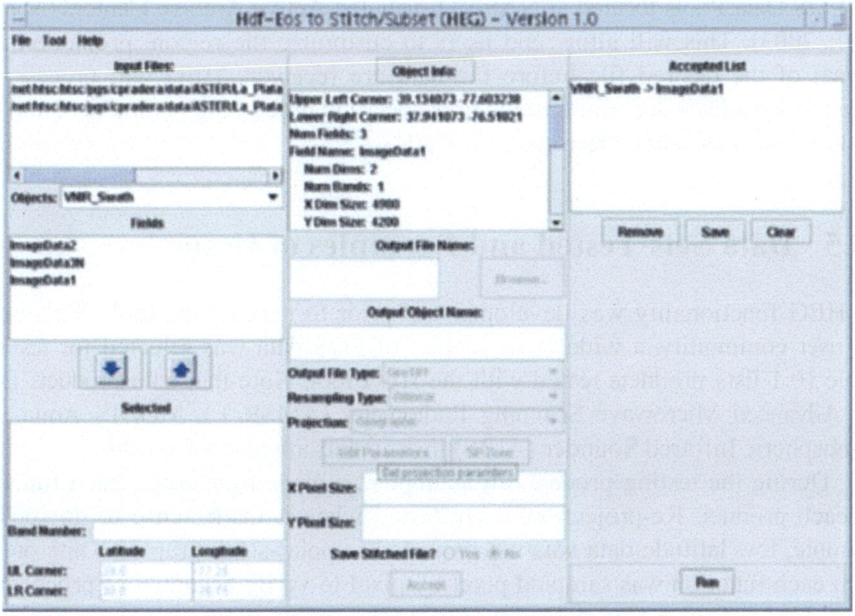

Figure 10.2 HEG Stitch/Subset Window

10.4 Access

The HEG Tool is available as an FTP file (ftp.east.hitc.com) for download onto Sun, SGI or Windows, Mac or Linux workstations. Instructions are as follows:

```
ftp edhs1.gsfc.nasa.gov
Name: anonymous
Password: <your email address>
ftp> quote site group sdptk
```

```
ftp> quote site gpass ecs-tkit
ftp> cd HEG_Too
```

In this directory are five files. These are as follows:

```
hegSUN.tar.Z - HEG for the Sun/Solaris system
hegSGI.tar.Z - HEG for the SGI system
hegLNX.tar.Z - HEG for the Linux system
hegWIN.zip - HEG for Windows
HEG_UsersGuide.doc.Z - HEG Users Guide in MS-Word format
```

HEG can also be obtained at:

http://newsroom.gsfc.nasa.gov/sdptoolkit/HEG/HEGHome.html. This site will have the lastest updates, platforms and products supported.

As described in Section 10.2, HEG functionality is also available through the EOS Data Pools located at NASA Distributed Active Archive Centers (Marley et al., 2003). This will allow end users to customize the region, projection and format of the desired file before the data are received. HEG will process the requested products according to the user-specified processing parameters, and the resultant product will be retrieved via FTP.

10.5 Data Sets Tested and Examples of Usage

As HEG functionality was developed and prior to making the tool available to the user community, a wide cross section of EOS data was selected for testing. Table 10.1 lists products tested with the HEG tool. Note that data products from the Advanced Microwave Scanning Radiometer (AMSR E), MODIS Aqua and Atmospheric Infrared Sounder (AIRS) instruments are also accessable.

During the testing process, an attempt was made to exercise each function on each product. Re-projections were tested when it made sense to do so. For example, low latitude data was not projected to polar-stereographic. Data output from each function was sampled pixel-by-pixel to verify accuracy of processing. Ingestibility of output products into GIS tools was verified. Tests were repeated on all supported platforms.

Several images follow which represent various invocations of HEG functionality during the testing process. We note that the data products from which these images were processed are different in structure, have different projections, some were in Swath format and some were in Grid format.

In Figs. 10.3 and 10.4, MISR and ASTER images, respectively, are co-registered over an eruption of Mt. Etna. The MISR image is a composite constructed with HEG by stitching several adjacent Grids, which are called "MISR blocks". MISR data sets are a special variation of the HDF-EOS Grid format called a "Stacked Grid" format. HEG stitches the adjacent Grids transparently to the user (the

Table 10.1 Data products tested with the HEG Tool

ASTER Products	L1B Calibrated Radiances (250 m)
Level-1B Registered Radiance at Sensor	L1A Geolocation Fields (5-Min Swath)
Level-2 Brightness Temperature	L2 Aerosol
Level-2 Emissivity Product	L2 Total Precipitable Water Vapor Test Results
Level-2 Decorrelation Stretch (VNIR)	L2 Cloud Product
Level-2 Decorrelation Stretch (SWIR)	L2 Temperature and Water Vapor Profile Product
Level-2 Decorrelation Stretch (TIR)	L2 Ocean Color Products
Level-2 Surface Reflectance (VNIR)	L2 Sea Surface Temperature Products
Level-2 Surface Reflectance (SWIR)	L2 Land Surface Reflectance (250 m)
Level-2 Surface Reflectance (TIR)	L2 Land Surface Reflectance (500 m)
Level-2 Surface Kinetic Temperature	L2 Land Surface Reflectance (1 km)
Level-2 Surface Radiance (VNIR)	L2 Thermal Anomalies/Fire
Level-2 Surface Radiance (SWIR)	L2 Snow Cover (500 m)
Level-2 Surface Radiance (TIR)	L2 Sea Ice Extent
Level-3 DEM (Absolute)	L2 Land Surface Temperature and Emissivity
Level-3 DEM (Relative)	L3 Gridded Surface Reflectances
MISR Products	L3 Gridded Snow Cover Product
L1B2 Ellipsoid Data	L3 Gridded Daily Land Surface Temp./Emissivity
L1B2 Terrain Data	L3 Gridded 96-Day Land Cover (1 km)
L2 Land Products	L3 Gridded Thermal Anomalies (1 km).
L2 Aerosol Products	L3 Vegetation Indices 16-Day Global (500 m)
L2 Cloud Products	L3 Vegetation Indices 16-Day Global (1 km)
Digital Elevation Model	L3 Leaf Area Index
1 km Global DTED	L3 Net Photosynthesis
MODIS Products	L3 Daily Gridded Sea Ice Extent
Level-1B Registered Radiance	L3 16-Day SemiEmpirical BRDF and Albedo
L1B Calibrated Radiances (1,000 m)	L3 BRDF Adjusted Nadir Surface Reflectance
L1B Calibrated Radiances (500 m)	L3 Ocean Color and SST
	L3 Vegetation Indices

stitching panel is not needed for MISR data sets). All that is required by the user is to specify the geographic area of interest. HEG will automatically determine which MISR blocks are to be used and stitched. The ASTER image was constructed by stitching, then subsetting adjacent scenes. Each image is a multi-band GeoTIFF. Three bands from the input data set were selected and stored in the output GeoTIFFs. These bands can be individually manipulated through the RGB planes in GIS software to accentuate various features in the image. In this case, the plumes of the volcano are highlighted to emphasize the eruption. The MISR image is in the UTM projection and the ASTER image is in a Geographic projection.

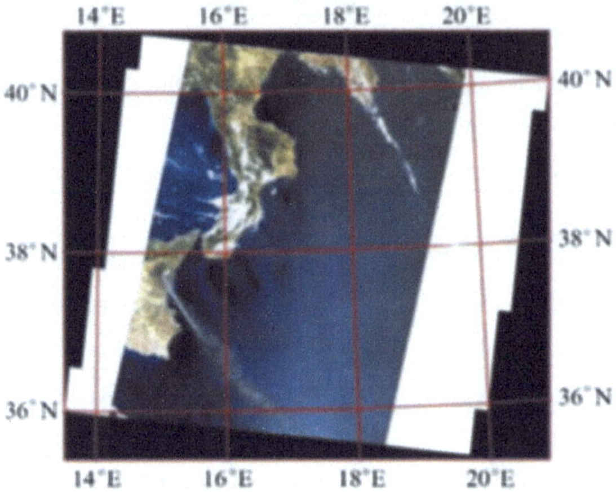

Figure 10.3 Mt. Etna eruption: MISR L1B Stacked Grids

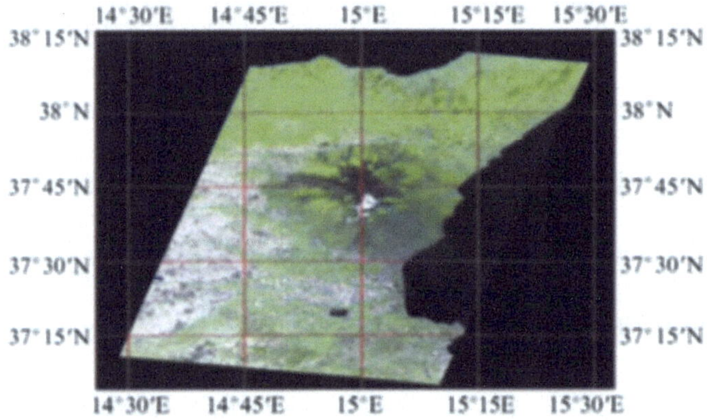

Figure 10.4 Mt. Etna: stitched, subsetted ASTER L1B

Figures 10.5 and 10.6 show the area in La Plata, Maryland that was struck by a Category 5 tornado in 2002. The dark strip running east/west in the center of the ASTER image is the ground damage track of the tornado. In both figures, the input datasets that produced these multi-band GeoTIFF images are HDF-EOS Swath Objects. Swath data is ungridded which a special situation that HEG can handle is. In this case, it has converted them to a Geographic projection. Three different bands from the input data set have been selected and stored in a multi-band GeoTIFF. When loaded into GIS software (i.e. ENVI), each band can be assigned one of the RGB colors. This accentuates various aspects of the images, for example, the clouds. Additionally, the area of interest in each image overlaps two contiguous data sets. Therefore, the stitching and subsetting features of HEG were needed for the multiple input data sets.

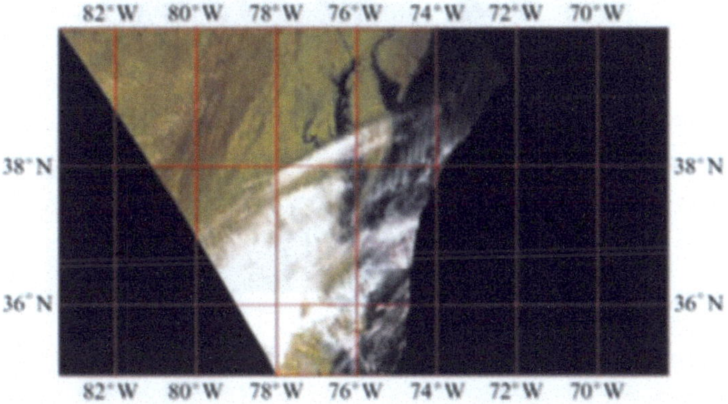

Figure 10.5 MODIS L1B: re-projected, co-registered with ASTER, La Plata, MD Tornado

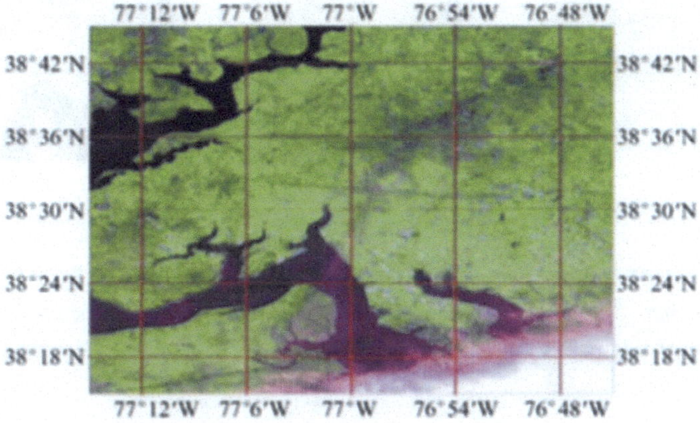

Figure 10.6 La Plata Tornado damage: ASTER L1B reprojected/stitched/subsetted

Figure 10.7 shows a composite made by HEG, subsetted from a global sea surface temperature map. The input data set is an HDF-EOS Grid that covers the entire globe. A smaller image was needed in the same projection. This image is a single band GeoTIFF. Color scaling (red is hot, green is cool) was assigned by using a false color table.

Figure 10.8 shows three bands from MODIS land product data, re-projected from the original projection and then co-registered. The MODIS images are in originally in the Sinusoidal projection and they have been re-projected to Universal Transverse Mercator (UTM).

Figure 10.9 shows MODIS brightness temperature. The input data set is a MODIS L1B HDF-EOS Swath. The image data set has been converted to a geographic projection.

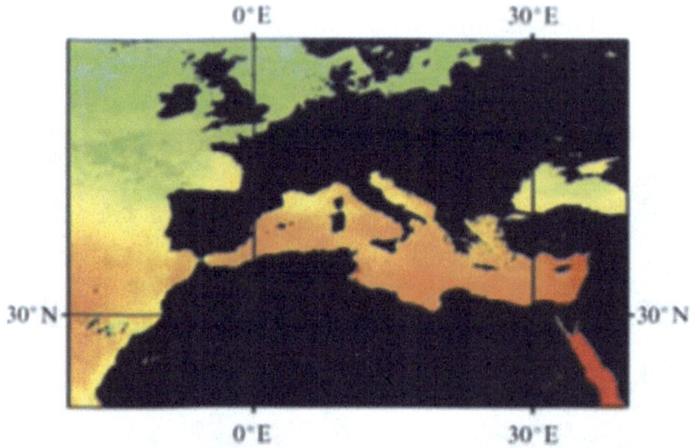

Figure 10.7 MODIS sea surface temperature subset from global map

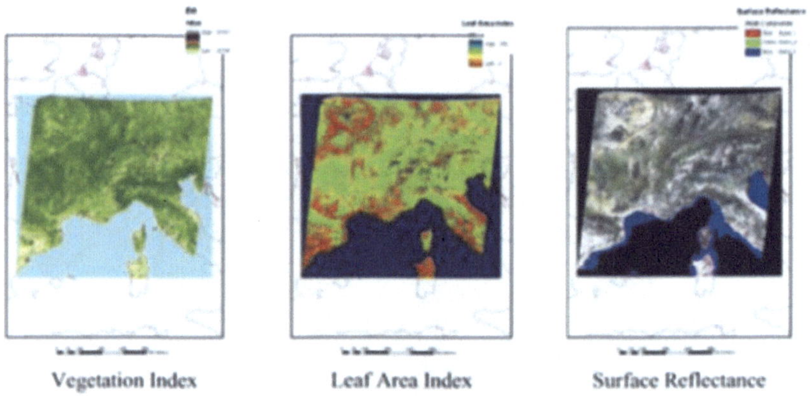

Figure 10.8 MODIS land products: reprojected, co-registered, processed in ArcInfo/ArcMap

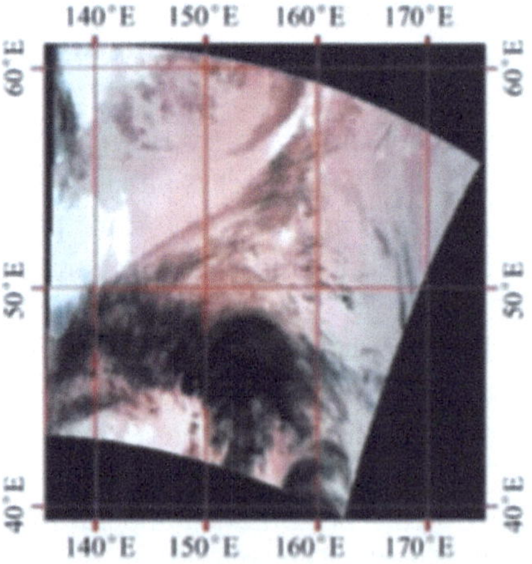

Figure 10.9 MODIS atmosphere product: brightness temperature

10.6 Conclusions

This Chapter describes a tool to be used for post-processing of commonly used EOS products. The tool, HEG, fills a gap in usability of the original archived data. By providing GIS formats, subsetting, re-projection, re-formatting, and other common functions, the data will become more assessable and more useful. Additionally, the tool removes the burden on the user of having to understand details of various data storage methods within HDF-EOS files. It is anticipated that the tool will expand the base of users of EOS data.

References

HDF 4.1R3 Users Guide (2003) http://hdf.ncsa.uiuc.edu/UG41r3_html/

Klein L (1996a) HDF-EOS Library Users Guide for the ECS Project, Volume I: Overview and Examples. 170-TP-006-001, Hughes Information Technology Systems. http://newsroom.gsfc.nasa.gov/sdptoolkit

Klein L (1996b) HDF-EOS Library Users Guide for the ECS Project, Volume II: Function Reference Guide. 170-TP-006-001, Hughes Information Technology Systems. http://newsroom.gsfc.nasa.gov/sdptoolkit

Marley S, Moore M, Clark B (2003) Building Cost-Effective Remote Data Storage Capabilities for NASA's EOSDIS. http://www.storageconference.com/2003/papers/04-Marley-Building.pdf

Taaheri A (2003) Release 6A.07 SDP Toolkit Users Guide for the ECS Project. Raytheon Information Systems. 333-CD-605-003.http://newsroom.gsfc.nasa.gov/sdptoolkit

11 HDFLook—Multifunctional HDF-EOS Tool for MODIS and AIRS Data Processing at GES DISC DAAC

Louis Gonzalez, Dimitar Ouzounov, Darryl Nickless and Gregory Leptoukh

11.1 Introduction

The Goddard Earth Sciences (GES) Data and Information Services Center (DISC) Distributed Active Archive Center (DAAC) plays a major role in enabling scientific research by providing scientific data and data support to the general remote sensing community. Several GES DISC DAAC Data Support Teams provide expert assistance to users in accessing data, including information on visualization tools and documentation for data products. To provide easy access, processing and visualization of the science data, the data support teams have additionally developed many desktop and online tools. In the early days of the Terra satellite's mission, the available tools were not able to meet the Moderate Resolution Imaging Spectroradiometer (MODIS) instrument's data validation and processing requirements for the Hierarchical Data Format (HDF)-Earth Observing System (HDF-EOS) data format. The MODIS Data Support Team's (MDST) goal was to unify the strengths of the current tools and add new features in a new tool intended to visualize, validate, and analyze MODIS data. The MDST started to develop a simple and quick MODIS data processing tool on the base of the existing HDFLook tool originally developed by the Laboratoire d'Optique Atmospherique (LOA) U. S. T. L., Lille, France.

11.2 HDFLook Main Features

The new HDFLook tool is an X-Motif based reader and viewer of the National Center for Supercomputing Application's (NCSA) HDF v.4.x and HDF-EOS 2.x science datasets. HDFlook was designed as a modular tool and allows easy additions of new features and works with different HDF v.4.x data sets simultaneously. It now provides a quick and reliable way to visualize, analyze and validate any MODIS data along with Level-1 data from the Atmospheric Infrared Sounder (AIRS) instrument (Ouzounov et al., 2003, 2004; Gonzales et al., 2003). It is based on the Motif Satellite Process Handling Images Under XWindows (MSPHINX) system, and works on all major UNIX, Linux, Macintosh, and Windows platforms.

The latest HDFLook functionalities are result of a joint collaboration between the NASA GES DISC DAAC, LOA U. S. T. L., Lille, France and the University of Maryland's MODIS Land team.

It is free and available from the NASA GES DISC DAAC website, http://daac.gsfc.nasa.gov/tools_services/HDFLook/ and LOA, http://www-loa.univ-lille1.fr/Hdflook/hdflook_gb.html

HDFLook works with any HDF v.4.x and HDF-EOS 2.x science data, and additional pre-defined features for MODIS and AIRS data products data sets were developed on top of the basic functionalities.

11.2.1 Common HDFLook HDF-EOS Features

(1) The main HDF-EOS visualization capabilities cover L1 radiometric and geolocation products, L2, L3 Land products, L2, L3, L4 atmospheric products; and L2, L3, L4 ocean mapped and binned MODIS/Terra, MODIS/Aqua, AIRS L1B, Cloud and the Earth's Radiant Energy System (CERES), and Polarization and Directionality of Earth's Reflectances (POLDER) data products (see Fig.11.1).

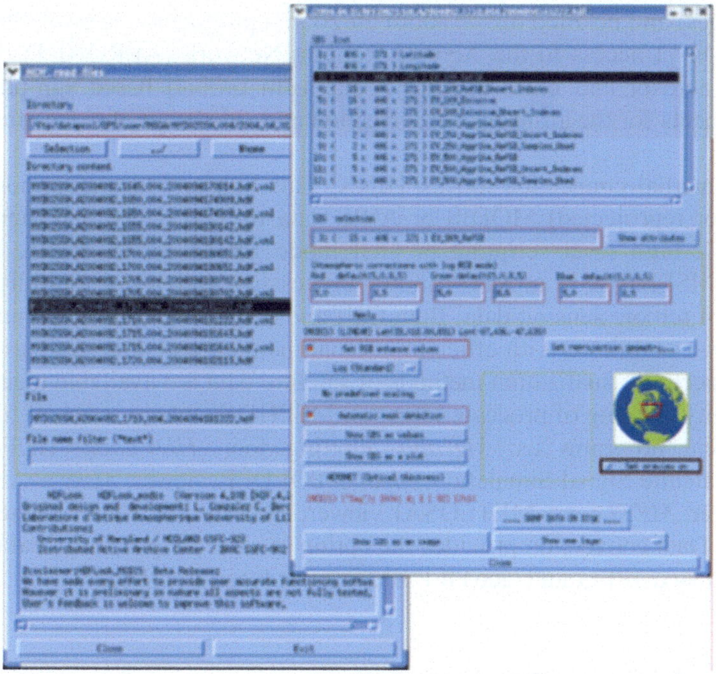

Figure 11.1 HDFLook main GUI Windows shown generic HDF and HDF-EOS MODIS functions as select SDS, select geo projection geometry, color enhancement and spatial subsetting features

(2) Mapping allows re-projection and conversion for L2, L3, and L4 products in fourteen different projections (Linear, Goode's, Mollweide, Hammer, Azimuthal, Lambert, Miller, Orthographic, Sinusoidal, Van Der Grinten, Wagner 7, ISIN, and POLDER) integrated with channel, parameter, and spatial subsetting options. Resampling is optional within any data level and is interactive with mosaic and stitching algorithms.

(3) Additional mapping features such as global grids, coastline, state borders, and global hydrology sets are available.

(4) All functionality can be executed in three different modes: interactive, command line, and operational mode. Multi granule processing is available only in command line and operational modes.

(5) Efficient multi-granule processing approach combined with optimal memory use makes it easy to build global MODIS or AIRS browse composites (mosaics) for red, green, and blue (RGB) (true color), single Science Data Set (SDS) (false color) (see Fig. 11.4). A multi-granule processing scheme is available for stitching in script mode and a customized processing scheme for ftp and other restricted computer environments version is available for operational mode.

(6) "On-demand" RGB or false color band composite. Any three-band sequence could be combined into a composite.

(7) The additional Graphical User Interface (GUI) options include single pixel validation feature, mapping of human activity information (showing towns and villages) and detailed image zoom/un-zoom optimization is available.

(8) Direct access to the Aerosol Robotic Network (AERONET) data bank provides aerosol and other ancillary data from ground based sun-photometer measurements for the atmospheric correction and validation of some atmospheric products.

(9) Internal connection to MSPHINX. This link provides an option to send all (raw or reprojected) MODIS or AIRS data to MSPHINX and then use the MSPHINX high level scripting or GUI mode for advanced science and data analysis. MSPHINX is an X11 Window Processing System for multidisciplinary analysis of remote sensing data. Both GUI and script modes proved multiple file processing capabilities with efficient internal memory and data structure. Several of multispectral visualization methods of satellite data are available together with high-quality printing of processed images. MSPHINX integrated shell for several independent programs as: *Mgraph*—Plotting tool; *HDFLook*—HDF viewer; *Msixs*—Satellite Signal Simulation; *Mpol*—POLDER viewer; *Msky*—AERONET data viewer; *METLook*—METEOSAT viewer; *TMLook*—LANDSAT viewer.

(10) Data conversion provides the option to save the processed data in HDF, ASCII, Binary, JPEG, and GeoTIFF file formats (see Fig. 11.6).

11.2.2 HDFLook MODIS Functions

The main challenges working with MODIS/Terra and MODIS/Aqua data are the large data volumes and the large variety of science products. The MODIS library includes more than 50 features targeting mostly L1 and L2 datasets. HDFLook MODIS is library of specific transformations and functionalities created to serve widely HDF-EOS data but also customized for the very specific and unique characteristics for MODIS data Main HDFLook MODIS features include as followed.

11.2.2.1 Atmospheric Correction

Rayleigh and Ozone corrections, on the base of 6S (second simulation of satellite signal in the solar spectrum) approach is available only for the purpose of imaging to minimize the atmospheric contribution and enhance the surface signal (Vermote and Tanre, 1992).

11.2.2.2 Color Enhancement

Some advanced color enhancement procedures (customized for dry and polar areas) have been used for true color imaging and integrated with the pixel validation capability. 250-m MODIS RGB is available also as combination of bands 1 (250 m), band 4 (500 m) and band 3 (500 m).

11.2.2.3 Cursor Values Capability

The cursor values preview values options is available in Interactive mode and allows viewing of the global position and cross SDS values for the selected point. For multiple parameter data products all SDS values (converted into geophysical parameters) are viewing together with navigation (lat, lon) values.

11.2.2.4 Daytime/Nighttime L1B Visualization

Depending on the time of observation, reconstruction of the spectral reflectance (a.k.a. true-color image) during the day (bands 1, 4, and 3) and brightness temperature (a.k.a. false-color image) during the night is selected. Nighttime browse images are composites of band 31 red (290 – 275 K), band 29 Green (280 – 270 K) and band 20 Blue (300 – 285 K). Clouds appear in white and the color scale shows the highest temperature is in the lowest intensity and the lowest temperature are in the highest intensity.

11.2.2.5 Thermal Anomaly RGB

The thermal anomaly RGB is using MODIS reflectance computed from band 20 (3.75 μm) in the red, band 6 (1.64 μm) corrected reflectance in the green, and band 7 (2.13 μm) corrected reflectance in the blue. For band 20, the reflectance is computed and same method is modified for MODIS by taking advantage of water vapor inversion using Band 18 and 19. Qualitatively, the RGB thermal composite is a very useful and simple way to locate fire activity, where the thermal energy released by fires is interpreted as an extra reflectance contribution at 3.75 μm and produces "red" spots on the image, the relative intensity of the red is directly proportional to the size and temperature of the fire (Petitcolin and Vermote, 2002).

11.2.3 HDFLook AIRS Functions

AIRS is a complex instrument, especially with regard to the downlink telemetry of science and engineering data.

The AIRS infrared level-1B data set contains AIRS infrared calibrated and geolocated radiances, including 2,378 infrared channels in the 3.74 to 15.4 μm region of the spectrum. HDFLook was customized to read L1B AIRS data including AIRS infrared geolocated radiances, AIRS Visible IR and Near IR geolocated radiances, Advanced Microwave Sounding Unit (AMSU)-A1 & AMSU-A2 combined, geolocated & calibrated brightness temperatures. More information can be found on the Atmospheric Dynamics Data Support Team's AIRS Web site, http://daac.gsfc.nasa.gov/atmodyn/airs/. All common HDFLook functionalities (in GUI, command line and operational modes) are available for AIRS data and some new functions were developed mostly for the GUI mode of operation (see Fig. 11.2).

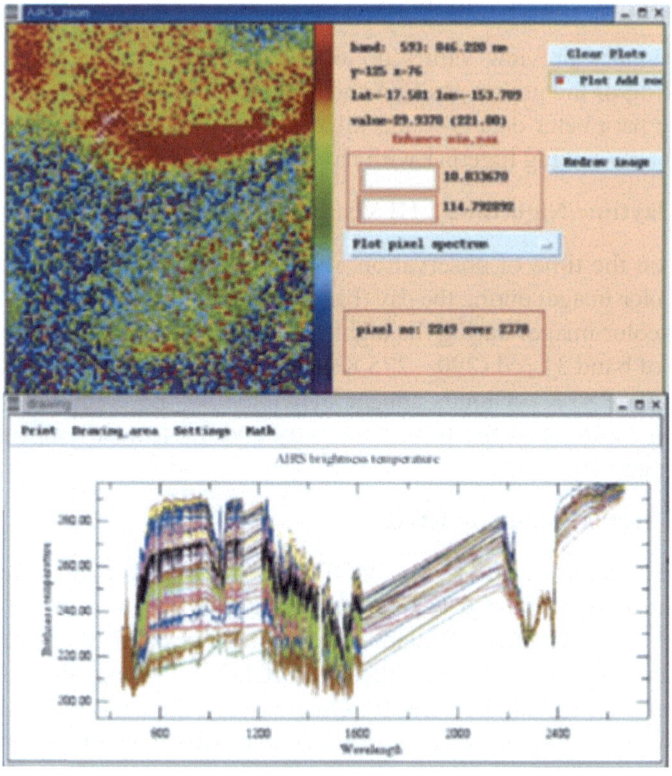

Figure 11.2 HDFLook AIRS functions: plot transect and plot add mode of AIRS L1B radiance

11.2.3.1 Display Raw Data Values (radiance) as well as Brightness from HDF-EOF Files (Swath and Grid)

The main image displays a mosaic of all the spectral bands, and subsetting with zoom/un-zoom enhancement is available. Users can define the region of interest and display transects and histogram plot functions. The display geometry of the images in the main window is the satellite viewing geometry. Additional re-mapping features are available to re-project in Linear, Orthographic and Sinusoidal projections.

11.2.3.2 Display Spectrum

This option allows a user to select a single pixel point and to draw the spectrum values for all 2,378 channels, with the capability of stacking the different spectra for comparison (see Fig. 11.2). All selected spectra can be downloaded as an ASCII file.

11.2.3.3 Display Bi-Dimensional Scatter Diagrams

This option is for a given location and for any two channels selected from the 2,378 spectral bands. The new functionality is the ability to combine image and plot data, and also to provide an option to select pixel range to directly visualize their position on the image area.

11.2.4 High-Level Script Features

HDFLook does provide script language for single and multiple HDF file processing. More than 50 functions are predefined for MODIS, AIRS and any HDF-EOS data types. Interactive and script language commands provide the same functionally with some exception of few functions (mosaic and stitching) defined only for script mode execution. Detail description of the script language with multiple examples could be found in HDFLook_Handbook delivered with the HDFLook package or on http://www-loa.univ-lille1.fr/Hdflook/E_HDF.html.

Script interpretation is very flexible and for single granule allow many functionalities could be executed in one run. In Fig. 11.3, from single granule L1B MODIS, a true color images has been created together with 8 bit single JPEG file form the near IR band 22.

One of the efficient ways to analyze MODIS L1 and L2 is building daily composite images which could be done very easy with HDFLook scripts. The predefined function "create_MODIS_RGB_Mosaic" allow in one step to aggregate 144 daily granules, stitch together, subset and build RGB image and save the output in binary, JPEG and GeoTIFF formats. In Fig. 11.4 the shown script was generated on the fly to produce daily global MODIS RGB image. For more details about the script and other options for multi granule processing referrer to the Global Browse GES Gateway described further in this Chapter.

- verbose
- clear_data
- set_map_directory /usr/modis/HDFLOOK/Maps
- set_output_directory /var/scratch/dimitar//ca
- set_input_directory
  /ftp/datapool/OPS/user/MOGT/MOD021KM.004/2002.11.13
- set_image_background color=White
- set_projection_to_geometry ProjectionTo="LINEAR" \
- LatitudeMinTo=41.0000 LatitudeMaxTo=47.0000 LongitudeMinTo=-85.0000
  LongitudeMaxTo=-75.0000
- set_misc_options CloudsRGBEnhance=Yes
- set_misc_options OnlyDay=Yes
- set_image_RGB_composite VISIBLE
- add_maps Coastlines=yes Rivers=no Boundaries=no \
- CoastlineColor=Black RiverColor=Blue BoundaryColor=Magenta
- set_input_hdf_file MOD021KM.A2002317.1855.004.2002318070109.hdf
- create_MODIS_JPEG_image
  FileName="MOD021KM.A2002317.1855.004.2002318070109.hdf.jpeg"
  RGBMode="LogStandard"
- ....
- select_SDS SDSName="EV_1KM_Emissive" index=2,height,width
- set_misc_options OnlyNight=YES
- create_MODIS_SDS_JPEG_imageFilename="MOD021KM.A2002317.1855.004.2002
  318070109.hdf.jpg"

Figure 11.3 Simple script of creating RGB true color JPEG image & 8-bit SDS JPEG images from single MODIS granule

```
verbose
clear_data
set_home_directory /var/scratch
set_tmp_directory /var/scratch
set_map_directory /usr/modis/HDFLOOK/Maps

set_output_directory          /ftp/data/6/browse_aqua
set_input_directory           /ftp/datapool/OPS/user/MOGA/MYD02SSH.003/2003.09.18

 set_projection_to_geometry ProjectionTo= "HAMMER" WidthTo=2000 HeightTo=1000 \
                    LatitudeMinTo=-90.     LatitudeMaxTo=90.0     LongitudeMinTo=-180.
LongitudeMaxTo=180.0

set_misc_options CloudsRGBEnhance=Yes
set_image_background          color=White
add_maps Coastlines=Yes CoastlineColor=Black
set_misc_options    OnlyDay=Yes
set_image_RGB_composite VISIBLE
set_input_hdf_file
ftp/datapool/OPS/user/MOGA/MYD02SSH.003/2003.09.18/MYD02SSH*

 create_MODIS_RGB_Mosaic      Filename="MYD02SSH_day_aqua_2003_09_018_day"
RGBMode="LogStandard" ClearFile=No \ BottomPixelFiller=2 CreateJPEG=yes
CreateGEOTIFF=yes
```

Figure 11.4 Example HDFLook command line scripts producing global true color image shown on Fig. 11.5

High resolution (250 m) MODIS RGB imaging is available as pre-define option and provide excellent performance in script mode. For single 250 m MODIS L1B granule additional 500 m and geolocation files are need in order to compute RGB. The third (Blue) channel of 250 m RGB is simulated from the 500 m corresponded file. The result of the high MODIS resolution browse for single granule is shown on Fig. 11.6. Example of this script generating this image is similar to the script shown on Fig. 11.3

Additionally to the high resolution browse a new option to build your own choice of RGB combining of any three bands from MODIS L1B radiances and is available in script an interactive modes. This on demand RGB imaging requires that user define own RGB band ratios (the standard colour optimization is off in this case) and in script mode user could import their own colour palette in order to perform SDS false colour mapping.

11.3 GES DISC DAAC HDFLook Applications

The versatility of HDFLook resulted in the integration of the tool in several GES DISC DAAC data production and distribution subsystem and has become a major tool for science validation for the MDST.

11.3.1 MODIS/Terra and MODIS/Aqua Browse Imagery

The L1B true-color for daytime and brightness temperature for nighttime is available for quick review through the Earth Observing System (EOS) Data Gateway (EDG) and the GES DISC DAAC Search and Order interfaces. These browse images are routinely generated by HDFLook as part of the GES DISC DAAC production system, and are created from Calibrated Radiance data (1 km resolution) granules, each covering 5-min of the satellite swath and with a frequency of 288 granules each day. 24-bit RGB images are composed from three bands extracted from these files. Browse images allow science data users to quickly evaluate the remote sensing data's geographic location or cloud coverage before ordering or downloading the large data files.

Each band (channel) is 8 bits and after assembling to one multi-component image, a 24-bit RGB true-color browse image is created. Color is normalized to maximum and minimum for each band separately. Subsampling of the output RGB image provides an option to obtain 1-km or 5-km resolution output images.

Depending on the time of observation, reconstruction of the true-color image during the day (Bands 1, 4, and 3) and false-color image during the night is selected. Following re-mapping process is essential to remove the scan-to-scan overlapping at scan angles greater than 25 degrees and minimizes the bow tie effect (Nishihama et al., 1997). Coastlines feature for 5-km resolution is added for the both daytime and nighttime observation. All data is projected in the Orthographic projection.

As in many multi-spectral images, some enhancement methods have been adapted for the de-correlation (noise reduction) and color enhancement processing (contrast improvement). Color enhancement is one simple way to normalize a true-color (bands 1, 4, and 3) image. Non-linear/linear stretch increases levels of saturation for each band without changing the color value. Four different schemes of color enhancements for drought areas, polar regions, mid-altitude areas and universal cross—correlated approach were designed for MODIS true-color imaging.

11.3.2 On-the-Fly Spatial Subsetting of Data from the GES DISC DAAC Data Pool

HDFLook is the backend engine for the MODIS L2 Ocean On-Demand Parameter Subsetting System. HDFLook is running in operational mode on the FTP server and performs parameter subsetting for multiple inputs to one output file. The GES DISC DAAC web interface to the Data Pool has the same look and feel as the regular GES DISC DAAC Search and Order Interface to the archived data, which includes on-the-fly spatial subsetting of the global gridded (Level-3 and -4) data using HDFLook.

11.3.3 MODIS L3 Atmospheric Products Online Visualization and Analysis System

This Web-based tool is designed to address the science users desire to study aerosols, water vapor and clouds on a large regional to global basis online and without ordering the HDF data. HDFLook is running in script mode as the back-end engine for this system and processes L3 MODIS monthly atmospheric products (1 degree latitude × 1 degree longitude resolution data available only from MODIS/Terra for April, 2000 through December, 2003) and converts to standard binary file with scaling the SDS to geophysical parameters, http://lake.nascom.nasa.gov/online_analysis/movas/.

11.4 Global MODIS Browse Imagery

HDFLook is used to produce daily global MODIS/Terra and MODIS/Aqua browse images with data stored on the GES DISC DAAC Data Pool FTP server, ftp://g0dps01u.ecs.nasa.gov.

Input data of 288 single granules of 5-km calibrated radiance data from MODIS/Terra and MODIS/Aqua data (MOD02SSH,MYD02SSH) is processed to create global daytime (RGB true-color) shown on Fig. 11.5, night time (false-color RGB composite from bands 31, 29 and 20), and brightness temperature (11 μm, both day and night).

Daily production (http://daac.gsfc.nasa.gov/gallery/global_browse/) includes Terra and Aqua MODIS true color images, night time false-color RGB and day and night time brightness temperature (11 μm). Additionally to MODIS data set also Aqua AIRS global images are available for the same parameter configuration.

Figure 11.5 Complete day from MODIS/Terra and MODIS/Aqua. This global 5-km pixel resolution image was generated using MODIS visible bands 1, 4 and 3 from both the Terra and Aqua platforms. While the MODIS/Terra data here is simply overlayed over MODIS/Aqua, the image demonstrates the unique daily coverage achieved by the Terra and Aqua formation. The image, created using the HDFLook tool, shows our planet on Sep. 18, 2003. On that day hurricane Isabel (Category 2) made a landfall on the Eastern Coast of the US (just south of Cape Hatteras of North Carolina) with winds of 100 mph

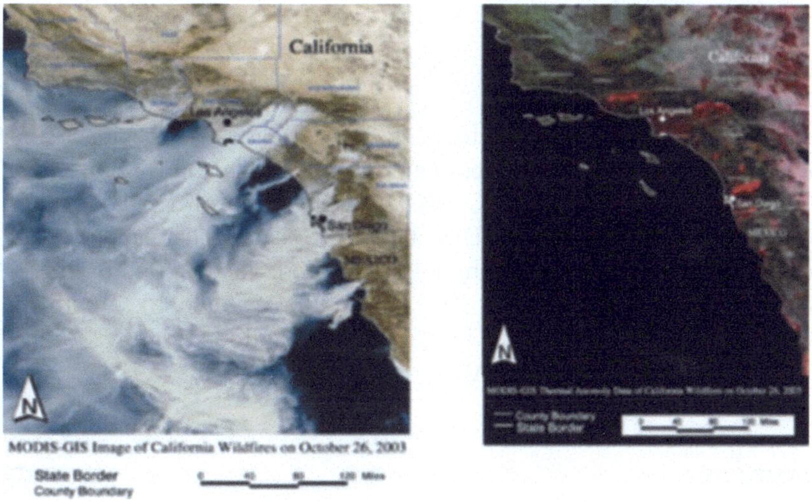

Figure 11.6 California wildfires for Oct. 26, 2003 has been shown in two MODIS/ Terra derived products using HDFLook GeoTIFF export features. On the left is a MODIS 250 m RGB image, while on the right is a MODIS 1 km thermal anomalies image with areas of very high temperature shown in red

11.5 HDFLook Releases and Distribution

HDFLook is an XMotif viewer and is available for all major UNIX operating systems such as SGI, HP, IBM-AIX, DEC, SUN, Linux and Macintosh. For Windows, a free Linux emulator called Cygwin needs to be installed first in order to run the binaries. The main package is distributed in two versions that include a separate desktop tool and as a part of the MSPHINX processing package.

HDFLook can be downloaded from the GES DISC DAAC HDFLook page, http://daac.gsfc.nasa.gov/tools_services/HDFLook/, and also from the LOA server, http://daac.gsfc.nasa.gov/MODIS/HDFLook/ ftp://loa.univ-lille1.fr/HDFLOOK/.

Full documentation, with an online tutorial, examples and detailed Handbook of all GUI and script features are available online at http://www-loa.univ-lille1.fr/Hdflook/hdflook_gb.html.

11.6 Conclusions

The new HDFLook was designed as a simple and quick MODIS and AIRS data processing tool to satisfy GES DISC DAAC needs and the need for wide range applications in the scientific community. This latest version reached the final goal as a modular tool, is flexible and easy to update, includes new features and, is free to obtain and use for HDF-EOS data processing. HDFLook "multi-dataset" philosophy provides new opportunities by working with different data sets simultaneously and can be adapted easily for any future data sets.

Acknowledgements

The authors would like to acknowledge the MODIS and AIRS Science Team for the Science Algorithms, the Processing Team for producing MODIS and AIRS data, and the GES DISC DAAC MODIS and Atmospheric Dynamics Data Support Teams for making MODIS and AIRS data, data support, and data services available to the user community. This work is supported by the NASA/GSFC Earth Science Enterprise. Also, the authors would like to thank Eric Vermote from the University of Maryland/GSFC, and also Suraiya Ahmad and Jason Li from the GES DISC DAAC for their contributions.

References

Gonzalez L, Ouzounov D, Leptoukh G, Savtchenko A, Li JY, Teng B, Deroo C (2003) HDF-EOS HDFLook data processing and visualization applications at GES DAAC. Eos. Trans. AGU 84(46): Fall Meet Suppl Abstract U41B-0008

Nishihama M et al. (1997) MODIS L1A Earth Location. ATBD MODIS-28, GSFC SBRC

Ouzounov D, Savtchenko A, Leptoukh G, Zhou B, Ostrenga D, Deroo C, Gonzalez L (2004) GES DAAC tools for accessing Terra and Aqua MODIS data. Adv. Space. Res. 37(7): 1,109 – 1,113

Ouzounov D, Gonzalez L, Deroo C, Ahmad S, Koziana J, Leptoukh G, Savtchenko A, Serafino G, Sharma AK, Qu J (2003) HDFLook—A New Multifunctional GES DAAC Data Processing and Visualization Tools for and, ocean and atmosphere MODIS data. Eos. Trans. Agu 82(47): Fall Meet Suppl Abstract A41B-0041

Petitcolin F, Vermote EF (2002) Land surface reflectance, emissivity and temperature from MODIS middle and thermal infrared data. Remote Sensing of Environment 83(1 – 2): 112 – 134

Vermote EF, Tanre D (1992) Analytical expressions for radiative properties of planar Rayleigh scattering media including polarization contribution. Journal of Quantitative Spectroscopy and Radiative Transfer 47(4): 305 – 314

12 Tropical Rainfall Measuring Mission Data and Access Tools

Long S. Chiu, Zhong Liu, Hualan Rui and William L. Teng

12.1 Introduction

The Tropical Rainfall Measuring Mission (TRMM), jointly sponsored by the National Aeronautics and Space Administration (NASA) of the United States and the Japanese Aerospace Exploration Agency (JAEA, formally National Space Development Agency (NASDA) of Japan), is the first coordinated international satellite mission to study tropical and sub-tropical rain systems. The TRMM provides visible, infrared, and microwave observations of tropical and subtropical rain systems, as well as lightning and cloud and radiation measurements. The satellite observations are complemented by ground radar and rain gauge measurements to validate satellite rain estimation algorithms (Simpson, 1988). Several field experiments have taken place for validating the algorithms and advancing the physics of precipitation systems.

Since the successful launch of the TRMM satellite on Nov. 27, 1997, over seven years of data have been collected. Originally planned for a three-year research mission, the mission goals were exceeded. The satellite was boosted on Aug., 2001 from its original 350 km altitude to 402 km, helping to extend the mission life. The data have gone through several reprocessing cycles as the science algorithms were validated and improved. These data are useful not only for hydrologists, atmospheric scientists, and climatologists, but also for the health community studying infectious diseases, the ocean research community, and the agricultural community.

12.1.1 TRMM Science

Tropical rainfall is important in many ways. First, it affects the lives and economies of a majority of the Earth's population. Tropical rain systems, such as hurricanes, typhoons, and monsoons, are crucial to sustaining the livelihoods of those living in the tropics. Excess rainfall can cause floods and great property and crop damage, whereas too little rainfall can cause drought and crop failure. Second, the latent heat released during the process of precipitation is a major source of energy that drives the atmospheric circulation. This latent heat can intensify weather systems, affecting weather thousands of kilometers away, thus making tropical rainfall an important indicator of atmospheric circulation and short-term climate change. Third, tropical forests and the underlying soils are major sources of many of the atmosphere's trace constituents. Together, the forests and the atmosphere act as a water-energy regulating system. Most of the rainfall is returned to the atmosphere through evaporation and transpiration, and the atmospheric trace constituents take part in the recycling process. Hence, the hydrological cycle provides a direct link between tropical rainfall and the global cycles of carbon, nitrogen, and sulfur, all important trace materials for the Earth's system.

Because rainfall is such an important component in the interactions between the ocean, atmosphere, land, and biosphere, accurate measurements of rainfall are crucial to understanding the workings of the Earth-atmosphere system. The large spatial and temporal variability of rainfall systems, however, poses a major challenge to estimating global rainfall. At present, the existing rain gauge networks are insufficient, especially over the oceans; network of ground radars are limited by mountain blockage, inter-radar calibration, and range attenuation; thus, satellite measurements provide a nice complement for global observation of rainfall.

There are four main science objectives of the TRMM (Simpson, 1988):

(1) To obtain and study multiyear science data sets of tropical and subtropical rainfall measurements.

(2) To understand how interactions between the sea, air, and land masses produce changes in global rainfall and climate.

(3) To improve modeling of tropical rainfall processes and their influence on global circulation in order to predict rainfall and variability at various periods of time.

(4) To test, evaluate, and improve satellite rainfall measurement techniques.

12.1.2 TRMM Orbit and Instruments

To satisfy the TRMM objectives, the TRMM satellite is flown in a low inclination (35 degree), non-sun-synchronous, and highly precessing orbit, which allows it to fly over each position on the Earth's surface at a different local time each day. This sampling strategy allows the examination of the diurnal cycle of precipitation. The orbit is maintained at approximately 350 km. The satellite was boosted from 350 to 402 km, completed in Aug. 24, 2001, to prolong satellite mission life. Table 12.1 summarizes the characteristics of the TRMM sensor package before and after boost, and their science applications. The three rain instruments are the Precipitation Radar (PR), the TRMM Microwave Imager (TMI), and the Visible and Infrared Scanner (VIRS). Data from these instruments are available from NASA/Goddard Space Flight Center, Data and Information Services Center,

Distributed Active Archive Center (GES DAAC). An assessment of the initial performance of the TRMM rain package was summarized by Kummerow et al. (1998).

Table 12.1 Characteristics of the TRMM sensor package

	Precipitation Radar	TRMM Microwave Imager	Visible and Infrared Scanner
Frequency/ Wavelength	Vertical Polarization: 13.8 GHz	Dual Polarization: 10.65, 19.35, 37, 85.5 GHz; Vertical: 21 GHz	0.63, 1.6, 3.75, 10.8 & 12 μm
Scanning Mode	Cross Track	Conical	Cross Track
Ground Resolution	4.3 km (*5.0 km) at nadir	4.4 km (*5.1 km) at 85.5 GHz	2.2 km (*2.4 km) at nadir
Swath Width	215 km (*247 km)	760 km (*878 km)	720 km (*833 km)
Science Applications	3-D rainfall distribution over both land and oceans, and latent heat release into the atmosphere	Surface rain rate, rain type, distribution, and structure; other atmospheric and oceanic parameters	Cloud parameters, fire, pollution

Note: The TRMM satellite was boosted from 350 to 402 km altitude in August 2001.

* Numbers in parentheses represent post boost values

In addition to the rain package, a Lightning Imaging Sensor (LIS) and a Clouds and Earth's Radiant Energy System (CERES) are carried on the TRMM satellite. The LIS is a calibrated optical sensor operating at 0.7774 μm and observes distribution and variability of lightning. The horizontal resolution of LIS at nadir is 5 km and the swath width is 590 km. The CERES is a visible/ infrared sensor which measures emitted and reflected radiative energy from the surface of the Earth and the atmosphere and its constituents. The TRMM CERES operates at 0.3 to 5.0 μm in the shortwave range and 8.0 to 12.0 μm in the longwave range. LIS and CERES data are available from the NASA Global Hydrology Resource Center (http://ghrc.msfc.nasa.gov) and the NASA Langley Research Center (http://asd-www.larc.nasa.gov/ceres/ASDceres.html) respectively.

12.1.3 TRMM Ground Validation Sites and Field Experiments

To support the validation of the TRMM satellite algorithms, a set of ground validation sites at climatologically significant areas were selected. The Ground

Validation (GV) radar sites include Darwin, Australia; Guam; Kwajalein; multiple sites in Florida; Houston, Texas and the TRMM Large Scale Biosphere-Atmosphere Experiment in Amazonia (TRMM-LBA) area in Brazil. Figure 12.1 shows the locations of the TRMM GV sites. Figure 12.2 shows the locations of the TRMM field experiments, which took place before and during TRMM, as indicated below:

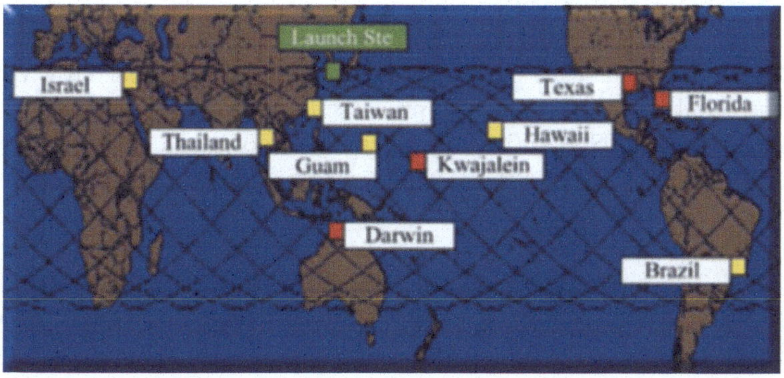

Figure 12.1 The TRMM Ground Validation sites and one day's TRMM satellite ground track. Florida, Texas, Kwajalein, and Darwin are Direct Data (DD) sites, where the raw radar data are processed by the TRMM satellite validation office. The other sites are Direct Product (DP) sites, where the raw radar data are processed by the site Principal Investigators. Florida and Texas are multiple radar sites

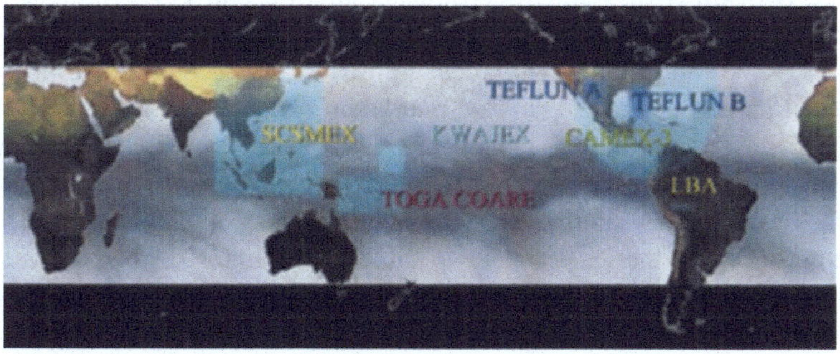

Figure 12.2 Location of the TRMM field experiments. The field experiments include TEFLUN A and B, SCSMEX, KWAJEX, TOGA COARE, CAMEX 3 and LBA. KAPP, not shown, took place in the Florida Keys

(1) Tropical Ocean Global Atmosphere Coupled Ocean Atmosphere Response Experiment (TOGA COARE), November 1992 – February 1993;

(2) Texas Florida Underflights Experiments (TEFLUN A and B), Apr. 1 – May 15, 1998 and Florida Aug. 1 – Sep. 30, 1998, respectively;

(3) South China Sea Monsoon Experiment (SCSMEX), May 1 – Aug. 31, 1998;

(4) Convection And Moisture Experiment (CAMEX-3), Aug. 6 – Sep. 23, 1998;

(5) TRMM Large Scale Biosphere-Atmosphere Experiment in Amazonia (TRMM-LBA), Jan. 1 – Feb. 28, 1999;

(6) Kwajalein Experiment (KWAJEX), Jul. 23 – Sep. 15, 1999;

(7) Key Area Precipitation Project (KAPP), Aug. – Sep. 2002.

12.2 TRMM Products

A set of the TRMM standard products has been developed by the TRMM Science Team. The satellite rain products are processed by the TRMM Science and Data Information System (TSDIS) (http://tsdis.gsfc.nasa.gov/) and the GV products are processed by the TRMM Satellite Validation Office (http://trmm-fc.gsfc.nasa.gov/trmm_gv/). All products are archived by the GES DAAC. TSDIS distributed TRMM and special products to the Science Team. Distribution to the general public is provided by GES DAAC. In addition, to facilitate data access, analyses and use, two general types of value-added TRMM subsets have been developed by the TSDIS and GES DAAC: coincidence subsets and gridded orbital subsets. These subsets have been developed with inputs from the TRMM Science Team and/or GES DAAC Users' Group.

12.2.1 TRMM Standard Products

There are two kinds of standard products: satellite and GV products. Level-1 products include VIRS calibrated radiances, TMI brightness temperatures, and PR return power and reflectivity measurements. Level-2 products are derived geophysical parameters (e.g. rain rate and latent heat) at the same resolution and location as those of the Level-1 data. Level-3 products are space-time-averaged parameters (Wharton and Myers, 1997). Figure 12.3 shows the processing flow of the TRMM standard satellite products. Characteristics of the TRMM standard satellite and GV products are listed in Tables 12.2 and 12.3, respectively. Each product is identified by a TSDIS reference number, which is useful for searching and identification.

12.2.2 TRMM Subsets

The main purposes of these value-added products are (1) to reduce the file sizes that users have to handle (on the order of hundreds of megabytes) and (2) in some cases, to convert from the Hierarchical Data Format (HDF) of the TRMM data to formats more familiar to users. These products have evolved as users' requirements change. Figure 12.4 shows the processing flow of the subsets.

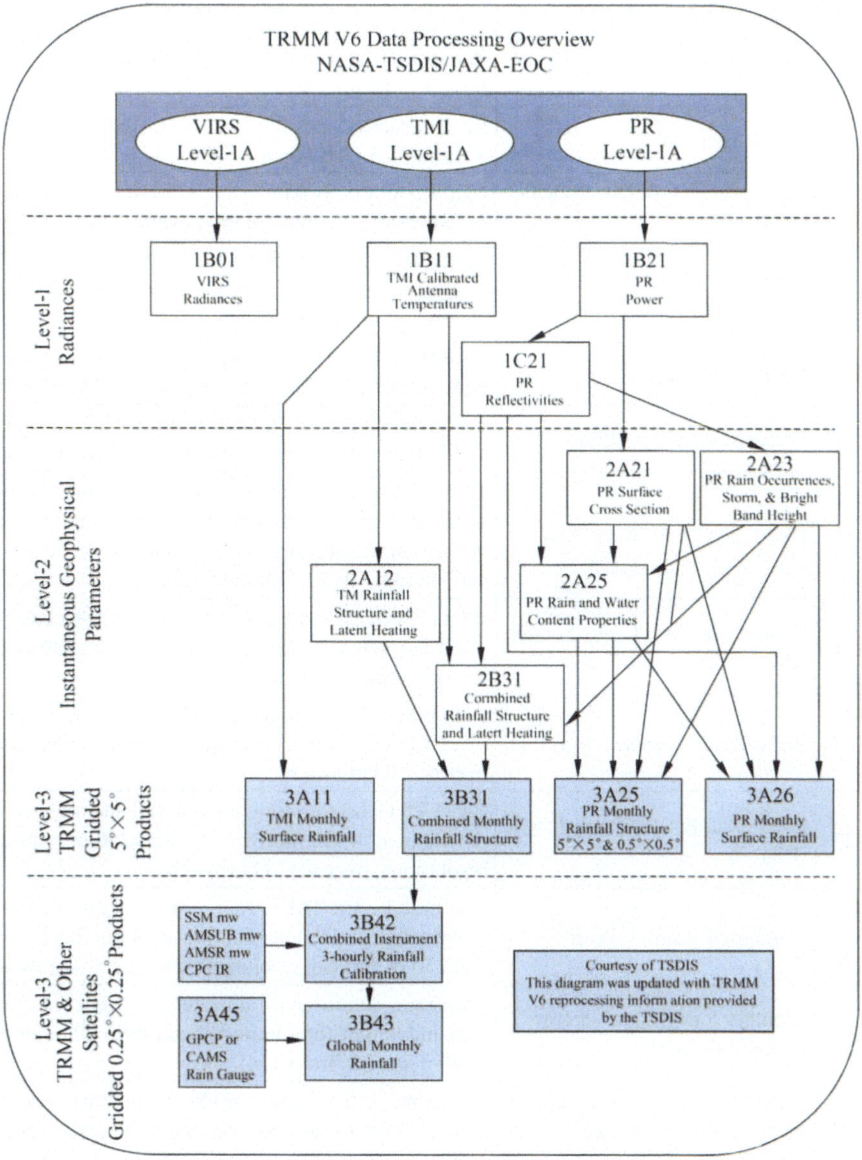

Figure 12.3 The TRMM standard satellite product processing flow. Starting with version 6 reprocessing, a new product, monthly TMI rain, latent heating and cloud liquid water (3A12), is added

Table 12.2 Characteristics of the TRMM Standard Satellite Products

TSDIS Ref.	Product Name	Product Description
1A01	VIRS Raw Data (VIRS)	Reconstructed, unprocessed VIRS (0.63, 1.6, 3.75, 10.8, and 12 μm) data
1A11	TMI Raw Data	Reconstructed, unprocessed TMI 10.65, 19.35, 21, 37, and 85.5 GHz) data
1B01	Visible and Infrared Radiance (VIRS)	Calibrated VIRS (0.63, 1.6, 3.75, 10.8, and 12 μm) radiances at 2.2 km resolution over a 720 km swath
1B11	Microwave Brightness Temperature (TMI)	Calibrated TMI (10.65, 19.35, 21, 37, and 85.5 GHz) brightness temperatures at 5 km to 45 km resolution over a 760 km swath
1B21	Radar Power (PR)	Calibrated PR (13.8 GHz) power at 4 km horizontal, and 250 m vertical, resolutions over a 220 km swath
1C21	Radar Reflectivity (PR)	Calibrated PR (13.8 GHz) reflectivity at 4 km horizontal, and 250 m vertical, resolutions over a 220 km swath
2A12	Hydrometeor Profile (TMI)	TMI Hydrometeor (cloud liquid water, prec. water, cloud ice, prec. ice) profiles in 14 layers at 5 km horizontal resolution, along with latent heat and surface rain, over a 760 km swath
2A21	Radar Surface Cross-Section (PR)	PR (13.8 GHz) normalized surface cross-section at 4 km horizontal resolution and path attenuation (in case of rain), over a 220 km swath
2A23	Radar Rain Characteristics (PR)	Rain type; storm, freezing, and bright band heights; from PR (13.8 GHz) at 4 km horizontal resolution over a 220 km swath
2A25	Radar Rainfall Rate and Profile (PR)	PR (13.8 GHz) rain rate, reflectivity, and attenuation profiles, at 4 km horizontal, and 250 m vertical, resolutions, over a 220 km swath
2B31	Combined Rainfall Profile	Combined PR/TMI rain rate, path-integrated attenuation and latent heating at 4 km horizontal, and 250 m vertical, resolutions, over a 220 km swath
3A11	Monthly 5 degree × 5 degree Oceanic Rainfall	Rain rate, conditional rain rate, rain frequency, and freezing height for a latitude band from 40 degree N to 40 degree S, from TMI
3A12	Monthly 0.5 degree × 0.5 degree mean 2A12, profile, and surface rainfall	0.5 degree × 0.5 degree gridded monthly product comprised of mean 2A12 data and calculated vertical hydrometeor profiles as well as mean surface rainfall
3A25	Monthly 5 degree × 5 degree and 0.5 degree × 0.5 degree Spaceborne Radar Rainfall	Total and conditional rain rate, radar reflectivity, path-integrated attenuation at 2, 4, 6, 10, 15 km for convective and stratiform rain, storm, freezing, and bright band heights, and snow-ice layer depth for a latitude band from 40 degree N to 40 degree S, from PR

Continued

TSDIS Ref.	Product Name	Product Description
3A26	Monthly 5 degree × 5 degree Surface Rain Total	Rain rate probability distribution at surface, 2 km, and 4 km for a latitude band from 40 degree N to 40 degree S, from PR
3B31	Monthly 5 degree × 5 degree Combined Rainfall	Rain rate, cloud liquid water, rain water, cloud ice, grauples at 14 levels for a latitude band from 40 degree N to 40 degree S, from PR and TMI
3B46	Monthly 1 degree × 1 degree SSM/I Rain	Global rain rate from SSM/I
3B42	3B42: 3-Hour 0.25 degree × 0.25 degree TRMM and Other-GPI, Calibration Rainfall	Calibrated geosynchronous IR rain rate using TRMM estimates
3B43	3B43: Monthly 0.25 degree × 0.25 degree TRMM and Other Sources Rainfall	Merged rain rate from TRMM, geosynchronous IR, SSM/I, rain gauges
CSH	CSH: Monthly 0.5 degree × 0.5 degree Convective & Stratiform Heating	TRMM Monthly 0.5 degree × 0.5 degree Convective/ Stratiform Heating (This is a TRMM experimental product)

Table 12.3 Characteristics of the TRMM GV Products

TSDIS Ref.	Product Name	Description
1B51	Radar Reflectivity	Volume scan of radar reflectivity, differential reflectivity, and mean velocity (if available), truncated at 230 km range, at original radar resolution, coordinate, and sampling
1C51	C51: Half-Hourly* Calibrated Radar Reflectivity	Volume scan of calibrated radar reflectivity and differential reflectivity (if available), and corresponding QC masks, truncated at 200 km range, at original radar resolution and coordinate
2A52	Half-Hourly* Rain Existence	Percent of rain in the radar volume scan
2A53	Half-Hourly* 2 km Radar Site Rain Map	Instantaneous rain rate over an area of 300 km × 300 km
2A54	Half-Hourly* 2 km Radar Site Rain Type Map	Instantaneous rain type classification (convective, stratiform) over an area of 300 km × 300 km
2A55	Half-Hourly* Radar Site 3-D Reflectivity	Instantaneous radar reflectivity and vertical profile statistics over an area of 300 km × 300 km, at 2 km horizontal and 1.5 km vertical resolution

Continued

TSDIS Ref.	Product Name	Description
2A56	1 Minute Average and Peak Rain Gauge Rain Rate	Time series of rain gauge rain rates over the radar site rain gauge network
3A53	5-Day 2 km Ground Radar Site Rain Map	Surface rain total from ground radar
3A54	Monthly 2 km Ground Radar Site Rain Map	Surface rain total from ground radar
3A55	Monthly 2 km Ground Radar Site 3-D Rain Map	Vertical profile of reflectivity and contoured frequency by altitude diagrams for stratiform, convective, and anvil rain over land and water

* Volume scans at original radar sampling when within half hour of TRMM satellite coincidence

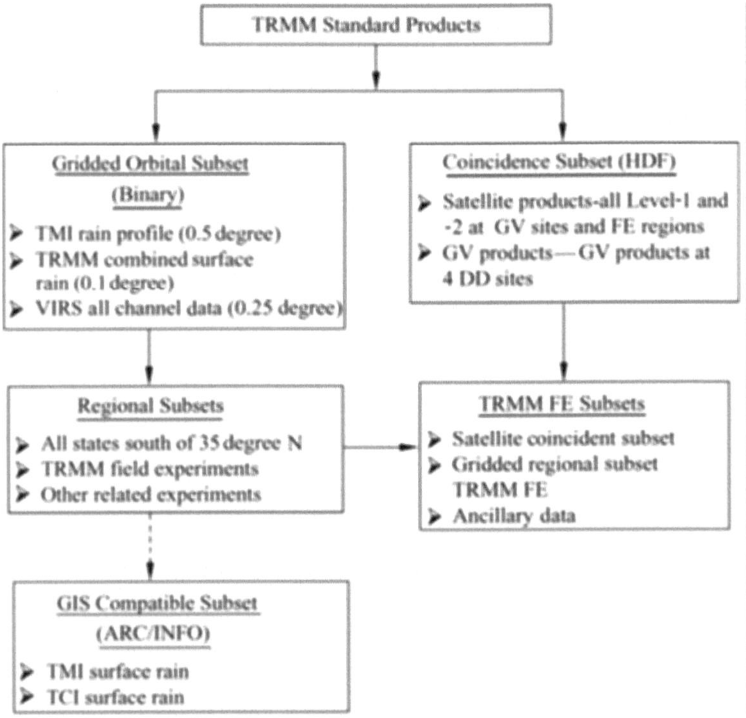

Figure 12.4 Processing flow of GES DAAC value-added TRMM subsets. The TRMM standard products from which these value-added products are produced include TRMM Microwave Imager, TRMM Combined Instrument (TCI), TRMM Precipitation Radar and Visible and Infrared Scanner, and Ground Validation products (see Tables 12.2 and 12.3)

12.2.2.1 Coincidence Subsets

There are two groups of Coincidence Subset Intermediate (CSI) products, both generated by TSDIS. A satellite CSI file is a subset (collection of instrument scan data) of a TRMM standard orbital product corresponding to a time period during which the sub-satellite point is within a specified distance of a given GV radar site. The distance is usually defined as the sum of the ground radar radius and the satellite sensor half-swath width. A GV CSI file is that ground radar volume scan corresponding to the time when the sub-satellite point is nearest the GV site.

The GES DAAC archives nine satellite CSI products at 27 sites and four GV CSI products at four sites. The 27 satellite sites include sites in Australia, Brazil LBA area, Bahamas, Borneo Island, Congo, West Africa, Kwajalein, Guam, Israel, Japan, Korea, South China and East China Seas, Thailand, Taiwan, and Florida, Oklahoma, Texas in the continental United States. The four GV sites are Darwin, Australia; Houston, Texas; Melbourne, Florida; and Kwajalein. The nine satellite CSI products are VIRS radiance (1B01_CSI), TMI brightness temperature (1B11_CSI), PR return power (1B21_CSI), calibrated PR reflectivity (1C21_CSI), TMI hydrometeor profile (2A12_CSI), PR radar surface cross section (2A21_CSI), PR rain qualitative (2A23_CSI), PR rain rate and profile (2A25_CSI), and combined PR/TMI rain profile (2B31_CSI). The four GV CSI products are calibrated GV reflectivity (1C51_CSI), GV site rain map (2A53_CSI), GV site convective/stratiform map (2A54_CSI), and GV site three-dimensional reflectivity (2A55_CSI). Where possible, the GES DAAC calculates the percent rain or rain existence information, inserts them as metadata into the database (not into the original files), and, thus, makes them available as data search criteria. All 13 CSI subset products are in HDF format.

12.2.2.2 Gridded Subsets

There are three gridded (G) subset products developed by the DAAC, and are derived from their corresponding TRMM standard products by sub- or re-sampling the orbital data onto Earth grids. For the VIRS radiance gridded product (G1B01), the sub-sampling is done on a 0.25 degree latitude by 0.25 degree longitude grid. The grid value is the value of the pixel closest to the center of the grid box for all VIRS channels. For the TMI hydrometeor profile gridded product (G2A12: surface rainfall and vertical profiles of cloud water at 14 layers), the re-sampling is done on a 0.5 degree latitude by 0.5 degree longitude grid. The grid value contains the total number of pixels, number of rainy pixels, and mean and standard deviation of conditional surface rain and cloud water at all 14 layers. For the combined PR/TMI rain profile gridded product (G2B31), the re-sampling is done on a 0.1 degree latitude by 0.1 degree longitude grid. The grid value contains the total number of pixels, number of rainy pixels, and mean and standard deviation of conditional surface rain rate. All three gridded subsets are produced in binary format.

Additionally, TSDIS produces daily files of the TRMM surface rain products (2A12, 2A25, and 2B31) at 0.5 degree latitude by 0.5 degree longitude resolution

(TSDIS reference number 3G68). The pixel value contains the number of satellite pixels, number of rainy pixels, number of rainy pixels determined to be convective, and mean rain rate. These files are in ASCII format.

12.2.2.3 Regional Gridded Subsets

For the three gridded subsets (previous section), the GES DAAC produces geographical region subsets for the US and selected regions (e.g. Lower Mississippi River valley, the US-Mexico border region) under TRMM coverage and for several TRMM field experiment areas (e.g. the South China Sea Monsoon Experiment, SCSMEX). All three regional gridded subset products are in binary format.

12.2.2.4 GIS-Compatible Subsets

Geographic Information System (GIS)-compatible TRMM subsets have been developed by the GES DAAC to extend the current user base beyond that of the science research communities. These GIS-compatible subsets are designed to attract users from fields such as water resources, agriculture, and land management (Chiu et al., 1999; Pollack et al., 2000). Regional gridded TRMM Combined Instrument rainfall products (RG2B31) are converted to Environmental Systems Research Institute (ESRI) GIS-compatible formats, allowing the use of these products in ESRI GIS packages (e.g. ARC/INFO, ArcView) or their importation to other GIS packages.

12.3 TRMM Field Experiment Data Sets

12.3.1 Field Experiment Data

Six field experiments were carried out for TRMM algorithm validation. The GES DAAC is the central archive for three of these experiments, TEFLUN-A and -B, TRMM-LBA, and KWAJEX; and the ancillary data archive for a fourth, SCSMEX. There are about 100 data sets collected during the TRMM FEs.

12.3.2 Ancillary Data

Ancillary data products, principally temporal and spatial subsets of other satellite data, are developed and/or stored by the GES DAAC for TRMM FEs and made available to users via ftp or web access (http://daac.gsfc.nasa.gov/hydrology/hd_data.shtml). The ancillary data include:

(1) Visible and IR channel data at original sensor resolution observed by geosynchronous meteorological satellites that include GOES, GMS and Meteosat.

The GOES 8 and GOES 10 data in HDF format are produced at the University of Hawaii and pumped over to the GES DAAC site. GMS data are acquired from the NASA/GSFC Remote Sensing Data Portal (http://rsd.gsfc.nasa.gov/). Meteosat data are collected from the Marshall Space Flight Center Global Hydrology and Climate Center (ftp://geo.msfc.nasa.gov/Weather/).

(2) The Special Sensor Microwave/Imager (SSM/I) GPROF 6.0 rain rates derived using the Goddard profiling algorithm (Kummerow et al., 2001). Both 0.5 and 0.25 degree gridded satellite orbit products are available.

(3) NOAA global 5 km merged infrared data set produced by J. Janowiak.

(4) Goddard Satellite derived Surface Radiation Budget (GSSRB) data developed by M-D. Chou.

(5) Goddard Satellite Sea Surface Turbulent Flux version 2 (GSSTF2) developed by S. H. Chou et al.

(6) Earth topography five-minute grid (ETOPO5).

12.4 Tools for Data Visualization and Analysis

A number of software tools are available to read and analyze TRMM data, which are stored in HDF. They include the TSDIS Toolkit developed by TSDIS, WebWinds, and the commercially available Noesys and IDL ENVI. The GES DISC provides access to two software tools for data visualization and analysis: the TSDIS Orbit Viewer and the TRMM On-Line Visualization and Analysis System (TOVAS).

12.4.1 TSDIS Orbit Viewer

The Orbit Viewer is based on the IDL software developed by Research Systems Inc. (RSI). It can be run in PC or UNIX modes. The Orbit Viewer displays most TRMM data formats described in Section 12.4.2, including the Satellite products, Ground Validation radar and rain gauge products, and GES DAAC subsets. The Orbit Viewer's interactive 2D and 3D capabilities are summarized below, as well as its automated mode. The basic image can be created with two mouse clicks. Figure 12.5(a) shows a 2D image of Hurricane Floyd. The orange circle in the middle is the intense rain at 2.5 km altitude, in the middle of the eye wall.

Additional visualization features include histograms, a statistical summary file for each variable, vertical cross sections of the satellite swath, and plots of navigation data stored as HDF Vdata structures. Figure 12.5(b) shows the vertical distribution of rainfall. At increasing altitudes, the eyewall of Hurricane Floyd slopes outward away from the eye, which is a well-known feature of hurricanes.

Some features of a storm can be identified more easily in the Orbit Viewer's 3D images. Figure 12.5(c) shows a narrow band of intense rainfall in the 3D image, connecting the eyewall with the outer region of rain.

Tar files and installation instructions of the Orbit Viewer can be downloaded from TSDIS or the GES DAAC at no charge (http://tsdis.gsfc.nasa.gov/). The installation includes the source code, making it possible for scientists to modify or reuse the code to create specialized images.

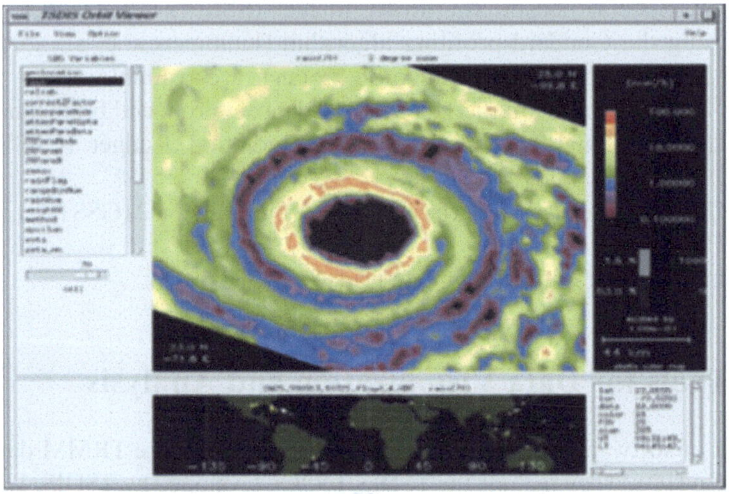

(a)

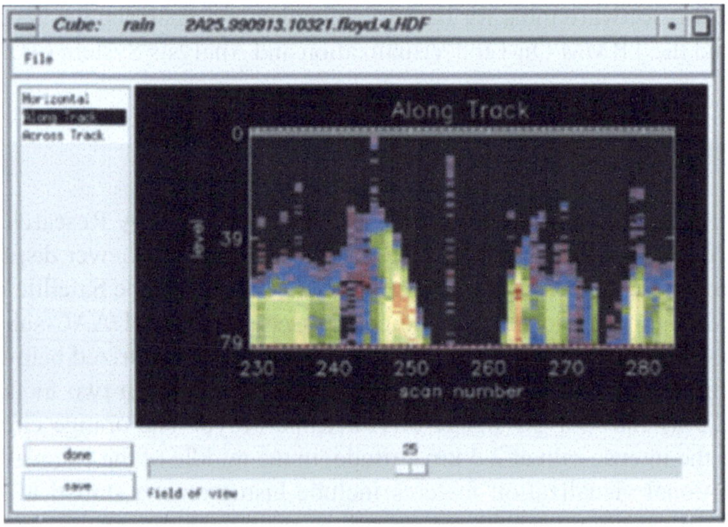

(b)

Figure 12.5 (a) Two dimensional view of Hurricane Floyd on Sep. 13, 1999. The image is an enlarged image of the white square shown in the lower panel. (b) Along track vertical cross section of the same Hurricane Floyd image shown in (a). (c) Three dimensional image of Hurricane Floyd, showing regions with rain rates in excess of 3 mm/h

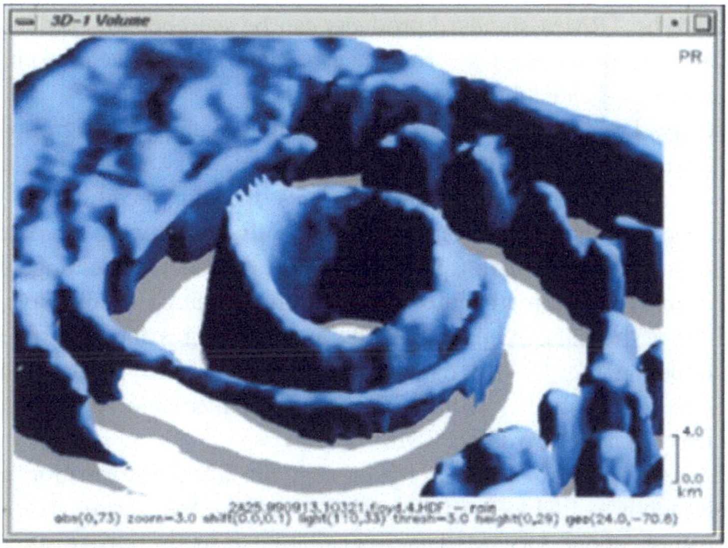

(c)

Figure 12.5 (Continued)

12.4.2 TOVAS

TOVAS, developed by the GES DAAC, provides fast and easy access for users to obtain precipitation information via the Internet (http://lake.nascom.nasa.gov/tovas/) (Liu et al., 2002). TOVAS currently supports area and time plots, animation, and display of time-longitude and time-latitude sections. Simple analysis functions include computing temporal average or accumulation maps, time series of area average rain rates over user-specified regions, and an ASCII output capability which allows users to download the computation results for their own analysis.

Several TRMM and related products are supported. The products include TRMM 3-hourly near-real-time (3B42RT), daily (3B42, Version 5 and 6), monthly (3B43, Version 5 and 6) products, and the TMI rain, latent heat and cloud liquid water profiles (3A12, Version 6) and the University of Delaware (Willmott and Matsuura, 1995) and Global Precipitation Climatology Center (GPCC) (Rudolf et al., 1994) gauge analyses.

12.5 TRMM Data Access and Usage

As of Dec. 2003, over 1,200 users from 38 countries have ordered TRMM data from the GES DAAC. In addition, the TOVAS has had over 30,000 hits by over 2,000 users within the first year of release.

Most of the TRMM standard products can be searched and ordered via the TRMM Data Search and Order System (Rui et al., 1999), http://lake.nascom.nasa.gov/data/dataset/TRMM/index.html.

Currently, TRMM data are organized into five data product groups: orbital, gridded, ground-based instrument, subsets, and ancillary. From a data product group, one can search for specific data products, then for specific years, months, and days. One can also search for data dynamically, by drawing a box around an area of interest on a global map, specifying the data products, time range of interest, and value range of certain attributes. Then, from the search results page, one can view and order the data. Available online are help pages for data ordering and documentation on the TRMM data products. Some TRMM data, such as the Gridded Subsets described in Section 12.2.2, are also available via anonymous FTP, ftp://lake.nascom.nasa.gov/data/TRMM/.

Data sets are distributed via ftp or sent on tapes. Information about the TRMM mission is available from the TRMM Web site http://trmm.gsfc.nasa.gov/ and information about TRMM data is available from the DAAC Precipitation Data and Information Service Center (PDISC) Web site, http://daac.gsfc.nasa.gov/www/precipitation.

As of Dec. 2003, about 579,000 files totaling 51 Terabytes (uncompressed) of TRMM standard products have been archived at the GES DAAC. Figure 12.6 shows the total data archive and distribution of TRMM satellite products. Level-1 satellite data are the largest by volume, constituting greater than 50%, followed

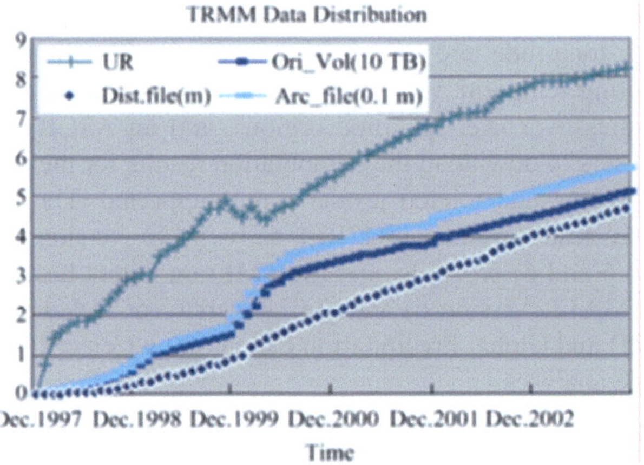

Figure 12.6 Summary statistics of TRMM satellite data archive and distribution. Units for the original volume are in 10 TB, the number of archived files in 0.1 million, and the number of distributed files in millions. The UR (Utilization Ratio) is the ratio of the number of distributed files to that archived

by Levels 0/1A and Level-2, which are approximately 15% each. GV data are about 5%. Level-3 and gridded products constitute less than 0.1% of total data volume, but are the most requested.

To monitor the usage of TRMM products, we define a utilization ratio as the ratio of the total number of files distributed to that archived. When the UR is constant during routine processing, the usage is constant. An increase in UR indicates increased usage. The UR for all products has been increasing during most of the mission life and is about 8.5 currently. To improve on the quality of the products, several reprocessings of TRMM products have occurred. During reprocessing periods, the entire data set was reprocessed at a rate of 3 – 5 times the steady processing rate. Hence, the accumulated files increased drastically and the UR experienced a dip. Reprocessing of TRMM Versions 1, 2, 3, and 4 began in March 1998, June 1998, September 1998, and November 1999, respectively. Reprocessing of version 4 took about six months. After the release of version 5 data, the improved algorithms resulted in more stable and better products. 3B42 and 3B43 are the two most requested products. Their URs were 10 and 14, respectively, in September 2000 (after rebound from Version 4 reprocessing), and increased to 57 and 68, respectively, in December 2003. The version 6 reprocessing started in April 2004.

TRMM data distributed via GES DAAC's anonymous FTP constitutes a fraction of the total distributed volume. Of these, the Gridded Orbital products (see Section 12.2.2) consistently are the most frequently requested, along with, at times, TRMM Field Experiment-related data.

12.6 TRMM Applications

TRMM data have been used in a number of research and applications. Application uses include a warning system for mosquito-borne diseases; rain events and diabetes relationship; supplementing existing gauge data in regions such as SE Asia, Africa, Afghanistan and Sri Lanka; flash flood warning and monitoring; crop yield models and monitoring; fire monitoring; range prediction of butterflies in America; monitoring rain events in Balkans; insect activities in the US, and initializing a soil moisture model for South Africa.

As an example of TRMM data use, Fig. 12.7 shows the use of TRMM data in monitoring crop yields in Africa by the United Nations World Food Program. The figure in the left panel shows the accumulated rainfall for the period Aug. 2002 – Mar. 2003, the growing period of maize in Africa. The figure in the right panel shows the percent departure from the same months in 2001 – 2002. The left panel is generated by the TOVAS. The information from this figure can complement field observations in assessing the severity of crop failure and provide a useful guide for the United Nations' food aid distribution.

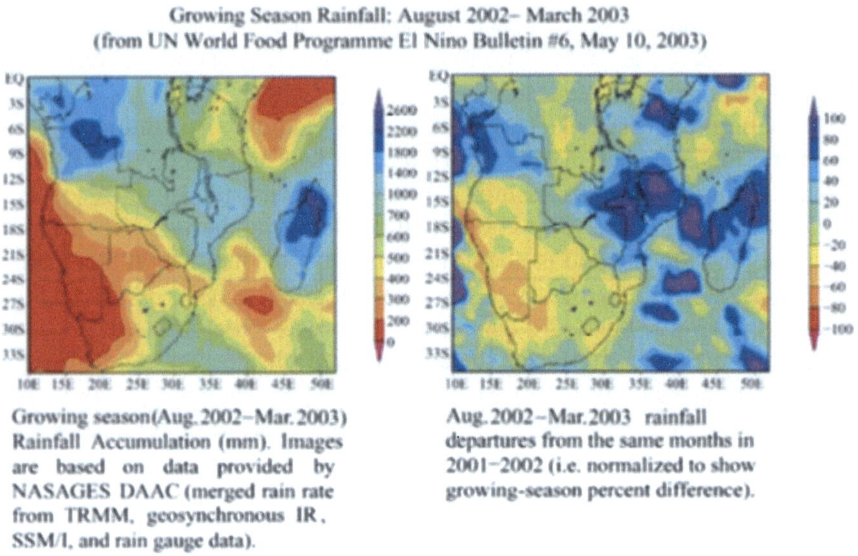

Figure 12.7 Use of TRMM data by the United Nation World Food Program in assessing crop yield and providing decision support in food aid distribution

Acknowledgements

This work is based on the collective work of the GES DAAC Hydrology Data Support Team. Additional members include John Bonk, Pat Hrubiak, Li Lu, currently and S. Ahmad, N. Pollack, A. K. Sharma, G. Serafino, G. Vicente and R. Yang, previously. Owen Kelley of TSDIS contributed to the Orbit Viewer Section and Dong-Bin Shin of GMU/CEOSR provided formatting support. The authors thank members of the TRMM Science Team, in particular, Drs. R. Adler and C. Kummerow, for comments and suggestions and members of the TSDIS, led by Erich Stocker, for processing the standard products and subsets. The processing, archiving, and distribution of TRMM and other hydrological data by the TSDIS and GES DAAC are funded by the NASA Earth Science Enterprise.

References

Chiu L, Pollack N, Serafino G, Teng W, Wong D (1999) Disseminating National Aeronautics and Space Administration (NASA) remotely sensed data to the agricultural and natural resource management communities. Third World Multiconference on Systemics, Cybernetics and Informatics and Fifth International Conference on Information System Analysis and Synthesis Orlando, FL 2: 151 – 155

Kummerow C, Hong Y, Olson WS, Yang S, Adler RF, McCollum J, Ferraro R, Petty G, Shin DB,

Wilheit TT (2001) The evolution of the Goddard profiling algorithm (GPROF) for rainfall estimation from passive microwave sensors. J Appl. Meteorol. 40(11): 1,801 – 1,820

Kummerow C, Barnes W, Kozu T, Shiue J, Simpson J (1998) The Tropical Rainfall Measuring Mission (TRMM) sensor package. J. Atmos. Ocean. Tech. 15: 809 – 817

Liu Z, Chiu L, Teng W, Serafino G (2002) A simple online analysis tool for visualization of TRMM and other precipitation data sets. Science Data Processing Workshop 2002, Greenbelt, Maryland

Pollack N, Teng W, Chiu L, Serafino G, Wong D (2000) Operational production and distribution of GIS-compatible remotely sensed data to facilitate their use. In: Proc., 2000 ASPRS Annual Conference, Washington, DC. Amer. Soc. Photogrammetry and Remote Sensing

Rudolf B, Hauschild H, Rueth W, Schneider U (1994) Terrestrial precipitation analysis: operational method and required density of point measurements. In: Desbois M and Desalmand F (eds) Global Precipitation and Climate Change NATO ASI Series 1(26): 173 – 186, Springer-Verlag

Rui H, Teng B, Sharma AK, Chiu L (1999) NASA/Goddard DAAC hierarchical search and order system for TRMM data: a web-based approach. 22nd General Assembly of the International Union of Geodesy and Geophysics, Birmingham, United Kingdom

Simpson J (ed) (1988) TRMM: a satellite mission to measure tropical rainfall. Report of the Science Steering Group, NASA p. 94

Wharton SW, Myers MF (1997) MTPE EOS Data Products Handbook, Volume I, TRMM and AM-1. (Available from Code 902, NASA/Goddard Space Flight Center Greenbelt MD 20771) p. 266

Willmott CJ, Matsuura K (1995) Smart interpolation of annually averaged air temperature in the United States. J. Appl. Meteorol. 34: 2,577 – 2,586

13 The Open GIS Web Service Specifications for Interoperable Access and Services of NASA EOS Data

Liping Di

13.1 Introduction

Data is a representation of objects or phenomena in a formalized manner subject to interpretation or to which meaning may be assigned (ISO/IEC 2382-1, 1993). Geospatial data is the data that can be associated with location on Earth. Geospatial data is the dominant form of data in terms of data volume. Geospatial data are widely used in many socio-economic activities, ranging from environmental management to military operations. Because of the importance of geospatial data, both public and private sectors have invested huge amounts of resources and money in collecting, managing, archiving, and distributing geospatial data. The total amount of geospatial data is approaching exabytes and is continuing to grow rapidly.

Remote sensing is one of the major methods for collecting geospatial data. As one of the contributions to the US national global change research, NASA has, since the late 1980s of the last century, been developing the Earth Observing System (EOS), an integrated, multi-satellite, long-term program designed for monitoring Earth's land, ocean, and atmosphere as an integrated system (Asrar and Dozier, 1994). Currently, EOS is generating more than 2 TB of remote sensing data per day. The data collected by EOS are processed, distributed, archived, and managed by the EOS Data and Information System (EOSDIS), and are distributed at nine Distributed Active Archival Centers (DAACs) (Asrar and Greenstone, 1995; King et al., 2003).

Although EOS data are intended for use in studying global climate and environmental change, they have also been increasingly used for local and regional applications. Both research and applications call for integrated analyses of data from multiple sources (Asrar and Dozier, 1994; King, 1999; Di, 2003). This requires the interoperability of data from multiple sensors and data producers. In order to achieve interoperability and integrated analysis, standards for remote sensing data and access have to be developed. In the past several years, individual data producers and many national and international organizations have worked on the issue of interoperability for remote sensing data and information and have set various levels of standards (Di et al., 2001, 2002a; Kresse and Di, 2002; Di, 2003). The most notable organizations setting geospatial standards include the US Federal Geographic Data Committee (FGDC) which sets the US federal standards on geographic information (FGDC, 2003), the Open Geospatial Consortium (OGC, formerly the open CGIS Consortium) which sets the industry specifications for interoperability of geo-processing software and services (OGC, 2003), and International Organization for Standar- dization (ISO) Technical Committee 211 (ISO TC 211) which sets the international standards on geographic information (ISO TC211, 2003). Among all those organizations, OGC is the only one that defines implementation standards for the interoperability of geospatial data, information, services, and systems. This paper discusses the NASA EOS data environment, summarizes OGC specifications applicable for accessing NASA EOS data in the web environment, and presents an approach for applying OGC specifications to NASA EOS data.

13.2 NASA EOSDIS Data Environment

In order to process, distribute, and archive data from the EOS satellites and to facilitate interdisciplinary research of the Earth system, NASA has developed the EOS Data and Information System (EOSDIS) and its Core System (ECS). EOSDIS is central to the EOS program. It provides the environment in which scientists, educators, policy makers, and other users access and use reliable large-scale data sets of geophysical and biological processes. It also facilitates the interaction and communication among EOSDIS users (Asrar and Greenstone, 1995; King et al., 2003).

Because the data EOSDIS processes can originate from field campaigns, aircraft and ship observations, satellite remote sensing, or the output of scientific models, they are in various forms. It is unrealistic for EOSDIS to deal with separate formats for the individual data sources. Therefore, common conceptual data models have to be established so that EOSDIS data products can be mapped into a limited number of data models. Then, encoding standards and software interfaces can be developed for the limited number of data models. In EOSDIS, data are grouped into three generalized models: grid, swath, and point (Di et al., 1996; Klein and Arrington et al., 2001):

(1) Grid: rectangular arrays or data structures for geocoded data.

(2) Swath: simple or complex remote-sensing data provided in sensor coordinates.

(3) Point: simple or indexed data of geo-located and/or time-tagged point observations or event data.

Each conceptual data model defines the data object (i.e., swath, grid, point), the components of the object, and the relationship between the components. In the swath and grid models, there are two required components: data and geolocation. The description of how to relate the geolocation component to the data components has to be packaged with these two components. In the point data model, the data collected at a point on Earth can be a record in a table and the geolocation information for the record is either contained in fields in the record or attached to the record through a pointer.

Hierarchical Data Format (HDF) is an object-oriented data format developed by the National Center for Supercomputing Applications (NCSA). Tools and libraries are available for creating and manipulating data in HDF. The detailed information about HDF can be found at http://hdf.ncsa.uiuc.edu. EOSDIS conceptual data models have been implemented both in HDF version 4 and HDF5 by mapping the conceptual models to HDF internal data models and structures, resulting in a profile of HDF called HDF-EOS. Libraries and many tools have been created to create and manipulate HDF-EOS data. Those tools and libraries can be downloaded from http://hdfeos.gsfc.nasa.gov.

There are two types of metadata in HDF-EOS—structure metadata and product metadata. Both metadata are stored as attributes in the HDF-EOS files. The structure metadata define the relationship between data components. The product metadata describe the data product. Both product and structure metadata are encoded in text format using Object Description Language (ODL) and are stored as global attributes in HDF-EOS files. The content standard for the structure metadata is defined in the EOSDIS conceptual data models. The structure metadata are generated, encoded, and stored automatically by the HDF-EOS software library when an HDF-EOS data object is created. For the product metadata, EOSDIS has developed its own metadata model and the metadata standard (Heller, 1994).

The EOSDIS data models and metadata standards are the basis for several national and international geospatial standards. The FGDC content standard for remote sensing swath data (Di and Carlisle, 1998; FGDC, 1999) is based on the EOSDIS swath data model with an extension for allowing more methods in the geolocation components. The FGDC standard is an important component in the data model part of ISO 19130 (Di et al., 2001). The FGDC Content Standard for Geospatial Metadata, Extensions for Remote Sensing Metadata (Di et al., 1999; FGDC, 2002) is based on ECS metadata models with additions for photogrammetric metadata. This FGDC standard is the foundation for the ISO 19115-2 Geographic Information-Metadata-Part 2: Extensions for imagery and gridded data.

NASA EOSDIS data are archived, managed, and distributed by nine Distributed Active Archive Centers (DAACs). In addition to searching and finding data through the individual DAAC's data systems, EOSDIS provides a portal called the EOS Data Gateway (EDG) for users to search and order EOS data. Currently, a typical process by which a data user obtains EOSDIS data for analysis and application involves the following steps:

(1) A data user goes to the EOS Data Gateway to search for and find the

data they want.

(2) The user places an order for the data. The minimum size of orderable data in EOSDIS is called a granule, which may cover a larger/or smaller geographic area than the user wants.

(3) The DAAC takes the order, pulls out granules from the archive, and copies the granules from on-line storage devices to distribution media such as CD-ROM and tapes or stages the data at an FTP site for retrieval. Normally, DAAC distributes the data in the same form that is in the archives. Services are not normally provided to convert data in archival form to the user-required form.

(4) The user obtains the data and then does preprocessing to convert the data to the user's form, so that the data will be in the format, map projection, spatial and temporal coverage and spatial and temporal resolution that is consistent with other datasets to be used in the analysis. This preprocessing can be called homogenization processing of multi-source data.

(5) The user performs the analysis of the homogenized multi-source datasets for scientific research and applications.

Although users can obtain EOSDIS data through the current method of data access, they suffer significantly in the following two ways. First, it takes a long time (e.g. weeks) for them to get the data. Only a small number of data users who have special privileges to make special arrangements with DAACs can do research and application projects that require timely data. Second, once users get the data, they need to invest a lot of time and resources to do homogenization processing that converts the data from the archival form to the user form. Many data users, although they know the usefulness of EOS data, do not have the resources and knowledge to do the processing. This prevents the EOS data from being used by broader user communities. However, in many cases, the homogenization processes employed by different users involve very similar process steps with the same matured processing algorithms. Hence, the homogenization processing should be fully automated, with the results automatically provided to users by the data system. Data users will then never need to know if the form of data they specify really exists in the data archive, because the advanced data system can generate the personalized data products on the fly if the data in the user-requested form don't exist. Such generation of personalized data products on demand is what most data users want.

13.3 The OGC Web-Based Interoperable Data Access Technology

The Open Geospatial Consortium, Inc. (OGC) is a not-for-profit international membership-based organization founded in 1994 to address the lack of interoperability among systems that process georeferenced data. The details about OGC can be found at http://www.openg.org.

OGC advances geospatial interoperability technology by developing interoperable interface specifications. Those specifications, tested through interoperability initiatives, are widely accepted by software vendors, the GIS community, and federal agencies in the US. They are also adopted by many different countries and international organizations. Based on the Memorandum of Understanding signed by ISO TC 211 and OGC, OGC will submit its specifications to ISO TC 211 for approval as international standards.

Since 1999, OGC has successfully implemented four web-based geospatial interoperability initiatives, including Web Mapping Testbed I (WMT-1) in 1999, WMT II in 2000, OGC Web Services Initiative-1.1 (OWS 1.1) in 2001, and OGC Web Services Initiative 1.2 (OWS 1.2) in 2002. The new OGC Web Services Initiative 2.0 (OWS 2.0) will be conducted in early 2004. Through those initiatives, OGC has produced a set of web-based interoperability specifications. The most important ones related to interoperable data access are the Web Coverage Services (WCS), Web Feature Services (WFS), Web Map Services (WMS), and Web Registries Services (WRS) specifications.

13.3.1 Web Coverage Service Implementation Specification

The OGC WCS specification defines the interfaces between web-based clients and servers for accessing on-line multi-dimensional, multi-temporal geospatial data in an interoperable way (Evans, 2003). Based on definitions by OGC and in ISO 19123, coverage data include all remote sensing images as well as gridded data such as DEM, land use classification, etc.

Three operations are defined in WCS, including:

(1) GetCapabilities: Client retrieves the XML-encoded capabilities document from a server. The document contains information about the data it serves, as well as about the server capabilities.

(2) GetCoverage: Client requests the server to send data based on client's requirements.

(3) DescribeCoverage (optional): Client retrieves the metadata for a specific coverage.

A typical session starts with the client issuing a GetCapabilities request, followed by one or multiple GetCoverage requests. The client formulates the GetCoverage request based on the information in the XML document retrieved in the GetCapabilities request. Two types of request encoding are defined in the specification: keyword-value pair encoding and XML encoding.

The WCS specification does not explicitly require a WCS server to provide data service functions. However, based on the options available to the client in the GetCoverage request, we can derive required service functions a server must provide in order to comply with the specification:

(1) Subsetting: Client can specify a subset of data with a bounding box and

a range subset. Subsetting based on geographic/map coordinates is not necessary because the server may declare it can only serve data in a native Spatial Reference System (SRS). The server has to be able to perform subsetting on both the spatial/ temporal domain and the range set.

(2) Resampling: Client can specify a resolution for the returned data that may not be the same as the native one. At least one resampling method must be implemented, although multiple methods can be implemented and declared in the capabilities XML for the client to select.

(3) Data Packing: The server should be able to pack the subset data in at-least one of the commonly used formats listed in the specification, including HDF-EOS, GeoTIFF, NITF, DTED, and GML.

In addition to the above mandatory data service capabilities, a WCS may implement other data service functions that can be explicitly or implicitly selected by clients through the GetCoverage request. Examples of such services include:

(1) Reprojection: On-the-fly reprojection allows a server to serve data in multiple SRSs. Such a capability can be declared as the availability of multiple response SRSs at the capabilities document.

(2) Georectification: Remote sensing swath data is un-georectified. Without the on-the-fly georectification, they can be served only in the native (row, column) reference system. Georectification will enable swath data to be served as georectified grid data.

(3) Reformatting: A server can serve data in multiple formats. This will allow data users to select their favorite format for the returned data.

13.3.2 Web Feature Service Specification

The OGC WFS specification defines the interfaces between web-based clients and servers for accessing feature-based geospatial data (Vretanos, 2002). Examples of geospatial feature data include transportation road networks, coastlines, political boundaries, utility lines, etc. The WCS and WFS together provide standardized, on-line access to all geospatial data. They form the foundation for the web-based interoperable access of geospatial data.

The WFS specification defines five operations for accessing and manipulation of feature data. The five operations are:

(1) GetCapabilities: A web feature server must be able to describe its capabilities. Specifically, it must indicate which feature types it can serve and what operations on each feature type are supported.

(2) GetFeature: A web feature server must be able to service a request to retrieve feature instances. In addition, the client should be able to specify which feature properties to fetch and should be able to constrain the query spatially and non-spatially.

(3) DescribeFeatureType: A web feature server must be able, upon request,

to describe the structure of any feature type it can serve.

(4) Transaction: A web feature server may be able to service transaction requests. A transaction request is composed of operations that modify features in the server. The transactions include "create", "update", and "delete" geographic features.

(5) LockFeature: A web feature server may be able to process a lock request on one or more instances of a feature type for the duration of transaction. This ensures that serializable transactions are supported.

Among the five operations, both the "Transaction" and "LockFeature" operations are optional. Therefore, a WFS-compliant web feature server must implement at least the GetCapabilities, GetFeature, and DescribeFeatureType operations.

Like WCS, WFS defines two methods to encode WFS operation requests, including XML encoding and keyword-value pair encoding (also known as URL encoding). The WFS specification also specifies some requirements a WFS server has to provide, including:

(1) At a minimum, a WFS server must be able to present features using GML.

(2) The predicate or filter language should be defined in XML and be derived from CQL as defined in the OpenGIS Catalogue Interface Implementation Specification.

(3) The datastore used to store geographic features should be opaque to client applications, whose only view of the data should be through the WFS interface.

(4) A subset of XPath expressions should be used for referencing properties. Because GML allows geographic features to have complex or aggregate non-geometric properties, the specification defines a subset of XPath expressions for providing references to those properties in the various places where property references are required.

13.3.3 Web Map Service Specification

The OGC WMS specification defines Web interfaces for dynamically assembling maps from multiple sources within a heterogeneous distributed computing environment over the Internet (de La Beaujardière, 2001). Maps are the visualization of data. A WMS server normally converts the data in its archive to a visualized form (map) based on the requirements from the client. In many cases, a WMS server may talk to a WCS or WFS server to obtain the needed data for making maps requested by a client. In this sense, a WMS server can be considered as a data visualization service for either WFS or WCS servers.

The WMS specification defines three operations:

(1) GetCapabilities (required): Obtain service-level metadata, which is a machine-readable and human-readable description of the WMS's information content and acceptable request parameters.

(2) GetMap (required): Obtain a map image whose geospatial and dimensional parameters are well defined.

(3) GetFeaturelnfo (optional): Ask for information about particular features shown on a map.

13.3.4 Web Registry Service Specification

OGC WRS defines the interfaces between web-based clients and servers for finding the user-required data or services from the registries (Reich, 2001). A data registry consists of data entries available at data archives that might be served by WCS, WFS, or WMS servers. WRS allows data users to find the requested data in vast archives. OGC WRS is designed to be a general interface that can be used not only for data discovery but also for service and resource discovery. Unlike WCS, WFS, and WMS, which are all official OGC interoperability specifications, WRS has appeared only in a discussion paper that needs to be further developed into an official specification. It is expected much work will be done in the OWS 2.0 initiative to further refine WRS.

The current draft WRS specification is based on the following usage scenarios:

(1) Adding, deleting, modifying and querying service offers;

(2) Adding, deleting, modifying and querying dataset descriptions;

(3) Adding, deleting, modifying and querying classification schemes;

(4) Classifying registry objects in a flexible manner;

(5) Associating registry objects in a flexible manner;

(6) Cooperation between registries.

The following basic WRS operations are defined:

(1) Publishing: This operation supports registration of the resources using standardized metadata and supports registration of any classification scheme. It is paired with a repository that stores resources.

(2) Classification: It supports flexible classification of registered items using user-defined classification schemes or taxonomies.

(3) Association: It supports the association of resources with other resources located both locally and externally.

(4) Querying: It supports queries by subject for registered items, using a common filter language. The server should be able to return registry metadata about the registered resources. It also should able to return repository items if the registry is paired with the repository, or a URL for repository items if the registry is not paired with the repository.

(5) Federated Registries: It provides a common distributed query mechanism to query federated registries.

The WRS specification supports queries of the "OGC-core" properties, including LatLonBoundingBox, Keyword, Title, Abstract, Format, DCPType, CRS, ServiceClass, ServiceType, and ContentType. It supports OGC Filter Encoding as a query language and XPath query language.

13.3.5 Results

With the above-mentioned four specifications, the OGC technology allows data users to discover where the data are located and then specify requirements for the retrieved data. An OGC-compliant WCS or WFS data server has to preprocess the data on demand on the server side, based on user requirements, and return the data back to users in the form they specified. At the end, the user will get the data that exactly match their requirements for both the content and the structure (e.g. format, projection, spatial and temporal coverage, etc.). Because OGC specifications are industry standards, many GIS vendors are implementing those specifications at servers or clients. The proliferation of OGC-compliant servers and the increased numbers of available OGC-compliant GIS clients enable a user to access data from multiple sources that are co-registered and in ready-for-analysis form. This will significantly reduce the time needed for users to acquire and preprocess the data before the data can be analyzed.

13.4 Applying OGC Technology to the NASA EOS Data Environment

With the OGC technology, we can implement a data access and service system on top of EOSDIS to provide seamless interoperable personalized on-demand access to NASA EOS data. The system will take care of steps 1 to 4 of current data access automatically. Data users need only provide their data specifications, and within seconds or minutes they will obtain the data on-line in ready-to-analyze form. The only requirement from this advanced data system is that the data have to be on-line.

HDF-EOS is the standard format for NASA EOSDIS. HDF-EOS has three data models, including swath, grid, and point. Both the swath and grid models store coverage data while the point model stores feature data. In order to serve these data to the user community through OGC interfaces, the system needs a web coverage server to handle swath and grid data, and a web feature server to handle point data. Because most EOSDIS data are in the swath and grid forms, a powerful web coverage server is essential to such a system. A web map server should be provided so that data users can view the data before they actually retrieve them. Because EOSDIS has a huge volume of data, it is practical to have a data catalog system for users to search and find the data they want. Therefore, a WRS server has to be part of the data system.

With such a data system in place, any OGC-compliant client can access data in EOSDIS. The data pre-processing is provided automatically by servers in the data system as the OGC specifications imply. With the new data system, typical access of EOSDIS data involves the following steps:

(1) User searches the WRS server through a WRS compliant client.

(2) The search results are fed to a WMS compliant client to preview the data and refine the selection.

(3) Depending on the type of data (feature or coverage), a WFS or WCS client will be used to retrieve the personalized data for analysis.

The above 3 steps can be performed on-line sequentially. However, these steps use different clients. They also use the result from the previous step as the input to the next step. An integrated intelligent client that can speak multiple protocols and handle all transactions and result feedback automatically will greatly ease users' burden. An example of such a client is the NWGISS multi-protocol geoinformation client (MPGC) (Deng et al., 2002; Di et al., 2002b). Such clients provide OGC interfaces for access not only to HDF-EOS servers, but also to any OGC-compliant data servers operated by other data providers. Therefore, it is also the natural platform for multi-source data integration and analysis.

From OGC point of view, interoperability at data access level is quite mature with the availability of WCS, WFS and WMS as the official interoperability specifications. The WCS, WFS, and WRS specifications form the common data environment on which interoperable data access and value added web service can be built. For example, a user can request a Web 3-D visualization service to produce a flight-through for Yellowstone National Park. The service finds, through the WRS interface, that digital elevation model (DEM) data is available at the US Geological Survey (USGS) WCS server and Landsat TM data is available from the NASA EOS WCS server. Then, the 3-D service can issue a GetCoverage request to the USGS server to request DEM data and to the NASA server to request the TM data. Both requests specify the same geographic region, map projection, format, and resolution. When the data return from both WCS servers, the two layers of data are ready for the 3-D services to produce a flyby simulation without needing further data integration processes. With service chaining, if OGC interfaces are implemented, powerful application services can be built on top of the interoperable common data environment of which the NASA EOSDIS could be part. Other applications that can be built on top of this common interoperable data environment include on-line data mining and interoperable simulation models.

13.5 The Current Implementation Status of the OGC Technology

All OGC Web service specifications mentioned in this paper were developed from the Draft Interoperability Program Reports (DIPR) or Interoperability Program

Reports (IPR) produced by the OGC interoperability initiatives. During those initiatives, participating organizations produced prototype software for testing interoperability. Currently, some of those software systems are already commercially available. OGC maintains a list of commercially available registered products at http://www.opengeospatial.org/resources/?page = products. Because of the power of the technologies shown in the OGC interoperability demonstrations, many organizations are developing in-house data servers by using those specifications, in addition to availability of commercial products.

NASA has been heavily involved in the OGC activities by sponsoring the OGC interoperability programs and defining the requirements for interoperability. The NASA Earth Science Enterprise (ESE) has funded several universities for developing OGC prototypes. An example of such prototypes is the NASA HDF-EOS Web GIS Software Suite (NWGISS) developed by George Mason University. NASA also maintains several operational Web map servers, including the globe Web map server. Currently, the ECS project is developing an operational OGC WCS server to be deployed at EOSDIS data pools running at DAACs. It is expected to be operational in 2004.

13.6 The Anticipated Impacts on End Users

With the deployment of current OGC interoperable data access technology in EOSDIS, any OGC-compliant GIS client can access the data in EOSDIS. On-the-fly reformatting, georectification, resampling, and subsetting (both spatial and parameters-based) are provided automatically by the servers in the data system. Users can assemble data from multiple data providers (not necessarily all from EOSDIS) on-line in a form that exactly matches their requirements, the interoperable, and personalized, on-demand data access and services (IPODAS):

(1) Interoperable means data users can obtain data from multiple sources and/or providers in the same form

(2) Personalized ensures that data services will meet the individual users' needs.

(3) On-demand means that data products must be created on-demand because the exact needs of data users are not known in advance.

Such access and services systems recognize that individual data users have their unique requirements on the data. IPODAS provides data to users in a form that exactly matches the user's requirement for format, spatial and temporal coverage, projection, and resolution. Therefore, end-users can significantly save time and resources for data access and preparation. In order for data users to access the IPODAS capability, they need to run OGC-compliant multi-protocol geoinformation clients at their local machines. A free copy of such a client can be obtained from http://laits.gmu.edu.

With further development of OGC technology in the geospatial services area, it is expected the geospatial services market will be formed based on OGC Web service technology. EOSDIS itself can provide certain types of geoinformation services directly to the end users. In addition, value-added services can access EOSDIS data through the OGC common data environment to derive user-specific information for the end-users on-line. Therefore, in the future the end-users will be able to access not only data but also information on-line from EOSDIS that tailors to the individual users' requirements.

References

Asrar G, Greenstone R (1995) MTPE EOS Reference Handbook. NASA Goddard Space Flight Center, Greenbelt, MD 20771

Asrar G, Dozier J (1994) EOS: Science Strategy for the Earth Observing System. American Institute of Physics Press, Woodbury, New York, 11797

de La Beaujardière J (ed) (2001) Web Map Service Implementation Specification, Version 1.1.1. OGC 01-068r2, Open Geospatial Consortium Inc. http://www.opengeospatial.org/docs/01-068r2.pdf

Deng D, Deng M, Zhao P, Di L (2002) A Thick Web Coverage Client (WCC) of NASA HDF-EOS Web GIS Software Suite (NWGISS). In: Chen and Jiang (eds) The International Archives of Photogrammetry, Remote Sensing, and Spatial Information Sciences. Vol. XXXIV, Part 2, Commission II, pp. 87 – 110

Di L (2003) The development of Remote-Sensing Related Standards at FGDC, OGC, and ISO TC 211. In: Proceedings of IEEE International Geoscience and Remote Sensing Symposium (IGARSS 2003), July 21 – 25, Toulouse, France, p. 4

Di L, Kresse W, Kobler B (2002a) The Progress and Current Status of ISO Standard Development Project 19130-Sensor and Data Models for Imagery and Gridded Data. In: Chen and Jiang (eds) The International Archives of Photogrammetry, Remote Sensing, and Spatial Information Sciences. Vol. XXXIV, Part 2, Commission II, pp. 107 – 110

Di L, Yang W, Deng, Deng M, McDonald K (2002b) Interoperable access of remote sensing data through NWGISS. In: Proceedings of IEEE International Geoscience and Remote Sensing Symposium (IGARSS 2002), Toronto, Canada, pp. 255 – 257

Di L, Kresse W, Kobler B (2001) The New ISO TC211 Standard Project on Sensor and Data Model for Imagery and Gridded Data. In: Proceedings of International Geoscience and Remote Sensing Symposium (IGARSS 2001). Sydney, Australia, pp. 1,188 – 1,190

Di L, Schlesinger B, Kobler B (1999) NASA-Sponsored US FGDC Remote Sensing Standards. In: Proceedings of the First International Symposium on Digital Earth, Vol. I. Science Press, Beijing, China, pp. 85 – 91

Di L, Carlisle C (1998) The Proposed FGDC Content Standard for Remote Sensing Swath data. In: Proceedings of the Annual ASPRS/RTI Conference, Tampa, FL, March 30 – April 3 (CD-ROM) p. 12

Di L, Suresh R, Ilg D, Meyer T (1996). EOSDIS data models and implementation. In: Proceedings, the Third International Conference/Workshop on Integrating GIS and

Environmental Modeling (CD-ROM and WWW), Santa Fe, New Mexico, Jan 21 – 25. National Center for Geographic Information Analysis (NCGIA), University of California, Santa Barbara, CA, p. 12

Evans JD (ed) (2003) Web Coverage Service (WCS), Version 1.0.0. OpenGIS© Implementation Specification. Open Geospatial Consortium Inc. http://www.opengeospatial.org/docs/ 03-065r6.pdf

FGDC (2003) The Federal Geographic Data Committee Homepage: http://www.fgdc.gov

FGDC (2002) Content Standard for Digital Geospatial Metadata: Extensions for Remote Sensing Metadata, FGDC-STD-012-2002. Reston, Virginia. http://www.fgdc.gov/standards/status/csdgm_rs_ex.html

FGDC (1999) Content Standard for Remote Sensing Swath Data, FGDC-STD-009-1999. Reston, Virginia http://www.fgdc.gov/standards/status/sub4_4.html

Heller D (1994) Proposed ECS Core Metadata Standard, Release 2.0. Hughes Applied Information System, Landover, MD: http://edhs1.gsfc.nasa.gov/

ISO/IEC 2382-1:1993. Information technology—Vocabulary—Part 1: Fundamental terms

ISO TC 211 (2003) ISO TC 211 homepage. http://www.isotc211.org

King MD, Closs J, Spangler S, Greenstone R (eds) (2003) EOS Data Products Handbook. NP-2003-4-544-GSFC, National Aeronautics and Space Administration, Washington DC

King MD (ed) (1999) EOS Science Plan. National Aeronautics and Space Administration, Washington DC NP-1998-12-069-GSFC

Klein L, Arrington D (2001), An HDF-EOS and Data Formatting Primer for the ECS Project. 175-WP-001-002. Raytheon Company, Upper Marlboro, Maryland

Kresse W, Di L (2002) Status Report of ISPRS WG II/4 Image Data Standards 2000 – 2002. In: Chen and Jiang (eds), The International Archives of Photogrammetry, Remote Sensing, and Spatial Information Sciences. Vol. XXXIV, Part 2, Commission II, pp. 229 – 230

OGC (2003) The Open Geospatial Consortium Homepage, http://www. opengeospatial.org

Reich L (ed) (2001) Web Registry Server Discussion Paper, OpenGIS Project Document 01-024r1. Open Geospatial Consortium Inc. http://www.opengeospatial.org/docs/01-024r1.pdf

Vretanos PA (ed) (2002) Web Feature Service Implementation Specification, Version 1.0.0. OGC 02-058. Open Geospatial Consortium Inc. http://www. opengeospatial.org/docs/02-058.pdf

14 Global MODIS Remote Sensing Data for Local Usage: Vaccess/MAGIC

John J. Qu, Menas Kafatos, Xianjun Hao, Swarvanu Dasgupta and Kwang-Su Yang

14.1 Introduction

The Moderate Resolution Imaging Spectroradiometer (MODIS), as a key research instrument, was successfully launched on Dec. 18, 1999 onboard the Terra (previously known as EOS AM-1) satellite. The second MODIS was launched onboard the Aqua (EOS PM-1) satellite in May 2002 (http://modis.gsfc.nasa.gov). Because the MODIS senses all the Earth's surface in 36 spectral bands spanning the visible (0.415 μm) to infrared (14.235 μm) spectrum at nadir spatial resolution of 1 km, 500 m and 250 m (Justice et al., 1998), the associated remote sensing applications are of interest not only to land, ocean, and atmosphere discipline researchers but also to application, interdisciplinary and environmental scientists (Salamonson et al., 1989, 2005).

MODIS products can be used for a large number of applications at regional and global levels. Although intended as an instrument focusing on global change research, MODIS, along with global Earth observing and remote sensing (RS) sensors, has been shown to be at least very useful for environmental applications. One such consortium that heavily utilizes Earth observing data has been assembled for the Mid-Atlantic region: Termed Virginia Access—Middle Atlantic Geospatial Information Consortium (MAGIC), hereinafter called, MAGIC, which is a consortium of Middle-Atlantic universities to develop a RS, applications-oriented, geospatial/temporal data and information system (http://cpgis.gmu.edu/vaccess/mission.htm). It focuses on two primary aspects: The usage of such data for national priorities, following the emphasis of NASA's Earth Science Applications; and the dissemination of such data through interoperable information systems, that couple to NASA's systems and promote open source solutions and standards. Therefore, the goal is to promote the usage of NASA Earth science data which can be used by broad communities. MAGIC prototypes focus on environmental applications for wetlands, agriculture, local impacts of global climate phenomena, forestry, public health and pollution for local, regional and state levels (Kafatos et al., 2002). It conducts applied research driven by significant issues identified by user communities, consistent with NASA's Earth Science Applications theme "to expand and accelerate the economic and societal benefits from Earth science, information and technology" (Earth Science Enterprise Applications Plan). The above two-fold aspects, data usage and dissemination, manifest in all MAGIC products: Environmentally-important applications are developed, the potential stakeholders are identified, relationships with partner federal agencies are developed to the extent possible and potential local, regional, and state relationships are identified. Applicable value-added data and information products are developed by the MAGIC team, adding to NASA's mission. Moreover, Decision Support Systems (DSS) and Decision Support Tools (DST) are identified applicable to these application areas.

With increasing numbers of Earth observing satellites in space, very large volumes of remote sensing data are being produced at rates exceeding several terabytes per day. Traditional remote sensing data processing and distribution methods may not be sufficient for different end user communities, which may include novice, intermediate and advanced users, to efficiently use such datasets for their applications. Tailored information and knowledge distribution development may help the data usage more efficiently.

Earth remote sensing data processing and distribution will face new challenges, namely maintaining the increasing volumes of data in forms that are readily accessible and that meet the needs of very diverse user communities. These intellectual challenges are at least equal to the challenges of building and launching advanced hardware into space. Information extraction (via, e.g. data mining) and distribution may be as important as data distribution. The following issues are considered crucial for wide usage of Earth observing/remote sensing datasets: (1) huge data volumes, exceeding many petabytes and rates exceeding many retabytes per day; (2) complex data formats, such as, Hierarchical Data Format (HDF) and Hierarchical Data Format Earth Observing System (HDF-EOS); (3) different map projections; (4) Geographic Information System (GIS) applications; (5) communication protocols and network capabilities; and (6) processing time. End users need to obtain Earth observing/remote sensing data in useful forms. Therefore, customized real-time remote sensing data with GIS/Web-GIS compatible formats are considered necessary for many end user communities. On the other hand, more widely distributed Earth observing/remote sensing data in different formats through diversified protocols will result in better usage of future Earth observing data.

To address the above issues, data compressing and pre-processing (sub-setting and sub-sampling), data format conversion (easy accessing data format such as GIS compatible formats), GIS and Open GIS applications, and simple real time data processing for future Earth observing satellite systems are discussed here. In this chapter, we emphasize some usage of MODIS products for the Mid-Atlantic region, which is the main focus of MAGIC.

14.2 MODIS Data Processing for Regional Use

GMU and NASA Goddard Earth Science (GES) Data and Information Service Center (DISC) have been jointly worked on new applications of Earth science remote sensing data through the Remote Sensing Information Partner (RSIP) Program. The purpose of the RSIP Program at the GES DISC is to provide participating institutions inexpensive access to the globally covered GES data holdings, and the ability to routinely acquire any/all of these data for local use or for future distribution (Qu et al., 2001, 2002; Serafino et al., 2002). The Goddard Earth Sciences Distributed Active Archive Center (GES DAAC) derives benefits from this arrangement by assuring that NASA Earth science remote sensing data are being more widely distributed thus affording better usage for NASA efforts. VAccess/MAGIC maintains the RSIP pipeline to redistribute value-added MODIS data.

We focused MODIS applications on hazards, specifically forest fire/smoke detection and drought monitoring. We have identified MODIS potential local users and products and also study the applications for combining Terra and Aqua, MODIS products, such as land surface evaporation product (MOD16). In addition, we provide customized MODIS 250 m Surface Reflectance and Vegetation Indices for the VAccess/MAGIC land and environmental scientists, in conjunction with local users, who define the requirements. In this chapter we only focus on some sample none real-time MODIS data, such as vegetation index and cloud mask, and real time data processing.

For regional use, MODIS data need to be processed in certain steps, such as mosaic, resampling, subsetting, reprojection et al. The MODIS Data Support Team (MDST) at GES DAAC has devoted much time to research, test, and enhance a variety of software packages to handle and manipulate MODIS data, these tools are freely available online (http://daac.gsfc.nasa.gov/MODIS/software.shtml). Some commercial software such as ENVI, IDL, Matlab et al., can also be used for MODIS data processing, but usually requires more efforts. We mainly focus on the usage of HDFLook-MODIS and MODIS Reprojection Tool (MRT), which are two important tools for MODIS data manipulation from MDST.

As a result of the joint collaboration between LOA USTL, France, and GES DAAC, HDFLook-MODIS is a powerful data processing and visualization tool designed for MODIS products. It can be run in either GUI mode or batch mode, and has integrated connectivity to MsPhinx, a comprehensive data analysis environment. Main functionalities of HDFLook-MODIS include:

(1) Accessing and visualization of all swath (level-1, and -2) and gridded (level-3 and -4) MODIS data products;

(2) Remapping and re-projecting;

(3) Subsetting and resampling;

(4) Extraction of meta data;

(5) Data format conversion;

(6) Atmospheric correction.

The MODIS Reprojection Tool is software designed to reproject MODIS gridded data products into more standard map projections. The heart of MRT is two executable programs, mrtmosaic and resampler, to mosaic and resample MODIS data respectively, and they can be run either from the command-line or from the MRT Graphical User Interface (GUI).

For Mid-Atlantic region, we processed major MODIS data products. In this paper, we only demonstrate the usage with MODIS Vegetation Index products, MODIS LAI products, and MODIS Cloud Mask products.

14.2.1 MODIS Vegetation Index and LAI Data Processing

MODIS vegetation index products contain two indices for vegetation, Normalized Difference Vegetation Index (NDVI), and Enhanced Vegetation Index (EVI). NDVI is a "continuity index"—the existing NOAA-AVHRR derived NDVI spanning two decades. EVI is a MODIS specific spectral index which was developed to improve vegetation signal sensitivity in high biomass regions and reduce the influences of canopy background signals and atmospheric effects. MODIS (Terra & Aqua) L3 Vegetation Indices Products provide NDVI and EVI on a 16-day basis at spatial resolutions of 250 m, 500 m, and 1 km on SIN grid. MODIS Leaf Area Index (LAI) is another important parameter providing information on the structural property of plant canopy. MODIS LAI product is provided on an 8-day basis at spatial resolution of 1 km on SIN grid. MODIS vegetation index and LAI products can be ordered from the EOS Data Gateway (http://delenn.gsfc.nasa.gov/~imswww/pub/imswelcome). To utilize these products for research and practical applications, some basic processing procedures are essential, namely: mosaic, reprojection, resampling, subsetting, and visualization.

1. Mosaic

A particular area of interest may be covered by several granules; mosaic is necessary and is the first step in data processing. Mosaic of MODIS Vegetation Indices (VI) and LAI products can be completed with MRT or HDFLook in batch mode.

2. Reprojection, Resampling and Subsetting

MODIS VI and LAI products are provided on SIN grid with specified spatial resolution. Reprojection, resampling and subsetting may be needed for visualization and analysis.

MRT supports reprojection, resampling, and subsetting in batch mode and GUI mode, the format of output files can be either HDF-EOS or binary or geotiff.

It is very convenient to reproject, resample and subset MODIS VI and LAI products with MRT. HDFLook also supports reprojection, resampling, and subsetting in batch mode, the format of output file can be HDF-EOS or binary. In this step, data for the area of interest is generated with specified map projection and spatial resolution.

3. Visualization and Analysis

For visualization and analysis of NDVI, EVI and LAI products, there are several issues that need to be considered:

(1) Scale factor and offset: values in data products are scaled. To get the real value, scale factors and offset must be applied.

(2) Valid range and fill value: visualize pixels with data in valid range and mark pixels with invalid data.

(3) Data quality flag: in many cases, data quality is critical, it may be necessary to filter data with specified data quality requirements.

(4) Color map: it is better to use color maps that can reflect the meanings of NDVI, EVI and LAI so as to make the images easier for interpretation and analysis.

Figure 14.1 shows three example images of NDVI, EVI, and LAI of VAccess/MAGIC area.

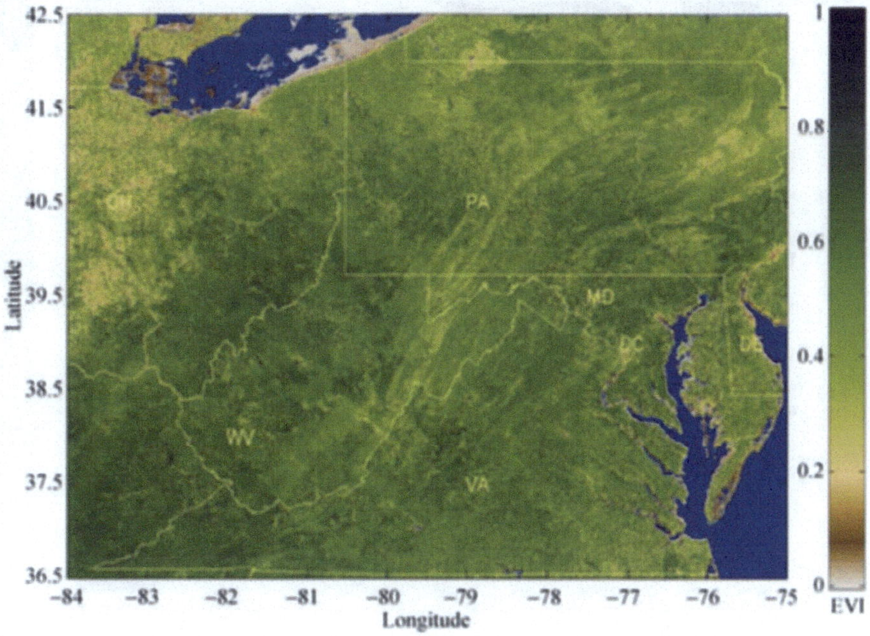

(a)

Figure 14.1 MODIS NDVI, EVI and LAI images over VAccess/MAGIC regions (a) is EVI images, (b) shows NDVI, and (c) is LAI image

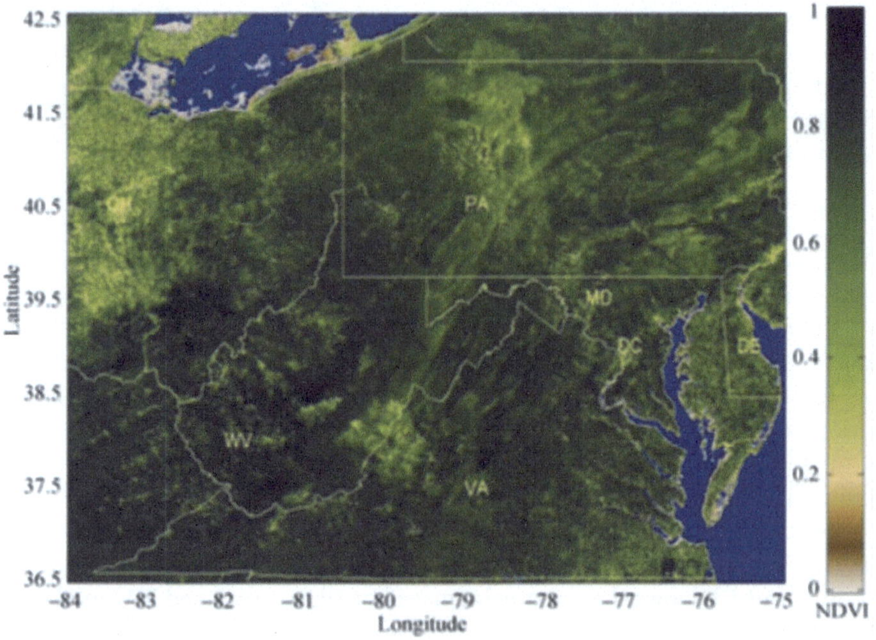

(b)

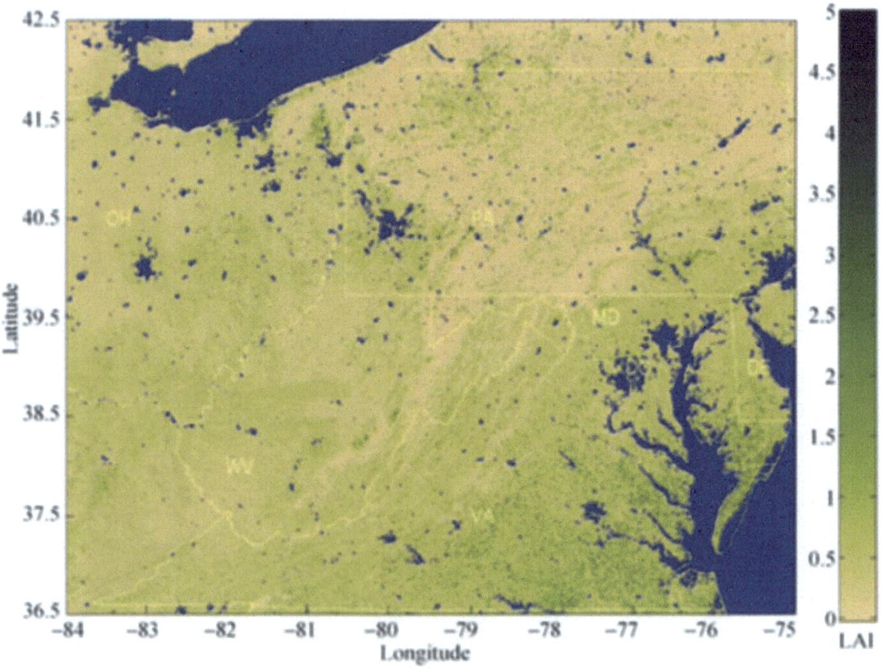

(c)

Figure 14.1 (Continued)

14.2.2 MODIS Cloud Mask Data Processing

MODIS Level-2 Cloud Mask Data Products (MOD35, MYD35) have been subsetted for the VAcess/MAGIC region by using HDFLook script. The spatial coverage for the region is 84 degree W to 75 degree W in longitude and 36 degree N to 41 degree N in latitude. The Cloud Mask Data Products contain the cloud mask information with eight bytes. The first byte provides the major information, as shown in Table 14.1, from which the major categories for classification can be extracted. The other seven bytes provide additional information in details. The relevant categories are cloud, snow/ice, land, desert, sun-glint, water, coast (shallow water), and undetermined. HDFLook script stitches granules and generates the binary file of cloud mask for the region on a daily basis. When HDFLook is stitching the granules, the most recent granule overlaps the other granules for purposes of duplicated coverage.

Table 14.1 Cloud Mask Bit-Field Interpretation of the First Byte (http://modis-atmos.gsfc.nasa.gov/MOD35_L2/format.html)

Bits	Field Description	Bit Interpretation
0	Cloud Mask Flag	0=Not Determined 1=Determined
1-2	Unobstructed FOV Quality Flag	0=Confident Cloudy 1=Probably Cloudy 2=Probably Clear 3=Confident Clear
3	Day/Night Flag	0=Night 1=Day
4	Sunglint Flag	0=Yes 1=No
5	Snow/Ice Background Flag	0=Yes 1=No
6-7	Land/Water Background Flag	0=Water 1=Coastal 2=Desert 3=Land

This may result in non-smoothing map for the overlapped areas. With the binary file, bit operations are applied to obtain the category for each pixel in a C program. Figure 14.2 shows the classification map for VAcess/MAGIC region for Aug. 25, 2004. The colors used are white for cloud, yellow for sun-glint, light pink for snow/ice, blue for water, medium aqua marine for coast (shallow water), dark khaki for desert, green for land and black for undetermined classes.

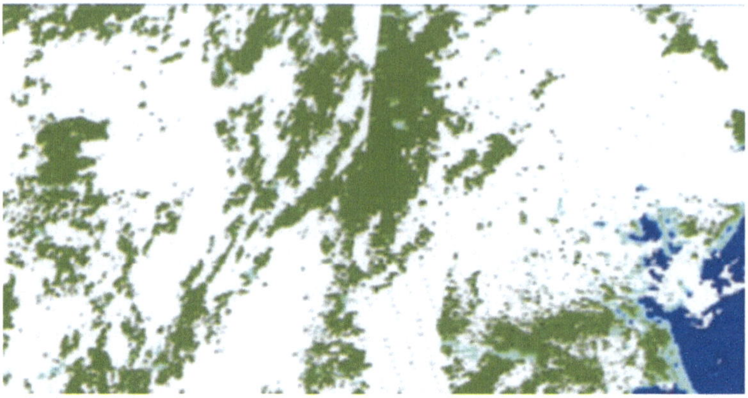

Figure 14.2 The classification map for VAcess/MAGIC region on Aug. 25, 2004

14.3 MODIS Real-Time Data Processing

Receiving data from earth observing satellites is normally accomplished by recording the instrument measurements onto an onboard tape recorder, which is transmitted later to a dedicated ground receiving station. The transmitted data is processed and then operationally distributed to broad user communities by agencies such as NASA DAAC centers or NOAA centers. Today Direct Broadcasting (DB) is a viable alternative to receiving satellite data products, which have been subjected to processing. To formally define the term, we can say that DB is the real-time transmission of satellite data to the ground. Normally a satellite would broadcast instrument data omni-directionally to the Earth continuously. Users having compatible ground receiving equipment and who are in a direct line of sight to the satellite, may receive or directly readout these transmissions. Direct Readout (DR) is thus the process of acquiring these freely transmitted live satellite data (Coronado, 2005). Currently NASA's EOS Terra and EOS Aqua satellites possess state of the art DB facilities. These stations broadcast MODIS sensor data continuously in the X band frequency of 8.2125 GHz at a data rate of 13.125 Mbps. Although establishing a direct readout station to receive MODIS broadcast data still remains an expensive proposition, primarily due to the cost of the X-band compatible antennae, public and private agencies are presently working to lower the associated costs. Presently, a more affordable alternative is to set up a few strategically-located stations for regional coverage and to share the data and derive geophysical products via the Internet. Following this approach, NASA has established four direct readout stations over the USA—Hawaii, University of South Florida, University of Wisconsin, and Goddard Space Flight Center, MD (GSFC). The DB service will be particularly useful to people who need satellite products in real time (sooner than the normal 24-hour availability through the EOS Data and Information System, EOSDIS) or who are unable to easily obtain data from the EOSDIS, which requires good connectivity.

Many interesting events, which affect societies occur with little or no warning. Such events require rapid response because of the dynamic and short-lived nature of the hazard involved. Forest fires, flash floods, hurricanes and volcanoes are such events. Direct broadcast is particularly useful in real time studies of such events. However DB suffers from one practical limitation—the coverage is regional. This limitation is, however, an actual advantage for regional users who are interested only in live local and regional environmental data. DB therefore is particularly useful in regional monitoring of the environment.

GMU/CEOSR is currently involved in research that would monitor the Middle-Atlantic Region of the USA using direct broadcast data from the MODIS sensors on board both Terra and Aqua satellites. The ongoing research under the VAcess/MAGIC project focuses on producing near real time data products for applications to hazard monitoring. Receiving the data from the satellite is only a fraction of the task involved. With the satellite revisit time on the order of 90 min, product generation cannot take an extended amount of time (Dodge, 1999). This calls for fast processing hardware and software. Modifying and tuning regular remote sensing algorithms in the direct broadcast mode is an area that CEOSR will focus on.

Software for Level-0 to Level-1 to Level-2 processing is provided by the Direct Readout Lab at GSFC and is available on their website. The Direct Readout Lab at GSFC algorithms come in two flavors: Direct Broadcast and Institutional. The DB version is designed to function in a stand-alone mode on a PC. The Institutional algorithms are heavily dependent on automated scripts and external libraries, which are shared among many other Level-2 algorithms, and therefore must operate in a very controlled environment (http://directreadout.gsfc.nasa.gov/). Currently, GSFC provides two Level-2 algorithms that have been converted from Institutional to Direct Broadcast version. They are NDVI/EVI and MOD14, which uses the 0.86, 3.9, and 11 μm radiances to identify actively-burning fires (http://directreadout.gsfc.nasa.gov/). Figure 14.3(a) is an example Level-1B true color image covering the VAccess/MAGIC region. Fig. 14.3(b) is the subset true color image for only this region. Figure 14.3(c) and 1d shows the NDVI data product generated using the direct broadcast algorithm provided by GSFC respectively over this region. Figure 14.3 represents the most favorable situation when a region under consideration lies within an overpass granule. However, in many cases a part of the region is outside the granule, in which case one has to be satisfied with the received data or wait for the next overpass in another 90 minutes. The granules received from the two overpasses can then be mosaiced to get the desired data products.

Other problems include obstruction by clouds. Cloud masking is usually done with the help of the MODIS optical bands. Note that in Fig. 14.3(c) the cloud covered region has valid values, which implies that further processing is necessary to mask out cloud covered regions.

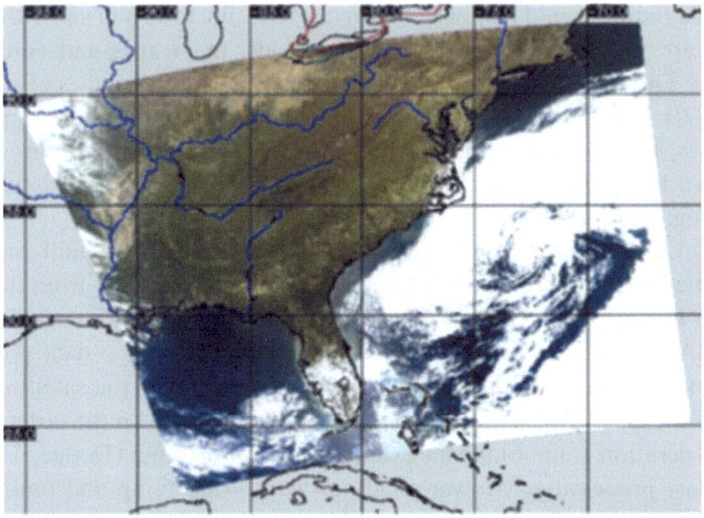

(a)

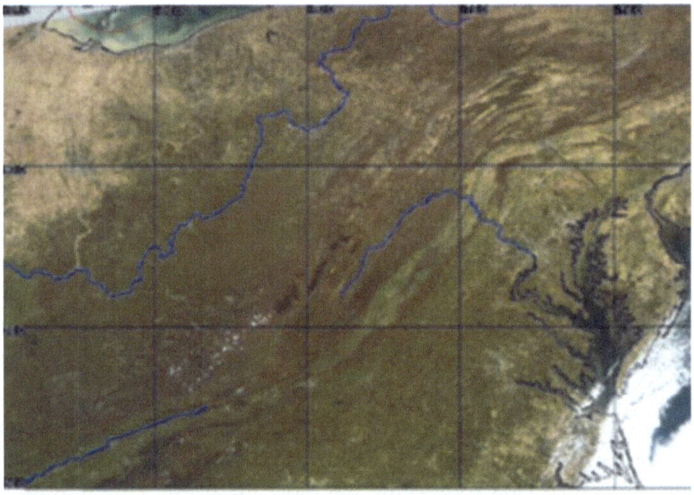

(b)

Figure 14.3 (a) MODIS Level-1B granule—true color image (MODIS Terra- 04-15-2005 1825-1830 UTC). (b) MODIS Level-1B subset true color image over the Middle Atlantic region (subsetting using HDFLook) (MODIS Terra- 04-15-2005 1825-1830 UTC). (c) NDVI data product over the Middle Atlantic region using GSFC direct broadcast application (MODIS Terra- 04-15-2005 1825-1830 UTC)

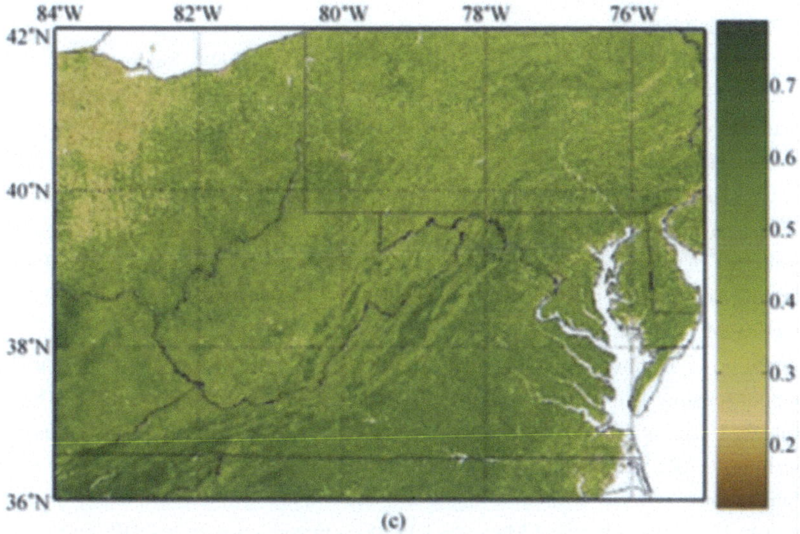

Figure 14.3 (Continued)

14.4 Summary and Discussions

MODIS data products have been shown to be of great use to regional applications. What is, however, required is processing to add value to standard EOS products as most user communities would face challenges that scientists, who are familiar with remote sensing, would not. In this chapter we showed several products for the VAccess/MAGIC region as well as the appropriate tools that user communities can employ. Direct Broadcasting is a particularly useful approach as an alternative to standard data access for rapid response.

References

Coronado P (2005) Direct Broadcast to Direct Readout—providing a bridge between NASA Direct Broadcast Earth Science Missions and the Broader User Community. In: Qu JJ et al. (eds) Earth Science Satellite Remote Sensing. Springer-Verlag and Tsinghua University Press (in press)

Dodge JC (1999) The Earth Observing System Direct Broadcast System. Earth Observation Magazine (EOM), March 1999

Kafatos M, El-Askary H, Chiu L, Gomez R, Hegazy M, Kinser J, Liu X, Liu Y, Liu Z,

McManus J, Nie Y, Qu J, Salem F, Sarkar S, Shen S, Taylor G, Wolf H, Wong D, Yang C, Yang K, Yang R (2002) Remote sensing and GIS for regional environmental applications. In: Proceeding of the International Society for Optical Engineering, SPIE, 23 – 27 September 2002 Crete, Greece

Justice C, Vermote E, Townshend JRG, Defries R, Roy DP, Hall DK, Salomonson VV, Privette J, Riggs G, Strahler A, Lucht W, Myneni R, Knjazihhin Y, Running S, Nemani R, Wan Z, Huete A, van Leeuwen W, Wolfe R (1998) The Moderate Resolution Imaging Spectroradiometer (MODIS): land remote sensing for global change research. IEEE Transactions on Geoscience and Remote Sensing, 36(4): 1,228 – 1,249

Qu J, Serafino G, Teng W, Ouzounov D, Pollack N, and Chiu L (2002) Responding to the challenge of producing and distributing MODIS data at the NASA/GES DISC via the Remote Sensing Information Partners (RSIP) Program. In: Proceeding of IGARSS' 2002 (CD-ROM Publication)

Qu J, Serafino G, Kafatos M, Lathrop R, Trimble J (2001) New application of Earth science remote sensing data at NASA/GDISC: Remote Sensing Information Partner (RSIP) with Rutgers University. In: Proceedings of International Geoscience and Remote Sensing Symposium, IGARSS 2001, July 9 – 13, at University New South Wales, Sydney (CD-ROM Publication)

Salomonson VV, Barnes WL and Masuoka EJ (2005) Introduction to MODIS and an overview of associated activities. In: Qu JJ et al. (eds) Earth Science Satellite Remote Sensing, Tsinghua University Press and Springer-Verlag (in press)

Salomonson VV, Barnes WL, Maymon PW, Montgomery HE, Ostrow H (1989) MODIS: advanced facility instrument for studies of the Earth. IEEE Transactions on Geoscience and Remote Sensing 27: 145 – 153

Serafino G, Teng W, Chiu L, Qu J, Pollack N, Liu Z, Maynard N (2002) Facilitating the access and utilization of satellite remote sensing data Via the Remote Sensing Information Partner (RSIP) Project at NASA/GSFC. In: Proceeding of Science Data Processing Workshop 2002, February 27 – 28, 2002 Greenbelt, Maryland

15 The NASA HDF-EOS Web GIS Software Suite

Liping Di and Kenneth R. McDonald

15.1 Introduction

The NASA Earth Science Enterprise (ESE) is generating a huge volume of remote sensing data to support Earth system science and applications research through its Earth Observing System (EOS) program. Most of the data are stored in HDF-EOS, the standard format for NASA's EOS Data and Information System (EOSDIS). EOS data are widely used not only for global change research but also for local and regional applications.

There are two typical groups of EOS data users. The first group of users is a small number of well-funded scientists who use large volumes of HDF-EOS data for their research and have significant resources in-house to process such data. For this group of users, EOSDIS provides special access interfaces with a dedicated fast network connection. The second group of users is a large number of small data users with limited resources. They typically use a Geographic Information System (GIS) to analyze data from EOS for a variety of applications. For most of them, obtaining the needed data from EOSDIS and preprocessing them into a form ready to be analyzed by their in-house GIS represents a major technical challenge that takes a significant amount of their time and resources. In addition to those two groups of existing users, there are many more potential users who would like to use EOS data for their applications but have no resources or expertise to derive information from the data provided by NASA data centers. Therefore, the development of capabilities to deliver GIS-ready EOS data directly to a variety of users' GIS systems through the Internet based on the users' requirements and to provide services that make user-specific products on demand will greatly enhance the interoperability and increase the public use of EOS data. This paper describes a NASA-funded project to develop such capabilities in a system called the NASA HDF-EOS Web GIS Software Suite (NWGISS), which provides interoperable access and services for HDF-EOS data using Open Geospatial Consortium (OGC) protocols. The paper first presents the current functionalities and components of the system and then discusses the current efforts to extend NWGISS, including the integration of NWGISS with Grid technology and the development of geospatial Web service middleware.

15.2 The Current NWGISS Components and Their Functionalities

Currently, NWGISS consists of following components: a web map server, a web coverage server, a catalog server, a web registries server, a multi-protocol geoinformation client (MPGC), and a toolbox. All NWGISS components can work both independently and collaboratively.

In order for as many users as possible to access data served by NWGISS servers, several Open Geospatial Consortium interface specifications have been imple- mented. The OGC interface specifications implemented in NWGISS servers include the Web Map Service (WMS) (de la Beaujardière, 2002), Web Coverage Service (WCS) (Evans, 2003), Web Registries Service (WRS) (Reich, 2001), and Catalog Service Specification (CAT) (Nebert, 2002). On the client side, WRS, WCS, WMS, and Web Feature Services (WFS) (Vretanos, 2002) specifications have been implemented. Figure 15.1 shows the architecture of NWGISS. The following paragraphs provide brief descriptions of each component. Additional information can be found in (Di et al., 2001).

The Web Map Server: The map server enables GIS clients to access HDF-EOS data as maps. Currently the NWGISS map server complies with OGC WMS Version 1.1.0 (de la Beaujardière, 2002). The OGC specification defines three operations, namely GetCapabilities, GetMap, and GetFeatureInfo. All three interfaces have been implemented and all three HDF-EOS data models (grid, point, and swath) are supported.

The Web Coverage Server: The OGC Web Coverage Service specification is designed to enable GIS clients to access multi-dimensional, multi-temporal geospatial data from WCS servers. WCS defines three interface operations, namely GetCapabilities, GetCoverage, and DescribeCoverage. The NWGISS coverage server has implemented Versions 0.5, 0.6, and 0.7 of the draft WCS specification. The WCS server works for both HDF grid and swath data. On-the-fly georectification is provided for swath data (Yang and Di, 2004). Reprojection is provided for grid data. Three return formats for coverage are supported in NWGISS, namely, HDF-EOS4 (NASA, 2002), GeoTIFF (Ritter and Ruth, 1999), and NITFF (NIMA, 1999). Currently we are implementing WCS Version 1.0, the first official version of the WCS specification.

The Catalog Server: Both WCS and WMS have the GetCapabilities operation that clients can use to find the geographic data/maps and services available at servers. This protocol works nicely when a server has a small data archive. If the server has a lot of data, the capabilities description, which basically is a data catalog, becomes very large. The catalog server allows GIS clients to search and find available geographic data and services at a NWGISS site based on the OGC catalog service specification Version 1.1.0 (Nebert, 2002). Both state-full and state-less OGC CAT have been implemented in the NWGISS catalog server, which

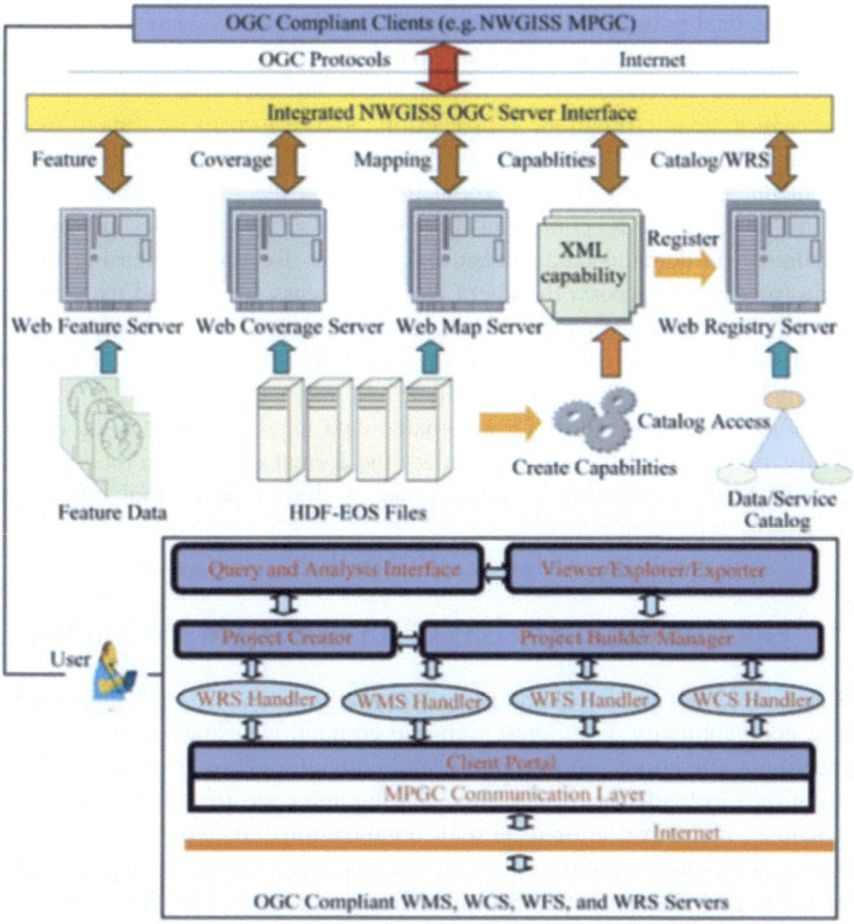

Figure 15.1 The current top-level architecture of NWGISS

reuses part of the catalog server for Data and Information Access Link (DIAL) (Di et al, 1999).

The Web Registries Server: The draft OGC Web Registries Service imple- mentation specification defines the interfaces for accessing both data and service catalogs (Reich, 2001). It also allows searching data based on available services and searching available services based on a given data type. This specification is one of the keys for service discovery in the Web Services environment. Currently, we have implemented a prototype WRS server as a part of NWGISS. The server uses an Oracle database at the backend.

The Multi-protocol Geoinformation Client (MPGC): MPGC is a comprehensive OGC client. Currently, OGC WRS, WMS, WFS, and WCS have been implemented in the client. For coverage access, the client can interactively communicate with all OGC-compliant coverage servers (not only with NWGISS) to access multidimensional geospatial data and handle the HDF-EOS, GeoTIFF and NITF coverage-encoding formats. Besides performing basic WCS client-server communication, coverage access, visualization, and user interaction, the client also provides georectification, reprojection, and reformatting functions. The execution of those functions is arranged automatically, based on the user's data requirements and the information about the data in the servers. The interaction between MPGC and OGC-compliant web coverage servers provides interoperable, personalized, on-demand data access and services for remote sensing data. In addition to coverage access, the client can also access feature and map data and overlay them with coverage for data analysis (Deng et al., 2002). All data accessed are stored and managed in a "project" folder for analysis later. Currently, we are implementing WCS 1.0 in the client.

Toolbox: It contains tools for automated data ingestion and catalog creation. Currently, two types of tools are provided: the format conversion tools (LAITS, 2002) and XML capabilities creation tools. A third type of tool, the catalog creation tools, will be provided later.

15.3 The Integration of NWGISS with Grid Technology

Grid is a rapid developing technology, originally motivated and supported for sharing geographically distributed high-end computing resources in science and engineering applications requiring high-end computing (Foster et al., 2001). The vision of the Grid is to enable resource sharing and coordinated problem solving in dynamic, multi-institutional virtual organizations. It provides on-demand, ubiquitous access to computing, data, and services, and it constructs new capabilities dynamically and transparently from distributed services. With Grid technology, new applications, such as distributed collaboration, distributed data access and analysis, and distributed computing, are enabled by the coordinated use of geographically distributed resources.

Currently, dozens of major Grid projects around the world in scientific and technical computing for research and education have been either deployed in operational use or demonstrated technically. Considerable consensus on key concepts and technologies has been reached. The key to Grid success is the open source middleware called the Globus Toolkit. It has become a de facto standard for all major Grid implementations. Although far from complete or perfect, the Grid technology is out there, evolving rapidly, and has a large tool/user base. The Global Grid Forum is a significant force that coordinates the development of the technology in the world.

OGC Web Service (OWS) specifications provide standards for implementing interoperable, distributed geospatial information processing software systems (e.g. GIS), by which a user can easily share geospatial data, information, and services with others. However, OGC technology consists mainly of interface specifications and content standards. It does not provide a mechanism for securely sharing the distributed computational resources. Meanwhile, because of the large volumes of Earth Observation (EO) data and the geographically scattered receiving and processing facilities, the EO data and associated computational resources are distributed. The multi-disciplinary nature of geospatial research and remote sensing applications requires the integrated analysis of a huge volume of multi-source data from multiple data centers. This requires sharing of both data and computing power among the data centers. Grid technology, because of its security and distributed resource sharing capabilities, is the ideal technology for filling the technology gap.

Currently, we are working on implementing Grid technology in NWGISS so that Grid managed geospatial data can be accessed by all OGC standards compliant GIS clients. The main task is to integrate Grid and OGC technologies so that OGC interfaces are provided in Globus. The integration is taking place between the backend of the NWGISS OGC servers and the front-end of data Grid services. The key is to make Grid-managed data accessible through NWGISS OGC servers. Additional information about the integration has been provided by Di et al. (2003).

15.4 The Development of Geospatial Web Services in NWGISS

Currently NWGISS servers provide the standard OGC interfaces for accessing EOS data. Combined with the multi-protocol geoinformation client MPGC, NWGISS allows data users to obtain data from data centers in the form that exactly matches the users' requirements. However, in many cases, the data products or information the users request may not be available at the data centers. The traditional way to solve this problem is for data users to obtain all necessary raw data from the data centers and then to process the raw data locally to obtain the information the users want. However, not all users have the capability or resources to do that. This problem has significantly limited the wider use of geospatial information. In order to solve this problem, we are developing interoperable geospatial services technologies. The development is based on the geo-object and geo-tree concept described by Di and McDonald (1999). The concepts are being implemented in the OGC web service environment described in Lieberman (2003). Figure 15.2 shows the architecture for the services extensions.

As we discussed in the previous section, currently NWGISS is a two-layer client-server system. The servers will directly talk to HDF-EOS data in the archives to provide on-demand data access. The extension will develop a middle service layer so that the NWGISS will become a three-tier system. The three tiers will be the interoperable data server tier, the middleware geospatial web services and knowledge management tier, and the integrated geoinformation client tier. In

Fig. 15.2, all standards-based system-wide interfaces are shown as white arrows. Any other systems using the same standard interfaces as NWGISS can be interoperable and federable with NWGISS. All other types of arrows represent either private/internal interfaces or user community defined interfaces. The size of arrows represents the size of data traffic. Since the system provides customized data products and information to the end-users by performing the data reduction and processing at the data sever and middleware tiers, the size of data traffic at the end users side is much smaller than at the data server side.

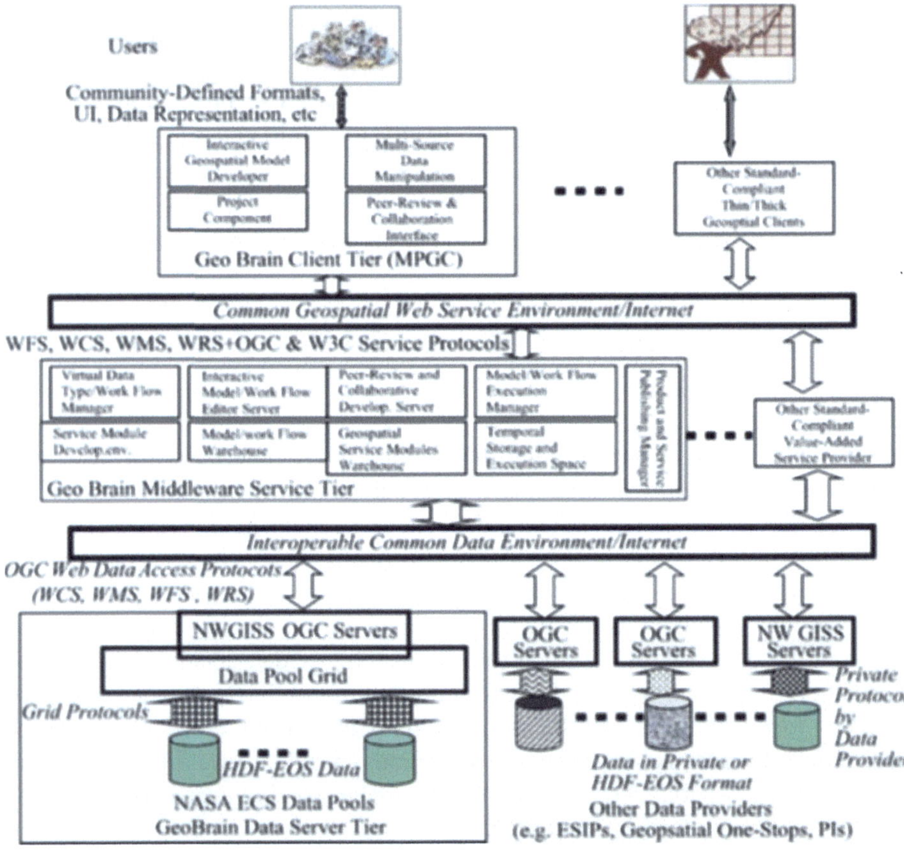

Figure 15.2 The Web Service Extensions of NWGISS

15.4.1 The Interoperable Data Provider Tier

This tier consists of data servers providing data to the geospatial services middleware, the application & data analysis systems, application clients, and human users (called requestors hereafter) through a common data environment. The NWGISS data servers are used in the data provider's tier. The common data environment is a set of standard interfaces for finding and accessing data in diverse data archives, ranging from small data providers to multiple-petabyte NASA EOS data archives. The environment allows geospatial services and value-added applications to access diverse data provided by different data providers in a standard way without worrying about their internal handling of data.

The interface standards for the common data environment are OGC Web Data Services Specifications, including Web Coverage Services, Web Feature Services, Web Map Services, and Web Registries Services. The specifications allow seamless access to geospatial data in a distributed environment, regardless of the format, projection, resolution, and the archive location.

15.4.2 The Middleware Geospatial Service and Knowledge Management Tier

The middleware tier consists of multiple components that perform geospatial data processing, information extraction, and knowledge management. This tier includes the following components:

(1) Geospatial service module warehouse contains individual geospatial web services executables.

(2) Geospatial model/workflow warehouse contains geo-trees that describe geospatial models encoded in a workflow language. The geo-trees capture the knowledge of the geospatial processes and modeling.

(3) Virtual data type/workflow manager serves two purposes. To external users, it is an OGC WRS server that helps requestors to find both data instances and data types. Internally, it manages the model warehouse, works as the client of WRS servers.

(4) Model/workflow execution manager acts as a WCS, WFS, or WMS server, depending on the type of geoinformation a requestor requests. If the requested geoinformation is non-virtual, the manager will retrieve the product from the data provider's server and deliver it to the requestor. If it is virtual, the manager will manage the execution of the workflow and deliver the materialized virtual geoinformation to the requestor.

(5) Interactive model/workflow editor server provides the interactive modeling environment that allows users to construct and test their models through the NWGISS client. Models will be constructed graphically by the client. The editor server will provide all available web service modules classified by their service categories, both virtual data types represented as workflows and the real data types to the client. The editor server will allow the user to instantiate, run, modify, debug, and save the model and submit it to a peer-review server for review.

(6) Service module development environment encourages end-users to develop additional service modules by providing a service module development environment.

The development environment includes a set of libraries for handling the interface protocols, data encoding and decoding, and general utilities functions. By using those libraries, web-service modules developed by users will be standards-compliant and interoperable.

(7) Peer-review and collaborative development server. Any user-developed web-service modules and the geospatial models are subject to peer review before being inserted into NWGISS as operational capabilities. The peer-review server will facilitate the processing by providing a common environment for reviewers to run and evaluate the submission through the Multiple-Protocol Geoinformation Client (MPGC).

(8) Product and service publishing interface. This component will publish the products and services available to external registries so that they also can be found though the registries.

15.4.3 The Integrated Multiple-Protocol Geoinformation Client Tier

The client provides not only accesses to all virtual and real data/information provided by NWGISS servers and all other OGC-compliant providers but also the geospatial modeling/workflow interfaces, peer-review interfaces, and the collaborative development interfaces.

15.5 Conclusions

Currently NWGISS has been used in many different testbeds and technology evaluation programs including the OGC Web Service Initiatives, the CEOS WGISS Test Facilities (WTF), and the NASA EOSDIS data pools. Those tests show that the NWGISS approach can significantly improve access to NASA EOSDIS data for the GIS user community. The current work on the further development of NWGISS will definitely enhance its functionality and applicability. The Grid extension will enable NWGISS to work in the Grid environment for construction of the geospatial information Grid. The Web service middleware development will enable many potential users, who are not able to use EOS data currently, to use EOS data in their applications.

Acknowledgements

This project has been supported by research grants from the NASA Earth Science Data and Information Project (ESDISP), NASA Earth Science Technology Office (ESTO), and Open Geospatial Consortium (OGC).

References

de la Beaujardière J (ed) (2001) Web Map Service Implementation Specification, Version 1.1.1. OGC 01-068r2. Open Geospatial Consortium Inc. http://www.opengis.org/docs/01-068r2.pdf

Deng D, Deng M, Zhao P, Di L (2002) A thick Web Coverage Client (WCC) of NASA HDF-EOS Web GIS Software Suite (NWGISS). In: Chen and Jiang (eds) The International Archives of Photogrammetry, Remote Sensing, and Spatial Information Sciences, Vol. XXXIV, Part 2. Commission II, pp. 87 – 110

Di L, Chen A, Yang W, Zhao P (2003) The integration of grid technology with OGC Web Services (OWS) in NWGISS for NASA EOS data. In: Proceedings of the Eighth Global Grid Forum (GGF8), June 24 – 27, Seattle, WA, USA p. 10

Di L, Yang W, Deng D, McDonald KR (2002) Interoperable, personalized, on-demand Geospatial Data Access and Services Based on OGC Web Coverage Service (OWS) Specification. In: Proceeding of NASA Earth Science Technology Conference, CDROM, Pasadena, California

Di L, Yang W, Deng M, Deng D, McDonald KR (2001) The Prototypical NASA HDF-EOS Web GIS Software Suite (NWGISS). In: Proceedings of NASA Earth Science Technology Conference, College Park, MD CD-ROM

Di L, Suresh R, Doan K, Ilg D, McDonald KR (1999) DIAL-an Interoperable Web-based Scientific Data Server. In: Goodchild M, Egenhofer M, Fegeas R, and Kottman C (eds) Interoperating Geographic Information Systems, Section 4. System Experiences Kluwer Academic Publishers, Norwell, MA pp. 301 – 312

Evans JD (ed) (2003) Web Coverage Service (WCS), Version 1.0.0. OpenGIS© Implementation Specification. Open Geospatial Consortium Inc. http://www.opengis.org/docs/03-065r6.pdf

Foster I, Kesselman C, Tuecke S (2001) The Anatomy of the Grid – Enabling Scalable Virtual Organizations. Intl J of High Performance Computing Applications 15(3): 200 – 222

LAITS (2002) HDF-EOS GIS translators. http://laits.gmu.edu/DownloadInterface. html

Lieberman J (ed) (2003) OpenGIS Web Services Architecture, OGC 03-025. Open Geospatial Consortium Inc. http://www.opengis.org/techno/discussions/03-025.pdf

NASA (2002) HDF-EOS4 (HE4), http://hdfeos.gsfc.nasa.gov/hdfeos/he4.cfm

Nebert D (ed) (2002) OpenGIS® Catalog Services Specification, Version: 1.1.1. OGC 02-087r3. Open Geospatial Consortium Inc. http://www.opengis.org/docs/02-087r3.pdf

NIMA (1999) National Imagery Transmission Format Version 2.1, National Imagery and Mapping Agency (NIMA), Bethesda, MD

OGC (2000) Catalog Interface Implementation Specification (version 1.0), Open Geospatial Consortium Inc. http://www.opengis.org/techno/specs/99-051.pdf

Reich L (ed) (2001) Web Registry Server Discussion Paper. OpenGIS Project Document 01-024r1. Open Geospatial Consortium Inc. http://www.opengis.org/techno/discussions/01-024r1.pdf

Ritter N, Ruth M (1999) GeoTIFF Format Specification GeoTIFF Revision 1.0, http://remotesensing.org/geotiff/spec/geotiffhome.html

Vretanos PA (ed) (2002) Web Feature Service Implementation Specification, Version 1.0.0 OGC 02-058. Open Geospatial Consortium Inc. http://www.opengis.org/techno/specs/02-058.pdf

Yang W, Di L (2004) An Accurate and Automated Approach to Georectification of HDF-EOS Swath Data. Photogrammetric Engineering and Remote Sensing (In press)

16 Network Geographic Information System

Chaowei (Phil) Yang, Menas Kafatos, David Wong and Ruixin Yang

16.1 Introduction

Since its interception in the 1960s (GIS World, 1996; Tomlinson, 1998), Geographic Information System (GIS) has been well established as an information system for acquiring, storing, processing, analyzing and visualizing spatial data. The success of GIS is reflected from the following developments: (1) numerous GIS software have been developed and are widely used; (2) A large amount of spatial datasets have been collected and used; (3) Many books and papers have been written to introduce and discuss different aspects of GIS. They range from providing an overview of GIS (Davis, 2003; Tomlinson, 1998), discussing selected advanced topics (Kreveld et al., 1997; O'Sullivan and Unwin, 2002), to recording the relevant history (Foresman, 1997).

However, relevant problems also arise: (1) Different software companies, such as ESRI (www.esri.com) and Intergraph (www.imgs.com), developed their own software with full GIS functions, and sold their software at a fully functional price. However, most customers just use the mapping function which accounts for no more than 30% of the full GIS functions, and almost no software are used all the time. (2) A large number of spatial databases are constructed from national level (such as NASA and USGS), state level (such as Virginia and Maryland), to county level (such as Fairfax and Loudoun). National applications (such as Homeland Security) call for full integration of these spatially dispersed databases. To save the investment on GIS nationally and internationally requires sharing and connecting spatial data and software. (3) While current literatures focus on using existing software for application research, GIS basic functions need to be revisited (Wise, 2002) for building new software platforms for sharing spatial data and processing (Yang et al., 2002).

Fortunately, the emerging and fast development of computer networks, especially WWW and the Internet, provide a communication framework to solve these problems. The marriage of GIS and computer networks has produced many types of GIS: such as On-line GIS (Plewe, 1997), Internet GIS (Peng, 1999; Huang et al., 2001; Peng and Tsou, 2003), Intranet GIS, Wireless GIS (Braun, 2003), and Distributed GIS (Peng and Tsou, 2003; Yang, 2000, etc.). To integrate GIS among networks for sharing the data and software, scholars propose many new architectures, such as agent-based GIS (Luo, 1999), distributed geospatial information services (e.g. Yang et al., 2002), and GridGIS (http://www.niees.ac.uk/events/GridGIS/). Although having different names, these different manifestations of GIS have the same characteristic of using computer networks as the communication infrastructure to implement GIS. Therefore, these different types of GISs can be grouped under the label of Network GIS.

These new manifestations reflect the rapid growth of Network GIS. however, they have no clear definition. What are the differences among Network GIS, Internet GIS, Intranet GIS, WebGIS, Mobile GIS, and Network GIS? What is the communication framework of a Network GIS and how would one build a Network GIS? How could we use Network GIS in disseminating remote sensing and other spatial data, and providing useful functions or services to users? What research aspects of Network GIS are needed? This chapter attempts to examine these questions based upon recent research on Network GIS (Yang, 2000; Kafatos et al., 2003; Peng and Tsou, 2003; Yang et al., 2003; Yang and Wong, 2003; etc). It is also hoped that this chapter could serve as a brief technical introduction of Network GIS to geospatial professionals who plan to use Network GIS or are interested in the technical issues of Network GIS.

To answer these questions and to provide a systematic introduction of Network GIS, the rest of this Chapter is organized in the following manner: Section 16.2 describes the communication foundation of Network GIS, i.e., the underlying network infrastructure. Section 16.3 reviews basic GIS functions, discusses another critical issue of Network GIS, that is, how to distribute these functions over a network. Consequently, a flexible Network GIS framework is proposed. Section 16.4 introduces Distributed GIS as a new envision of Network GIS. Section 16.5 proposes taxonomy of Network GIS and discusses the usage of different types of Network GIS. Two examples of Network GIS are given in Section 16.6 to illustrate how Network GIS can be used in disseminating data/information and providing services. Finally, in Section 16.7, we present several research areas of Network GIS.

16.2 Network Infrastructure

Building a Network GIS depends on two critical issues: (1) the underlying network infrastructure on which the GIS functions are built, and (2) the distribution or allocation of GIS functions among different computers within a network. This section provides an overview of the first issue and describes the procedure of Network GIS communication.

A computer network connects computers located in different places by providing electromagnetic channels (wired or wireless) to exchange information. The information exchange process via the channels is very complex. The International Standard Organization (ISO) realized the complexity and the importance of developing an open network model to ensure the interoperability of different network devices and computers. Consequently, ISO proposed a 7-layer network model in the 1980s to formalize this complex process. This 7-layer network model is widely used in introducing network infrastructure (Jain and Agrawala, 1993; Tanenbaum, 1996). However, the most widely used network model is the TCP/IP 5-layer model (Murhammer et al., 1998; Stallings, 2000); TCP/IP stands for Transportation Control Protocol/Internetworking Protocol.

Figure 16.1 shows a typical Internet TCP/IP 5-layer model. The 5-layer model includes a physical layer, a network layer, an internet layer, a transport layer, and an application layer. The internet layer is not the Internet that refers to the global network. For simplicity, some scholars (Komar, 2002) also combine the lowest two layers as the network access layer. In general, an upper layer depends on the immediately lower layer by calling the interface provided by the lower layer to perform its tasks. This dependency is illustrated in Fig. 16.1 by the interfacing service provided to the upper layers by the lower layers. Each layer has its own protocol (a set of communication rules) for communicating with a peer layer on another computer, i.e. two layers at the same level in two computers use the same protocol. Because the two layers are at the same level, their communications are also labeled as peer or peer-to-peer communication. In this type of communication, the two layers do not communicate directly but through the lower layers or other devices except the directly connected physical layers. Most communications between two computers on the Internet will go though many other network devices, such as routers, which select routes for data to travel from a source to a destination. Routers route data units with specific volume size called packets from computer to computer. In this manner, a simplified network model is comprised of computers, network cables (or wireless devices) for exchanging electromagnetic signals, and routers for routing packets. Peng and Tsou (2003) provide a description of the ISO 7-layer and the TCP/IP 5-layer models. The following example illustrates how a Network GIS utilizes the network infrastructure and illustrates the functions of each layer using the 5-layer model.

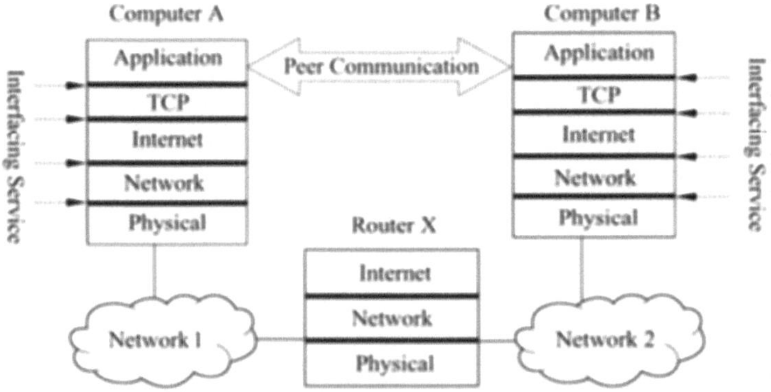

Figure 16.1 The Internet TCP/IP 5 layer model

If GIS α on computer A needs more data from GIS β which manages the spatial database on computer B, and the application layer protocol is HyperText Transfer Protocol (HTTP), α can call the data selection function of β through HTTP and will receive the response data. However, the above seemingly straightforward and simple communication procedure between α and β is actually a complex process as illustrated in Fig. 16.2 and involves the following steps:

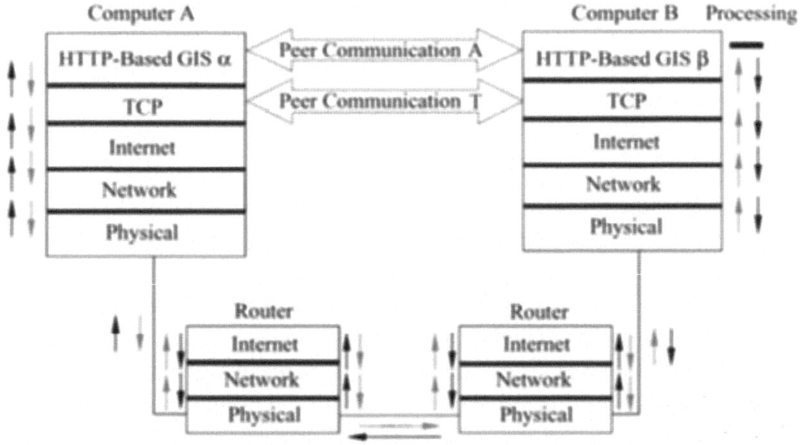

Figure 16.2 The communication procedure between two GISs in the Internet: The gray arrow shows the information flow of request from α to β, and the black arrow shows the response from β to α

(1) Data request information will be generated by α according to a GIS communication protocol, which is understandable to β. The data request information will be passed on to the application (HTTP) layer, which defines certain data format, interprets the data request information and transforms it to a data stream for passing to the TCP layer.

(2) The data stream is passed on to TCP layer by calling a TCP layer function provided to the application (HTTP) layer. The TCP layer ensures the end to end correct transmission of the data stream, i.e. the TCP layer on computer A will ensure the complete data stream reaches the TCP layer on computer B. The TCP layer will also divide the data stream into packets. Some priority and loss control functions can be added to this layer to ensure quality and level of service.

(3) The TCP layer will call the internet layer and pass the packets to the internet layer. According to the IP addresses of A and B, the internet layer will choose routes for the packets. The packets belonging to the same data stream may take different routes to travel from A to B because routes may involve various numbers of routers and other network devices, such as a firewall. Two routers are illustrated in this example; the internet layer may also divide packets into smaller packets called sub-packets, according to the data size restriction of

the internet protocol.

(4) The internet layer will call the network layer to send the packets/sub-packets on computer A. The network layer will ensure error free exchange of packets/sub-packets between any two directly connected computers or routers.

(5) The network layer on computer A will call the physical layer to send electromagnetic signals, which have the digital data encoded. The physical layer on router X will receive the signals.

(6) The physical layer defines the electrical standards, such as number of pins for a port and electric voltages. These standards will ensure the direct communication between the two physical devices for transmitting signals with digital data encoded.

(7) The physical layer of router X will decode the digital data and pass them onto the network layer on router X for checking correctness. Router X's internet layer will reassemble the data into packets from sub-packets, and may divide the packets into sub-packets using different sizes according to different internet layer protocols.

(8) The 4th, 5th, 6th, and 7th steps will be repeated among A&X, X&Y, Y&B en route until packets reach computer B.

(9) The physical layer of computer B will receive signals from the physical layer of router Y and pass decoded digital data to the network layer.

(10) The network layer of computer B will check the correctness of the digital data and request the data again from router Y if the data are incorrect. Received digital data will be passed to the internet layer on computer B.

(11) The internet layer on computer B will assemble sub-packets into packets according to the same protocol used by the internet layer on router Y, and passes the packets to the TCP layer.

(12) The TCP layer will check the correctness and sequence of the packets, and assemble the packets into data stream, which will be passed on to the application layer. The application layer will interpret the data stream according to formats defined by HTTP. The interpreted data request information will be passed on to β.

(13) β will perform the data selection process on computer B as requested and generate datasets accordingly.

(14) The output datasets will be sent back to α in a similar procedure from step (1) to step (12).

Among the layers of the TCP/IP 5-layer model, the most upper two layers (TCP and application layers) are most frequently used in developing application software, such as GIS. For instance, HTTP is widely used in developing Web-based software. A better performance can be achieved by programming on TCP by calling Winsock or NetBIOS (Jones and Ohlund, 1999) directly (Yang, 2000). However, this method requires the developers to possess in-depth network knowledge; it is often used by system developers in developing commercial or system level software, such as ArcGIS. In general, the computer sending requests is termed a client while the computer responding is called a server in a network environment. In the research, design, and development of a Network GIS, only the two upper most layers, the server, and the client, are considered. All other network devices and functions will be provided by lower-level software and hardware, such as the operating systems.

16.3 Distributing GIS Functions

Another critical issue of building a Network GIS is how to distribute GIS functions over a network. This section introduces the GIS functions to be distributed, discusses how to distribute these functions, and proposes a flexible framework for Network GIS.

According to most widely accepted definitions of GIS (Worboys, 1995), a traditional GIS includes functions of capturing, storing, analyzing, and displaying spatial data. Practitioners' definitions of GIS (Tomlinson RH and Tomlinson R, 2003) may also include procedures, operating personnel, and spatial data. Although the procedures and operating personnel are not GIS functions, Network GIS is transforming some processes of procedures and capabilities of operating personnel into GIS functions. For example, if a driver needs a map from the mapping office to drive from LA to DC, the procedure could be:

(1) Call the mapping office to submit a map request.

(2) The mapping office will generate a route map according to the request.

(3) Someone will ship the map to the driver.

If the mapping office has a map server to generate the route map and the driver has a network connection to access the corresponding server, this procedure and the functions of calling and sending map could be automated and built into one system. Therefore, the procedure processes and the sending capabilities of personnel are transformed to the functions of a Network GIS.

By incorporating these transformed GIS functions, the functions of a GIS can be classified into six general technical areas according to the spatial data processing procedure (as illustrated in Fig. 16.3): Geospatial data collection, data input and editing, geospatial database, geospatial data access, geospatial analysis, and visualization/interaction. Data collection and input is collecting spatial data using manual, automatic, or semi-automatic methods, such as remote sensing, ground surveying, or digitizing from paper maps. Data editing includes checking data error, validating attribute data. Data management is performed through spatial databases, such as ArcSDE from ESRI or SpatialWare from MapInfo. Data access provides interfaces for a caller program to access and operate the spatial database. For example, Intergraph uses an Open DataBase Connectivity (ODBC) link to Microsoft Access database from GeoMedia. ESRI provides Spatial Data Engine to access its ArcSDE database (ESRI, 1997). GIS is distinguished from other information systems through its spatial analysis capabilities, such as selecting the shortest path, choosing a location to build a cell phone tower to serve a large area. The visualization and interaction functions are built to provide users with a displayed map, and users can perform selection, zoom, query on the map.

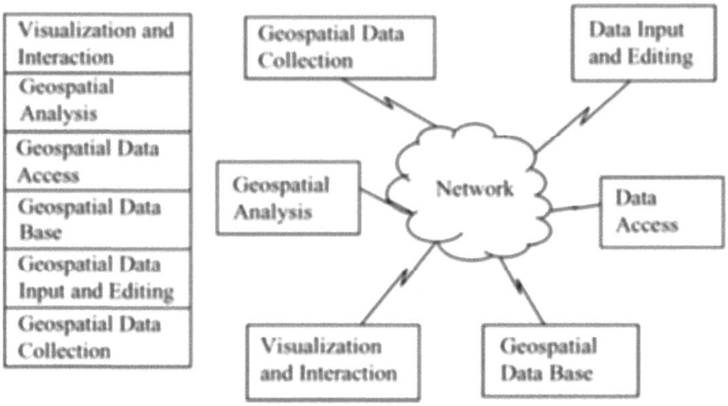

Figure 16.3 A traditional monolithic GIS (left) and a networked GIS (right) with different GIS functions distributed in a computer network. These functions can be allocated on two or more than two computers

In a traditional GIS architecture, these functions are all located within one computer. In a Network GIS, these functions will be allocated to different computers in a network to take advantage of the computing resources, and comply with the fact that datasets and users are widely distributed. The strategy of distributing GIS functions can be quite different according to specific project requirements and the long-term objectives of implementing a Network GIS. However, there are some basic principles in formulating these strategies:

(1) The data visualization interface and the system interaction should be accessed by users directly; therefore, the client (s) should host the data visualization interface and the user-interacting functions.

(2) The data management and databases should be allocated to a place that provides easy access to all clients and users with an acceptable performance; therefore, the server would be the ideal place to host the databases and data management.

(3) The data processing component has the greatest flexibility to be distributed to either the server side or the client side according to a specific application.

(4) Data collection often happens in fields, such as ground surveys. A mobile device could host this function and be equipped with a wireless connection to the server, so client can automatically update the database.

(5) The data access and integration function will be the connection between server and client or just inside the server according to the strategy taken in 3).

When data and data processing are all residing on the server, the data access and integration for linking the data and data processing will reside on server. If data and data processing are distributed on two computers, the data access and integration will be implemented on the two computers. This variation in data exchanged resulted from the data access and the integration strategies requires different protocols in Network GIS.

(6) The editing function resides mostly on computers which reside within the intranet or local network with the server. Network GIS could also automate this function by provide mobile or online update.

These general principles could help in strategic plan to allocate these six functions, but the implementation of Network GIS could be very complex. One aspect of this complexity is that these functions also include series of sub functions, which can also be allocated to different computers or be implemented all on one computer. For example, data visualization includes preparing data (coordinate transform, etc), visualizing data by symbolizing features, drawing image based on the symbolization, and displaying a drawn image. The display function of the visualized image can reside on the client side and all other functions can reside on the server side. This smaller client strategy is adopted by Mapquest (www.mapquest.com). On the other hand, we could put all visualization functions on the client side, and make the client bigger. This latter strategy is adopted by some commercial software, such as the Java client of ArcIMS (http://www.esri.com/software/arcims/index.html), and the ActiveCGM of GeoMedia WebMap (http://imgs.intergraph.com/products/default.asp?id=gmw&Submit=Go).

The smaller client strategy is ideal to implement simple functions or services with great speed, while the bigger client strategy can provide complex functions or more demanding services. The smaller or bigger client strategies are also referred as thin and thick client strategies respectively (Foote et al., 1998).

The adoption of thick or thin clients is dependent on the requirements, performance, and system-supporting infrastructure of a specific application. Mapquest, for instance, uses a thin client to let users submit queries and send only an image back to the client, through a web browser, while the Java version of ArcIMS includes a series of basic GIS functions, such as feature selection, feature querying, and display functions. According to the principles for distributing GIS functions, some GIS functions must be located on the server side but not on the client side, vice versa. But some functions can be located on either side. According to this principle, a flexible framework for building a Network GIS is illustrated in Fig. 16.4.

These strategies are based on the traditional client-server model. The rapid development of computing technology and computer networks also enable us to implement GIS among a number of computers in a network. This provides us another option to implement Network GIS. Instead of using client-server architecture, GIS functions can be implemented on different computers, and these computers can collaborate by serving as both server and client for a specific GIS

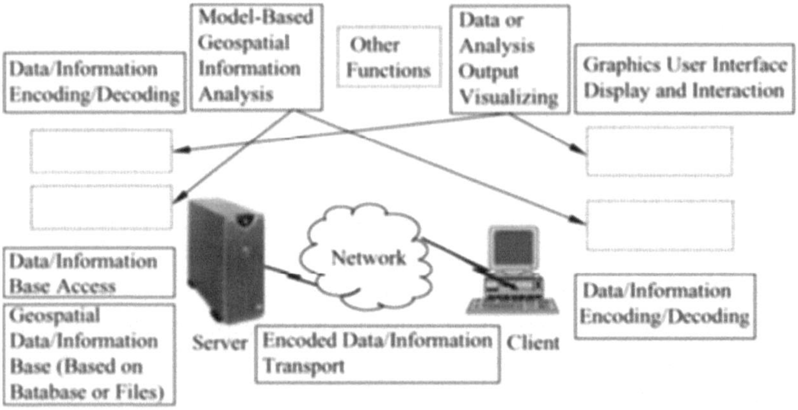

Figure 16.4 A flexible distributed GIS framework: the server side hosts data, provides data access interface; the client side displays the visualized data/output and interacts with users; data/information transmitted through the network will be decoded/encoded at both client and server side; model-based analysis, data visualizing, and other functions can be hosted at either server side or client side

task, which may require massive computing capability or data. Therefore, this implementation takes advantage of the distributed data and network computing power. The concept of agent-based GIS, peer-to-peer GIS (Li, 2003) and GridGIS are new research directions along this line.

16.4 Distributed GIS

By implementing GIS functions on two or more computers, the concept of Distributed GIS (DGIS) is introduced. This section provides an overview of this manifestation.

The basic idea of a Network GIS is to distribute GIS functions among different computers within a network. This idea of "distribution" gives Network GIS another name of DGIS. As its name implies, DGIS can be regarded as a GIS with its functions distributed at different places or on different computers. This definition emphasizes the distribution of all of the general six functions. According to this broad definition, almost all GISs are DGIS. Spatial data capturing, for instance, will be separated from the location where a computer hosts the processing of spatial data. This general definition can help us to manage a DGIS project by including the required personnel and procedures.

A more technical definition of DGIS similar to Network GIS would be a GIS that has functions distributed among different computers connected by a computer network. The functions could be hosting spatial data, accessing spatial data, analyzing spatial data, visualizing spatial data and providing interactions to users. In this definition, only the functions that can be implemented on computers or networks are considered, therefore, this definition is used for technical design, and implementation of DGIS (Yang and Wong, 2003).

In computer science, a distributed system is defined as "a collection of (probably heterogeneous) automata whose distribution is transparent to the user so that the system appears as one local machine". (http://www.ditionary.com/) According to this definition, DGIS can be defined as an information system that allows the storing, accessing, analyzing, visualizing, and user interacting distributed within a network; while the system internal processing is transparent to users. For example, computer users can interact with a DGIS in selecting routes without knowing where the data are, who selects the routes, which computer did the visualization, but feel like all data and operations are in the computer in front of them.

This definition envisions a highly transparent DGIS, which requires implementing functions to replace some capabilities of professionals, and building each of the processing functions into a uniformly accessible component over the network (Li, 2000). There is no such type of DGIS yet. However, conceptually, this definition would help us to implement the general defined functions of GIS, including professional capabilities and procedures, into DGIS. This definition could also help us to envision a future distributed geospatial information service platform, such as peer-to-peer GIS or grid GIS.

16.5 Network GIS Taxonomy

Different manifestations of Network GIS are based on two facts: (1) the types of computer networks, and (2) the distribution of GIS functions. Network GIS could also be classified in these two ways.

In the first classification, if a Network GIS is implemented on the Internet, e.g., the mapquest system, then, the Network GIS can be regarded as an Internet GIS. If the network is an Intranet inside a company, university or department, the Network GIS can then be labeled as an Intranet GIS. If the client side of the GIS is implemented inside a Web browser, the Network GIS can be regarded as a WebGIS and a WebGIS can be an Intranet GIS or Internet GIS. However, some Internet GIS, such as ArcExplorer, are not WebGIS. As illustrated in Fig. 16.5, different types of Network GIS overlap. Mobile GIS (Braun, 2003; Spencer, 2000) has emerged and is regarded as Network GIS. ESRI, Trimble, and Tele Atlas (2003) defined mobile GIS as the movable devices with GIS and GPS functions. From this definition, a mobile GIS is not necessarily a Network GIS. However, if equipped with a wireless connection, a mobile GIS could become a wireless Network GIS.

With different functions distributed between servers and clients, the second classification divides Network GIS into two types (Foote, 1998). (1) a thin-client Network GIS with only a few functions, such as the display and interaction

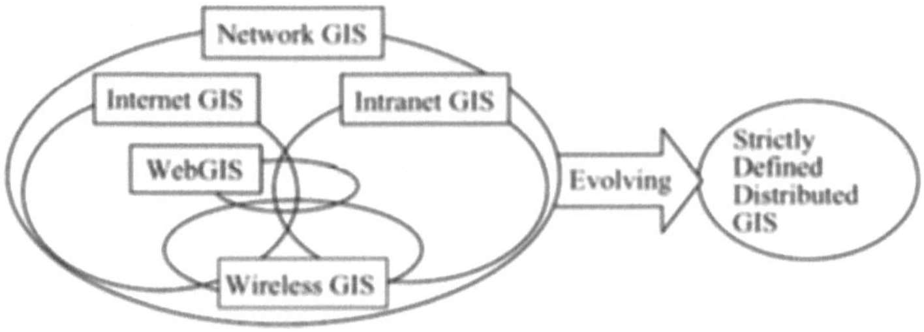

Figure 16.5 Network GIS taxonomy: different terms of network GIS refer to different implementations of distributed GIS: Network GIS is implemented in a network, which includes all kinds of GIS built on computer network; Intranet GIS is implemented in the Intranet. Internet GIS is implemented within an Internet. WebGIS is implemented inside a web browser. Distributed GIS is a future Network GIS with transparent characteristics that require more research and development

implemented on the client side. In this case, the server will handle functions related to data storage, processing, accessing, visualizing, and most interaction responses. (2) A thick-client Network GIS has most GIS functions implemented on the client computer, where the client can also communicate with the server for data and massive computing capability. In reality, we cannot just simply choose one from these two types. For example, we may need half of the functions to be implemented on the client computer while putting the other half on the server, or duplicate some functions on both server and client to improve performance. In this scenario, it would be flexible to implement GIS functions inside a component and make the internal system transparent to the component user (Brown, 1996; Li, 2000). This implementation of components could produce transparent components to help us in building the strictly defined DGIS.

Different applications require different types of Network GIS. For a department to share spatial data and GIS software, an Intranet GIS would be proper. Most GIS companies produce the Intranet GIS products, such as ArcGIS from ESRI and GeoMedia from Intergraph. If some data and software are used to build services for the public through the Internet, an Internet GIS is then proper. As such, ArcIMS from ESRI, GeoMedia Webmap from Intergraph would be proper to build Internet GIS. If in-field data update, collection, or validation is needed, mobile GIS will be a good choice. ArcPad from ESRI and Intelliwhere from Intergraph would be proper for this mobile GIS need. Wireless devices can be installed on the mobile devices to update the databases automatically if needed.

16.6 Examples of Network GIS

The Center for Earth Observing and Space Research (CEOSR) received funding from NASA through the Virginia Access-Middle Atlantic Geospatial Information

Consortium (VA-MAGIC) project (http://cpgis.gmu.edu/vaccess/). In this project, we develop RS-oriented applications for the Mid-Atlantic region using datasets from NASA and other agencies. One of the applications is the hurricane Isabel web service, which collected and continues to disseminate satellite data and information about the hurricane. When hurricane Isabel hit the mid-Atlantic region in September, 2003, over 3,000 individual users accessed the website in one day to obtain information about this natural hazard. Another application is the Mid-Atlantic regional portal of geospatial information service that collects information from various government agencies, and provides high-resolution geospatial information services. The first example is a thin-client Network GIS because it provides only preprocessed static spatial data. The second example resembles more to a thick-client Network GIS because it provides more complex GIS functions. Both of them are Internet GIS and WebGIS since they all provide information within web browser and through the Internet.

The hurricane Isabel website is located at http://cpgis.gmu.edu/vaccess/news/IsabelIndex.htm, and is a project integrating spatially referenced images, maps, and information from different Earth observing systems and/or sensors, such as Tropical Rainful Mesuring Mission (TRMM), Geostationary Operational Environmental Satellite (GOES), Advanced Very High Resolution Radiometer (AVHRR), and Moderater Resolution Imaging Spectroradiometer (MODIS). Data from the National Hurricane Center, the Storm Center Inc., and a few other organizations are accessed through this web service to provide the hurricane information. The basic approach in developing this website service is shown in Fig. 16.6.

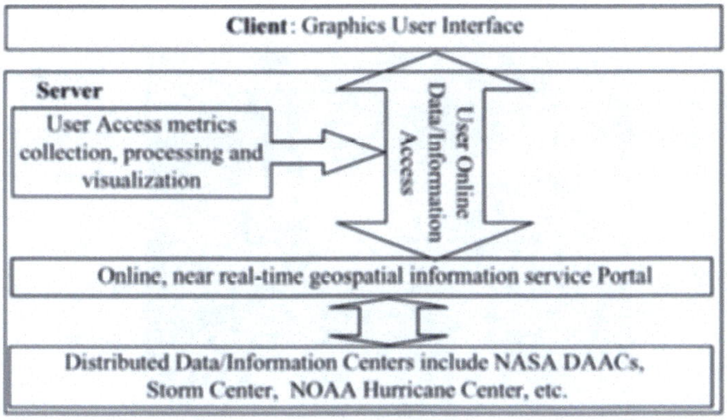

Figure 16.6 System architecture of the Hurricane Isabel Web Service

The Mid-Atlantic Regional Geospatial Information Service is a project supported by NASA, Intergraph, and USGS. The objective is to share data and information collected through VAccess-MAGIC (http://cpgis.gmu.edu/vaccess/) and the SIESIP project (http://esip.gmu.edu/). The geographical coverage of these datasets ranges from the global level to the regional and local levels. Other spatial data provided by the Commonwealth of Virginia, NASA, NOAA, USGS, are utilized in the portal to build the services. GeoMedia WebMap (http://www.intergraph.com/gis/), a Network GIS software package from Intergraph, is adopted to provide GIS functions. The software framework is illustrated in Fig. 16.7. The dash-lined

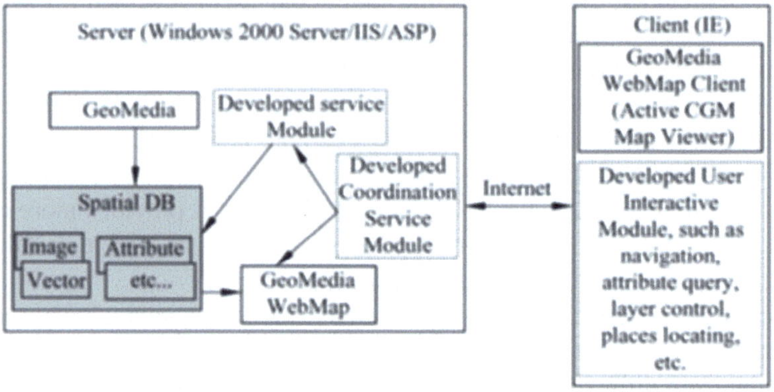

Figure 16.7 Software Framework for Mid-Atlantic Regional Geospatial Information Service System

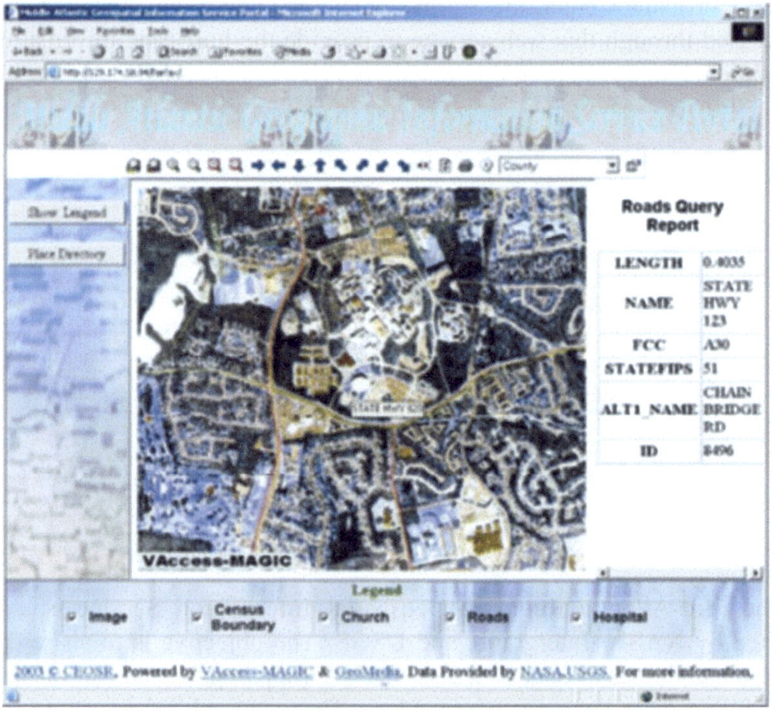

Figure 16.8 The GUI of Middle Atlantic Geospatial Information Service Portal

modules are developed to serve two functions: (1) to coordinate interactions between the GeoMedia products and the developed modules, and (2) to provide customized functions or services. The solid-lined white modules are software modules from Intergraph that are used for data management and web mapping. The gray modules refer to the data and information that CEOSR has collected through its various projects. The server side provides data, information, and services through the GeoMedia software and our developed modules. The client side provides visualization, interaction, and analysis functions to the users.

The portal provides users information about Fairfax county and Northern Virginia with online geographic information services. Users can explore the geographic environment of Northern Virginia by using the GIS functions provided online. Users can also use their background knowledge to design a customized map and print it out or use it in a document. The physical network supporting portal includes a DELL PowerEdge 4600 Server and a 100 Mbps network connection to the Internet. Users connected to the Internet through a 100 kbps+ connection can access the server with a good performance. The client-side interface is illustrated in Fig. 16.8.

16.7 Research Topics in Network GIS

Over the past ten years, Network GIS has been accepted and widely used in different disciplines from natural resource management (Kearns, 2003) to transportation (www.mapquest.com). However, to fully utilize the capability of Network GIS and build a completely transparent DGIS, the following issues should be addressed:

(1) The distribution strategy of geospatial information processing;

(2) Computational techniques for improving the overall performance;

(3) Proper protocols for effective transmitting of geospatial information;

(4) Interoperability of the data and software to enhance the transparency of DGIS;

(5) Adoption the peer to peer and grid computing infrastructures;

(6) Establishing a series of standards for measuring the quality and levels of services; and

(7) Geospatial knowledge discovery, management, and integration.

How should the geospatial information processing functions be distributed? Where should the sub functions be allocated? Where the data should be stored? Although answers to these questions will change according to specific applications, principles for guiding application developments should be established based on some criteria, such as performance and computing infrastructure.

The performance of distributed geospatial information processing is critical to both simple data dissemination and more complex functional geospatial information services. It requires more investigations on information extracting, compression, transmission, pre-fetching, pre-processing (Yang et al., 2004) and other computing techniques.

As illustrated in Fig. 16.5, where one puts the processing functions will also affect the data/information needs for transmitting. The variety in data/information transformed requires a set of protocols dealing with all levels of data from raw data to drawn images. These application level protocols (with respect to the network architecture) could be fundamental (with respect to Network GIS).

Interoperability research and implementation have been conducted for about ten years. Successes have been achieved in some projects, such as NSF funded Digital Library, ESRI's Geography Network, and NASA earth science data interoperability. However, there is a long way to go before we can fully integrate the data and sharing the processing module from different companies as envisioned by Buehler and McKeel (1996). Both government (such as GeoSpatial One-Stop), non-government organizations (such as OGC), and company efforts are needed.

Agent-based GIS (Tsou and Buttenfield, 1998; Luo, 1999) and Grid GIS (http://www.niees.ac.uk/events/GridGIS/) have been suggested as solutions for utilizing the power of massive networked computers. The implementation of agent-based GIS or Grid GIS will require a completely new design of the GIS according to computing environment of the agent or grid.

Online geospatial information service or Web Service becomes practical ways to gradually build DGIS (Tao and Yang, 2004). Levels of service, quality of service, and metrics for recording service performance (Stallings, 2002) are needed for providing effective and efficient web service and utilizing the power of massive networked computers.

The geospatial decision support system is based on geospatial information systems and disciplinary-based knowledge systems (Jankowski and Nyerges, 2001). The seamless integration of geospatial information and the decision support knowledge require investigations in spatial data mining and spatial knowledge discovery/management.

As an evolving field, Network GIS needs the contribution of various research and development related to computer networks and GIS. This chapter is only a general introduction. More detailed information can be found in related references. We believe that more literature will appear on this fast developing topic.

Acknowledgements

Research and developments discussed in this chapter are supported by NASA-funded Virginia Access—Middle Atlantic Geospatial Information Consortium (NASA NAG13-01009) and USGS-funded Building an Interoperable Web Mapping Portal for the Mid-Atlantic Region (03HQAG0146).

References

Braun P (2003) Primer on Wireless GIS, Urban and Regional Information Systems Association, http://www.urisa.org/store/primer_on_wireless_gis_qs.htm

Brown AW (1996) Foundations for Component-Based Software Engineering. In: Brown AW (ed) Component-Based Software Engineering. Los Alamitos, IEEE Computer Society Press, pp. 3 – 6

Buehler K and McKeel (eds) (1996) The OpenGIS Guide: Introduction to Interoperable Geo-Processing, Wayland, MA, OpenGIS Consortium, Inc.

Davis DE (2003) GIS for Everyone, 3rd edition. ESRI Press

ESRI, Trimble, and Tele Atlas (2003) What is Mobile GIS? http://www.esri.com/events/mobilegis/

ESRI (1997) Getting started with SDE, A ESRI white paper

Foresman T (1997) The History of GIS (Geographic Information Systems), Upper Saddle River, NJ: Prentice Hall

Foote, Kenneth E, Anthony PK (1997) WebGIS, NCGIA Core Curriculum in GIScience, http://www.ncgia.ucsb.edu/giscc/units/u133/u133.html, posted July 13, 1998

GIS World (1996) Roger Tomlinson: The Father of GIS, GIS WORLD Interview, http://www.geoplace.com/gw/1996/0496/0496feat2.asp

Healey R, Dowers S, Gittings B, Mineter M (1998) Parallel Processing Algorithms for GIS. Bristol: PA: Taylor & Francis

Huang B, Jiang B, Lin H (2001) An integration of GIS, virtual reality and the Internet for Visualization, analysis and exploration of spatial data. International Journal of Geographic Information Science 15: 439 – 456

Jain B, Agrawala (1993) Open systems Interconnection. New York: McGraw-Hill

Jankowski P, Nyerges TL (2001) Geographic Information Systems for Group Decision Making: Towards a Participatory, Geographic Information Science. New York: Taylor & Francis

Jones A, Ohlund (1999) Network Programming for Microsoft Windows. Redmond, Washington: Microsoft Press

Kafatos M, Yang C, Wong D, Yang R (2003) Mid-Atlantic Geographic Information Service Center, VAGIS Conference 2003, Richmond, Oct. 27 – 28

Kearns FR, Kelly M, Tuxen KA (2003) Everything happens somewhere: using webGIS as a tool for sustainable natural resource management. Front Ecol. Environ. 1(10): 541 – 548

Komar B (2002) Teach Yourself TCP/IP Networking in 21 Days. Indianapolis, Indiana: SAMS pp. 39 – 40

Kreveld M, Nievergelt J, Roos T, Widmayer P (1997a) Algorithmic Foundations of Geographic Information Systems. Springer: New York

Murhammer M, et al. (1998) TCP/IP: Tutorial and Technical Overview. Upper Saddle River, NJ: Prentice Hall

Li B (2003) Peer-to-Peer sharing of cartographic data and software. In: Michael Peterson (eds) Maps and the Internet. Elsevier Science pp. 157 – 168

Li B (2000) A component perspective on geographic information services. Cartography and Geographic Information Sciences 27: 75 – 86

Luo Y, Wang X, Xu Z (2003) Extension of spatial metadata and agent-based spatial Data navigation mechanism. In: Proceedings of the eleventh ACM international symposium on Advances in geographic information systems, pp. 102 – 109

Luo Y (1999) Study on Agent-based Distributed GIS. Unpublished PhD dissertation (in Chinese) (Beijing: Peking University)

OGC (1999) The OpenGIS Abstract Specification Topic 12: OpenGIS Service Architecture Version 4

O'Sullivan D, Unwin D (2002) Geographic Information Analysis. John Wiley & Sons

Longley PA, Goodchild MF, Maguire DJ, Rhind DW (1999) Geographical Information Systems. New York, John Wiley & Sons: pp. 1 – 20

Peng ZR, Tsou MH (2003) Internet GIS: Distributed Geographic Information Services for the Internet and Wireless Networks. Hoboken, New Jersey: John Wiley & Sons, Inc. pp. 2 – 5

Peng ZR (1999) An assessment framework for the development of Internet GIS. Environment and Planning B-Planning & Design 26: 117 – 132

Plewe B (1997) GIS Online: Information retrieval, Mapping, and the Internet. Santa Fe: OnWord Press, pp. 6 – 14

Rogul D (2003) Geographical Information (GIS) Market Trends. Faulkner Information Services http://www.faulkner.com/products/faulknerlibrary/pdf/00016293.pdf

Spencer B (2000) Mobile GIS. Geo Solutions http://www.geoplace.com/asiapac/2000/0700/0700sol.asp

Stallings W (2002) High-speed Networks and Internets. Upper Saddle River, NJ: Prentice Hall

Stallings W (2000) Data and Computer Communications, 6th edition. Upper Saddle River, NJ: Prentice Hall

Sui DZ, Goodchild MF (2001) GIS as a media? International Journal of Geographic Information Science 15: 387 – 390

Tanenbaum A (1996) Computer Networks. Upper Saddle River, NJ: Prentice Hall

Tao V, Yang C (2004) Distributed Geographic Information Services. In: Rana S and Sharma J (eds) Frontiers of Geographic Information Technology. London: Springer

Tomlinson RF, Tomlinson R (2003) Thinking About GIS: Geographic Information System Planning for Managers. ESRI Press

Tomlinson R (1998) The Canada Geographic Information System. In: Foresman T W (ed) The History of Geographic Information Systems. Upper Saddle River, Prentice Hall: pp. 21 – 32

Tsou MH, Buttenfield BP (1998) An Agent-based, Global User Interface for Distributed Geographic Information Services. In: Proceedings of the 8th International Symposium on Spatial Data Handling, Vancouver, Canada, July, 1998 IGU: pp. 603 – 612

Wise S (2002) GIS Basics. Taylor & Francis: New York

Worboys M (1995) GIS: A Computer Science Perspective. Taylor & Francis

Yang C, Wong D, Yang RX, Li Q, Tao V, Kafatos M (2004) Performance improving techniques in WebGIS. International Journal of Geographic Information Sciences (in press)

Yang C, Wong D (2003) Distributed GIS, A course developed at George Mason University, website: http://cpgis.gmu.edu/courses/04fall/

Yang C, Wong D, Kafatos M, Yang R (2003) PgGIS: A high performance WebGIS. Next Generation of Geographic Information, Boston, Oct. 19 – 21, 2003. A patent application of this technology has been submitted and PgGIS is in the process of commercializing

Yang C, Kafatos M, Yang R, Qu J, and Wong D (2002) A geographic information service environment. In: Proceeding of GeoInformatics 2002 Nanjing, China, June 1 – 3, 2002, (CD-ROM publication)

Yang C (2000) Theory, Techniques and Implementation Methods of WebGIS. Unpublished PhD dissertation (in Chinese) (Beijing: Peking University)

17 A Content-Based Search Method and Its Application for EOS

Ruixin Yang, Kwang-Su Yang, Jiang Tang, John Qu and Menas Kafatos

17.1 Introduction

Advances in Earth observing from space and associated Earth systems numerical simulations have resulted in rapidly growing data volumes for Earth science communities. The NASA's Earth Observing System (EOS) program alone is producing massive data products with total rates of more than 1.5 Terabytes/day (King and Greenstone, 1999). To find interesting data, data users need an effective and efficient way to search through the data. Traditionally, metadata are provided in a database which support data searching by commonly used criteria such as spatial coverage, temporal coverage, spatial resolution, temporal resolution, etc. (Yang et al., 2002; NASA (GCMD), 2003; NASA, 2003). However, metadata search itself may still result in large amounts of data which data users need to know about before determining the usefulness of the underlying data. Content-based data search, that is, searching data based on not only metadata but also actual data content will help data users to narrow down selected data. In fact, this may be the only way to find useful data in the future.

There are two types of value-range queries. A simple one, which we term Type I query in this article, is to find areas on which a given geophysical parameter assumes values in a specific range. A type II query, on the other hand, is an extension of type I query by adding area size and missing value ratio constraints. To study a phenomenon in the real world, data values in isolated locations may not make significant contribution even though the values are in the appropriate range. On the other hand, due to limits of instrumentation and corresponding remote sensing algorithms, Earth observing data may not be available in certain cases. The above two factors require the more meaningful type II content-based queries. Note that in this article, "content-based" query and "value-range" query are used interchangeably because the data sets used here contain values of geophysical variables, i.e. values are the main content of the data.

To answer value range queries by directly scanning data values is obviously unrealistic for very large data sets. Approximations based on aggregation are a natural choice. Approximate answers to queries based on aggregation are the basis of On-Line Analytical Processing (OLAP) (Han and Kamber, 2001), which has been studied quite extensively in database communities. The aggregation values are usually created by using the data cube concept (Gary et al., 1996). Histograms, as one kind of aggregation, have been used in databases for efficient estimation of query result size (Poosala et al., 1996). Poosala and Ganti (1999) even use histograms to summarize the aggregation values (contents of data cube) to approximately answer aggregation queries. Here, a data pyramid model and histograms based on the model are described. The histograms are clustered as indices to the database to answer the type I queries (Li et al., 1998; Yang et al., 2001b).

Type II queries can be answered based on Type I query result and extra condition checking. As the first step, areas are constructed under the area size constraint. Instead of a simple KD-tree (Samet, 1990), which may result in searching for unnecessary areas which result from the rectangular ranges, a convex hull algorithm is used to generate the smallest convex domain containing selected data points (Preparata and Shamos, 1985). Furthermore, a point-in-polygon algorithm (Preparata and Shamos, 1985; O'Rourke, 1994) is used to determine whether a point is inside a convex hull. Since the algorithm is computationally intensive, a modified version is used to make the overall computation more efficient (Yang et al., 2001a). The missing value ratio condition is then checked for points inside the constructed convex hulls.

Based on the algorithms for both query types, a prototype system has been developed (Tang et al., 2003), and sample EOS data are used in this system. This prototype is a web-based system. The displaying component has been built using Java and the Scalable Vector Graphics (SVG) (Ferraiolo, 2001), a W3C-recommended standard. Web users can use this system to interactively perform the content-based queries on the sample data. Of course, at this time, SVG viewer plug-in must be installed with a regular web browser for using this system.

The rest of this article is organized as follows. In the next section, Section 17.2, the details of the query algorithms are described. The prototype system is described in Section 17.3. Experiment results are given in Section 17.4. Finally, discussion and future potential work are presented in Section 17.5.

17.2 Method

The purpose of a value range search is to find regions on which parameter values fall in certain ranges with or without additional area size constraints. The speed for answering such a query and the accuracy of the result due to aggregation inversely affect each other. The goal of this work is to rapidly answer value range queries with reasonable accuracies. Instead of finding *all* regions satisfying present conditions, we want to quickly find out not less than a certain percentage (say, 95%) of the total qualified regions. These kinds of approximate queries are termed statistical range queries (Li et al., 1998).

The statistical range queries are answered based on a pyramid data model and aggregated data on that model. The aggregation is achieved by using histograms and clustering procedures. Additional area size conditions in type II queries are satisfied by implementing additional feasible checking algorithms.

17.2.1 Pyramid Model

A data pyramid model (Li et al., 1998) was proposed in the past to answer above statistical range queries. In this data model, the problem space is divided into cells at different levels. At each level, all cells are disjointed and exhaustively cover the entire study space. Each higher level cell covers a space which is the union of a number of lower level cells. Each lower level cell is associated with one and only one upper level cell. The construction of a data pyramid based on this model is bottom-up. Values associated with a level are computed from the values of the immediately lower layer. To answer a statistical range query, the procedure, on the other hand, is a drilling-down (Li et al., 1998). In other words, the procedure is to search the highest level first to find regions in the immediate lower resolution, then, to drill one layer down to search within the selected regions only. By limiting the lower level searches within the selected regions only, one may save a significant amount of computing time.

Following the same strategy, a two-level pyramid is used to avoid the accuracy problem with the existence of multiple levels of the data pyramid (Yang et al., 2001b). In this work, only one lower resolution layer is created above the original higher resolution layer.

17.2.2 Histograms

Earth science data can generally be considered as data with five dimensions, namely three spatial dimensions, one temporal dimension, and one variable dimension. For a specific application, one can define low resolution cells in any selected dimensions. In this work, cells are defined only on 2-D spatial regions, that is, in longitude and latitude. The spatial domain with the original spatial resolution of the given data sets is divided into lower resolution cells. Each of the higher level (lower resolution) cells contains a fixed number of lower level cells.

All higher resolution data values are associated with one lower resolution cell. Aggregation values on the cell can be computed via the higher resolution values. One method for aggregation is creating histograms. One histogram can be created by binning all higher resolution data values corresponding to the same lower resolution cell together. To obtain a histogram with high accuracy and a fixed number of bins, the best way is to use adaptive bin sizes according to the minimum and maximum values associated with this histogram. However, in this application, histograms are clustered based on their similarity, which in turn, requires histograms covering the same value range with the same bin sizes for distance computations. Therefore, global minimum and maximum values for a given data set are chosen to define a universal histogram frame.

17.2.3 Clustering and Type I Query

Although histograms may reduce the data volume, scanning individual histograms for value ranges is computationally expensive. Therefore, histograms are clustered together to further aggregate the data values. Clustering techniques have been used quite extensively in various applications (Everitt, 1993; Jain et al., 1999). This technique is mostly used for data exploration such as pattern classification, image segmentation, etc. The focus of this kind of work is to explore relations among different groups of data (clusters), that is, data mining. Recently, Jain et al. (1999) review clustering methods extensively from a perspective of statistical pattern recognition, providing taxonomy of clustering techniques and their applications such as image segmentation, object recognition, information retrieval, and data mining.

In order to cluster histograms, one may consider a histogram as a one-dimensional array, or a vector of components N where N is the number of bins of the histogram, or an N-dimensional point. The distance between two vectors can be measured with a specific metric. After the distances between all vectors are computed, a clustering algorithm can be used to perform the clustering. Here for simplicity and prototyping purpose, the agglomerative hierarchical method is used. There are no general optimal procedures for clustering, such as dissimilarity metric selection, clustering algorithm, number of clusters, etc. The actual result depends on value distribution of underneath data and the purpose of the clustering. The detailed algorithm for this work and discussion on the clustering parameters are discussed in later sections with the achieved results.

With clusters of cells, each cell belongs to one and only one cluster. The individual histograms in a cluster are summed up to form a representative histogram of the group. The representative histograms record the total counts of data values being in the given bin ranges and index the lower resolution cells. Figure 17.1 demonstrates the processes described above including the pyramid building-up, histogram creation, and clustering. From bottom to top, higher resolution data are compressed by using histograms associating with lower resolution cells. The histograms are clustered together based on their relative "distance" and then the histograms in the same clusters are summed up to give representative histograms, which index the lower resolution cells.

When a type I query is issued, the system computes the count of bottom cells with data values in the particular defined range. The clusters are ordered according to their counts in the representative histograms. For any defined accuracy, a certain number of groups are included in the search. The answer on the higher

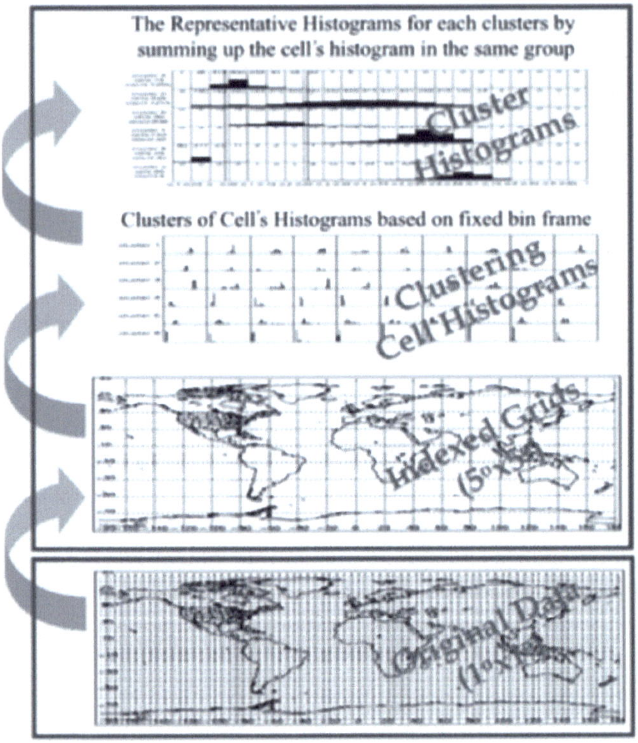

Figure 17.1 Diagrams of the data preprocessing steps

resolution level could be achieved by scanning cells in the selected clusters only. Therefore, the searching time is reduced by searching only a fraction of the total number of cells. The accuracy is defined as the ratio of the selected number of lower level cells to the total number of cells in the given value range. As more and more groups are included in the search, higher and higher accuracy is, obviously, reached. The advantage of this clustering method is that the clusters are ordered. Therefore, clusters of high contribution to the result are included first. Consequently, the accuracy increases quickly with the first few clusters.

17.2.4 Type II Query Algorithms

A type II query can be defined as: To find areas such that on those areas the following conditions is satisfied:

v_{min}„ mean (value)„ v_{max},
size (area)...given size, and
ratio of missing value points„ given percentage.

In this work, the number of grid data points is used as a surrogate for the size of an area. For example, the size of an area is constrained such that it is not smaller than a certain number of grid data points. This substitution is valid only for data sets with equal-area map projections.

For gridded data, the area size condition is actually a combinatorial problem. For example, let us consider a 1 degree × 1 degree resolution global data set which has 64,800 grid data points. If one tries to find areas covering more than 500 data points, the number of candidate areas is ${}_{64,800}C_{500} + \cdots + {}_{64,800}C_{64,800}$. It is impossible to consider all feasible areas in a real computation. This motivates us to present a reliable, feasible method to find areas satisfying the given constraints.

The algorithm of the method is based on depth-first hierarchical clustering method, in which areas are constructed by using a convex hull. The data points within the value range for a given parameter are initially selected. The hierarchical agglomerative clustering technique is applied for the selected data points with the Euclidean distance measure and the complete linkage method. Convex hull, which is a candidate area, is constructed for them. Point-in-polygon algorithm identifies the data points inside the convex hull. The area consists of data points within the given value range, those outside the range, and those with missing values. The constraints for missing ratio, area size, and average value are checked. When the size constraint is not met, the area is discarded. When the size constraint is met but at least one of missing ratio constraint and average value constraint is not met, the procedure is repeated for the selected data points inside the area. When all three constraints are met, the area is finally selected.

It is suggested that the number of clusters at the first level should be larger than that of the rest of the levels. Many feasible areas are found at the first level and a small number of clusters such as two or three is enough to find the areas at the rest of the levels. The convex hull is the smallest area to cover the data points within the range in a cluster. The area seems to be more reasonable than that generated by KD-tree algorithm (Samet, 1990) which may generate unnecessarily large rectangular areas.

Figure 17.2 illustrates how the algorithm is working with four levels. The number of clusters at the first level is large and one of the clusters is chosen here for the illustration. The number of clusters at the second, the third and the fourth level is 3, 2, and 2, respectively. There are four types of data points shown in Fig. 17.2. The symbol • denotes a data point within the value range, the symbol a data point above the maximum of the value range, the symbol Δ a data point below the minimum of the value range, the symbol ∇ a data point beblow the minimum of the value range, and the symbol × a data point with missing value. The hierarchical agglomerative clustering technique is applied to data points within the value range. The convex hull is drawn with a thicker line showing one cluster at the first level. The first number in the sequence identifying a cluster starting from the left denotes the cluster at the first level, the second number from the left denotes the cluster at the second level, the third number from the left denotes the cluster at the third level, and the fourth number from the left denotes the cluster at the fourth level.

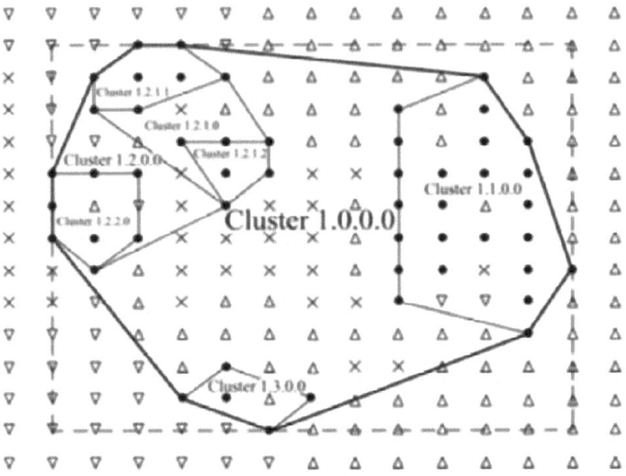

Figure 17.2 Illustration of how the algorithm is working

Assuming that the size constraint for data points is 7, that the missing proportion constraint is 0.2, and that the average value constraint of an area is specified, Fig. 17.2 shows the resulting clusters up to four levels. The area of the first cluster at the first level, Cluster 1.0.0.0, shown here is not supposed to be satisfying the average constraint. Three clusters at the second level are generated using the data points within the value range, • , and they are Cluster 1.1.0.0, Cluster 1.2.0.0, and Cluster 1.3.0.0. Since the data points, • , are considered, the areas of the three clusters at the second level exclude most of data points outside the value range but inside the convex hull of Cluster 1.0.0.0. The three candidate areas appear to be reliable because the areas consist of many data points within the given value range. The area of Cluster 1.1.0.0 is satisfying the first two constraints. If the area meets the average constraint, the area is chosen. Next, the convex hull of Cluster 1.2.0.0 is constructed. The area is satisfying the first two constraints. However, if it turns out not to meet the average constraint, the clustering technique is applied to data points within the value range inside this convex hull. Two clusters, which are Cluster 1.2.1.0 and Cluster 1.2.2.0, are then generated. The area of Cluster 1.2.1.0 is satisfying the first two constraints. If it does not meet the average constraint, the clustering technique is applied to the data points inside this convex hull and clusters at the next level are then generated. They are Cluster 1.2.1.1 and Cluster 1.2.1.2. The area of Cluster 1.2.1.1 is satisfying all constraints because the number of data points is 8 and the values of all data points in this convex hull are within the value range. The area of Cluster 1.2.1.2 does not meet the size constraint so that it is discarded. Next, Cluster 1.2.2.0 is considered. The area of Cluster 1.2.2.0 is satisfying the first two constraints, since the number of data points is ten and there are no missing data points. If the area is satisfying the average constraint, the area is chosen. Finally, Cluster 1.3.0.0 is considered. Since the area of this cluster does not meet the size constraint, the area is discarded. All clusters generated have been considered so that the algorithm stops at this point.

17.3 Prototype System

A web-based prototype system is developed based on the algorithms described in Section 17.2. Figure 17.3 shows the architecture of the prototype system. Standard protocols and implementation are considered in this design so that third party components could be easily integrated when they are available. The prototype involves software developments on both clients as well as on servers. In other words, there are data servers and application servers on the server side. Presently, only selected client and server components are developed as denoted by clean boxes in the architecture figure and will be explained below.

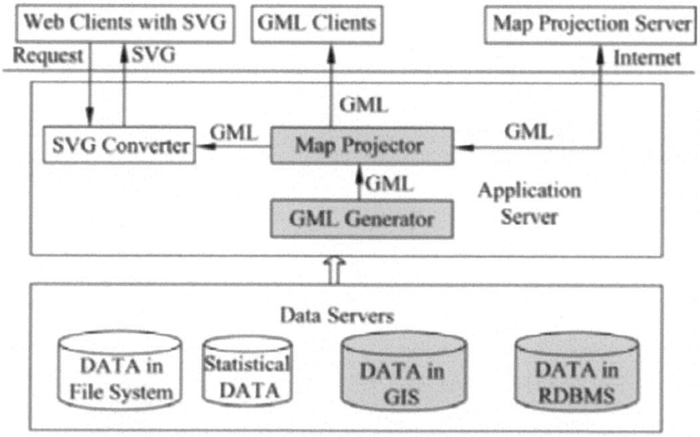

Figure 17.3 Architecture of the prototype system

The data servers provide original data and aggregation data created for answering content-based queries. The data also include information for cell division, representative histograms and the corresponding indices. The original data could be in a traditional relational database management system (RDBMS) or in a Geographic Information System (GIS). In both cases, the aggregation data should be stored inside the corresponding systems along with the original data. In the current implementation, original data are in a data file system. The corresponding aggregation data are also stored in the same file system.

The application server consists of two major parts. The query processor is for receiving queries from the clients, parsing the queries, and invoking the data servers for data, and processing the data. After the data are created, the SVG converter will convert the output result directly into SVG documents so that users with an SVG-enabled web browser can browse the results easily. Although not implemented, a Geography Markup Language (GML); (Lake and Cuthbert, 2000) generator has been conceptualized in the system because third party applications could be used to enhance the functionality users to receive browse data in specific projection but the reprojection will be realized by using third party standard products. Right now, only a simple version of Map Projector has been implemented and it supports a linear scalar projection as default as well as the Plate-Carree Projection (Pearson, 1990). This Map Projector converts the retrieved data from searching engine to the designated projection, which later is used by the SVG converter to generate an SVG-formatted file.

A special web-based client with Java applet is also developed to support the content-based queries. With this client shown in Fig. 17.4, users can issue queries

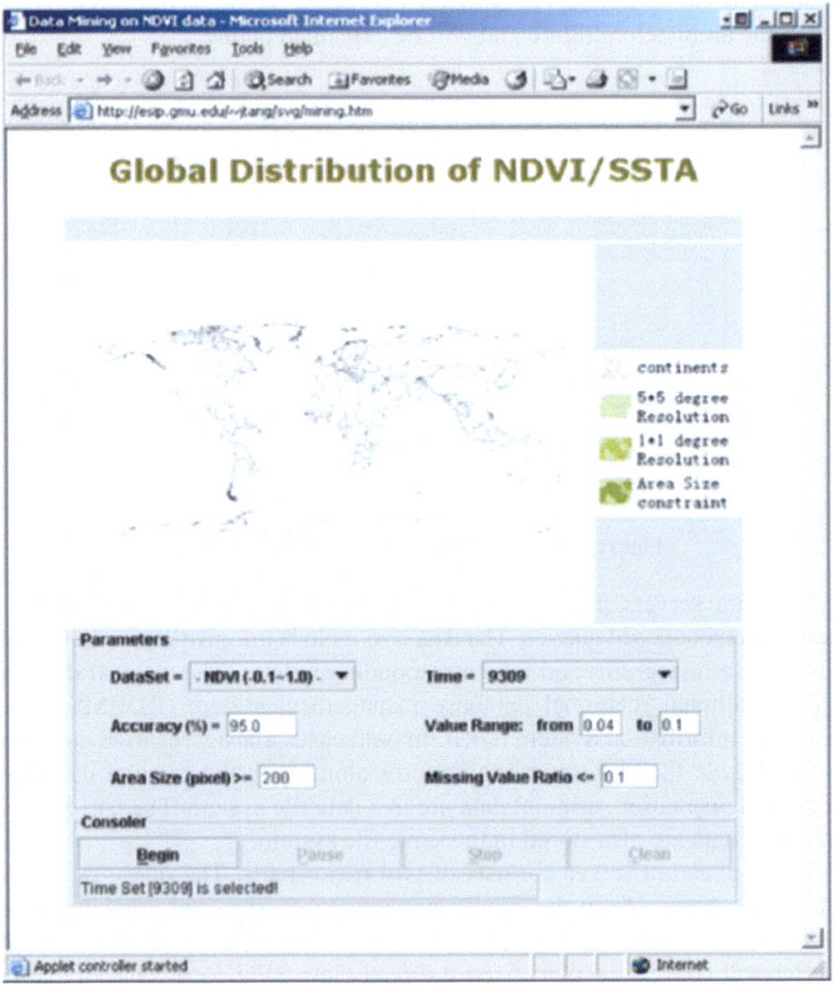

Figure 17.4 Client interface of the prototype system

by selecting a geophysical variable and inputting control parameters. The Java and SVG-enabled client will also communicate with the application server for query requests and results and will display the query results. In order to illustrate how the accuracy limit has been satisfied, this prototype system returns these candidate cells per cluster as intermediate results back to the client side. As the searching algorithm continues, the candidate pixels selected are returned to the client side cluster by cluster. If the query is a type II query, the third party algorithm is called after the pixels are selected and it outputs convex polygons with satisfying area size and missing value proportion constraints. The above steps can be viewed through the Java and SVG-enabled client phase by phase, leaving to users the ability to start, pause, resume, and terminate the query session as well.

17.4 Results

Although for this study a workable system is our temporary goal, many experiments must be carried out to archive "optimal" system performance. Therefore, before the results of the prototype are illustrated, more detailed information on the data and algorithm and parameter selections is presented.

17.4.1 Data and Pyramid Structure

A well-known Earth science data set, the gridded global monthly Normalized Difference Vegetation Index (NDVI) data set (Kyle et al., 1998) is chosen as an example. NDVI values are derived from the National Oceanographic and Atmospheric Administration (NOAA) polar orbiter AVHRR instruments by normalizing the difference of infrared and red reflectance measurements (Cracknell, 1997). The NDVI values provide a strong vegetation signal and are used widely for classification of vegetation. Though NDVI at a given time alone is not sufficient to provide vegetation classification, temporal profiles of NDVI do help for carrying out the classification (Cracknell, 1997). In the classification process, finding areas in which NDVI values fall in a certain range, in particular when associated with a sufficient large area, is in itself useful. At present, a data pyramid is created in spatial dimensions only, therefore, to a fixed time, and August, 1981, was randomly chosen.

The NDVI data we are using are of 1 degree × 1 degree spatial resolution in both longitude and latitude. The spatial coverage of the data set is global land. The original resolution areas of 1 degree × 1 degree are taken as the higher resolution (lower level) cells. Above this layer, 5 degree × 5 degree cells are created as the higher level (low resolution). For each higher level cell, there are 25 associated cells in the lower level. The 25 NDVI data values on each lower level cell are mapped into a fixed bin frame to form a histogram. In this case, we use 22 bins of width 0.05 covering the value range from – 0.1 to 1.0. The mapping result is a histogram for each higher level cell. Since the NDVI data cover only global land, there are many lower resolution cells which are not associated with any valid NDVI value such as over large area of water. Therefore, no histogram is built up for such an empty cell. Even so, due to the same reason, some histograms are composed of less than 25 values because the corresponding higher level cells cover both land and water.

Other data sets are also investigated in this study. For such data, the spatial and temporal resolutions are the same but the valid spatial areas are different for evaluating the usefulness of the algorithms.

17.4.2 Clustering Criteria

The hierarchical clustering method is performed because of its simplicity and its intuition. The clustering process is based on a dendrogram built on the distances between all the histograms as shown in Fig. 17.5. By viewing this dendrogram, one may think it is easy to divide the data into groups. Each group contains histograms which are close to each other. As a matter of fact, there is no practical optimal method for choosing the number of clusters. The dendrogram works only as a reference for one to identify the number of clusters. The choice of cluster number is problem dependent. For the problem here, clustering as a data compression (aggregation) mechanism is used to reduce the time required in answering value-range queries with assured accuracy. For that reason, a quantitative definition of time saving ratios should be defined first.

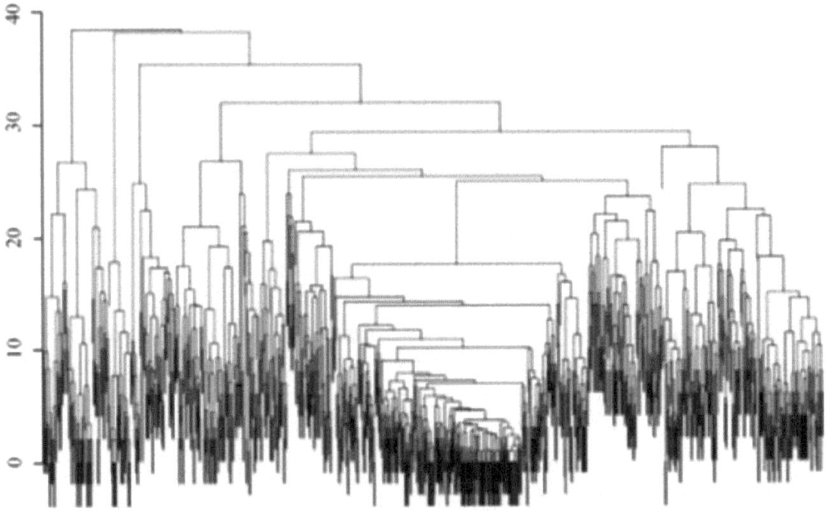

Figure 17.5 Dendrogram illustrating a tree structure

17.4.2.1 Time Saving Ratios

Time savings result from two effects. One is from the number of empty cells. Since NDVI covers only land, most of the lower level cells are not associated with any valid NDVI values (oceans). Such cells are termed empty cells. Other cells are then non-empty cells. For this example, there are a total of $360 \times 180 = 64,800$ lower level cells. However, there are only 15,722 non-empty cells. Similarly, there are $72 \times 36 = 2,592$ high level cells, only 887 of which are non-empty. When the index system is built, the empty cells are dropped from the problem. Therefore, searching the cells by indices will automatically ignore the empty cells and, correspondingly, save the searching time. The time saving ratio is defined as the ratio of the number of the selected cells to the total cell number in percentage. If the empty cells are included in the total, the time saving ratio is termed "saving type 1." If empty cells are eliminated in the counting, the saving ratio is termed "saving type 2," which should be smaller than "saving type 1" when empty cells appear in the problem.

17.4.2.2 Effect of Number of Clusters

An example query, to find all high resolution cells on which the NDVI values are between 0.04 and 0.26 with 95% accuracy, is used to study the influence of cluster number on the query result. Table 17.1 illustrates the results of this test. The first column is the number of total clusters chosen for the problem. The second column gives the number of selected clusters to obtain at least 95% accuracy. Since the accuracy is not improved continuously, the real accuracy we achieve by including the selected number of clusters should be higher than 95%. These real accuracies are listed in column 3 in percentage. The last two columns register the time savings with/without including empty cells, respectively.

Table 17.1 Dependence on number of clusters

Total	Selected	Accuracy	Saving 1	Saving 2
3	2	99.5	66.8	2.93
4	3	99.5	66.8	2.93
5	3	99.3	70.5	13.6
6	3	98.8	77.5	34.2
7	3	98.8	77.5	34.2
8	3	98.6	78.7	37.7
9	3	98.6	78.7	37.7
10	4	98.6	78.7	37.7

It is interesting to note that the results are not very sensitive to the number of clusters. Moreover, for some number ranges, say (3,4), (6,7) and (8,9,10), the time saving result is the same. The reason is that the change of cluster number does not affect the contents of clusters with values concentrated in the given query value range. Another point is that the time saving, in particular, the time saving without including the empty cells increases with the cluster number. This is not a surprising result because more clusters would result in more accurate distinction of value ranges. Therefore, fewer high level cells need to be scanned for a given value range. Of course, since more clusters would consume more resources in a real system for answering given statistical range queries, only a few clusters for a real problem should be included.

17.4.3 Type I Query Processing Procedure

Based on the results from above experiments, six clusters and the Manhattan metric are chosen in building up the representative histograms and indices. The query execution procedures are detailed here to explain the advantage of the algorithms. The same example, to find all high resolution cells on which the NDVI values are between 0.04 and 0.26 with 95% accuracy, is used again. The processing summary is listed in Table 17.2.

Table 17.2 Query result. All values except the first two columns are percentage values

ID	Cells	Count	Accuracy	Saving 1	Saving 2
4	76	40.84	40.84	97.07	91.43
2	458	32.48	73.32	79.4	39.8
5	50	25.46	98.78	77.47	34.16
1	182	0.56	99.34	70.45	13.64
6	26	0.5	99.84	69.44	10.71
3	95	0.16	100	65.78	0

First, the contribution from each cluster is computed based on its representative histogram and the clusters are sorted based on the resulting contributions. In this case, the descending order is clusters 4, 2, 5, 1, 6, and 3. The order is associated with the count of values in the given range in descending order. The contribution from each cluster, that is, the percentage of values in the given range for the cluster, is listed in column 3 as "count." The contribution from the current cluster and those from previous clusters are summed up in the order, and the result is the accuracy. From Table 17.2, one can find that after including only 3 clusters, one has obtained accuracy 98.78%. In other words, all cells in the first 3 cluster are associated with more than 98.78% NDVI values in the given range. By scanning on those selected cells, one is guaranteed to have more than 95% searched values. The numbers of involved lower resolution cells with each cluster is listed as "cells" in column 2, and the time saving ratios are computed based on data in this column.

It is obvious that the accuracy increases with the number of clusters included in the search. If we propose a real statistical range query, say with 95% accuracy, we should reach the desired answer after including only three clusters. The contribution from the last three clusters is of very small percentage. As a result, the computing time is saved by working on only the first three clusters of cells. As more and more clusters are included, the time saving from clustering drops very quickly. Of course, when all clusters are considered, no time saving at all results from the clustering method. The key point is that only a few clusters would usually give a very high accuracy. The last column in the table indicates that the time saving is significant even without including empty cells.

Another query on the same parameter, NDVI, but with a different value range, [0.56, 0.78], is performed to see the dependence on value ranges. The result is demonstrated in Table 17.3. Since the searching order is computed according to the value range, this order is of course different from the order in Table 17.2. However, the efficiency is almost the same. After including three clusters of cells in the search, the accuracy is already more than 99%. If the given accuracy is 95%, the time saving should be 71% and 17% for the situation with and without including empty cells, respectively.

Table 17.3 Query result for NDVI values between 0.56 and 0.78

ID	Cells	Count	Accuracy	Saving 1	Saving 2
1	182	47.35	47.35	92.98	79.48
3	95	37.99	85.35	89.31	68.77
2	458	14.44	99.79	71.64	17.14
5	50	0.19	99.98	69.71	11.5
4	76	0.02	100	66.78	2.93
6	26	0	100	65.78	0

As discussed before, ignoring empty cells contributes to the time saving for answering statistical range queries along with the histogram data aggregation. To constrain the contribution of empty cells, one should use algorithms developed here to a data set without empty cells (except for a few missing values). One such data set is a 1×1 global monthly mean surface skin temperature data (Kyle et al., 1998). The data format and structure are the same as those of NDVI data we used. Table 17.4 shows an example result for a query range between 250 and 280 K. Here, time saving is fully from the clustering method itself, and the amount is very significant. As with the parameter NDVI, the effect of the cluster number was also studied. The general trend is the same.

The experiment results on type I content-based queries are used in determining system parameters. As the result, in the operative prototype system, representative

Table 17.4 Result for data without empty cells

ID	Cells	Count	Accuracy	Saving 1	Saving 2
4	481	36.86	36.86	81.44	81.44
5	290	35.93	72.79	70.25	70.25
1	541	26.94	99.73	49.38	49.38
2	318	0.23	99.95	37.11	37.11
3	670	0.05	100	11.27	11.27
6	292	0	100	0	0

histograms are built with six clusters for each variable. The Manhattan metric is applied in the distance computation. With all aggregation data and parameters ready, the prototype system is ready to accept and to process content-based queries, essentially, type II queries.

Although the current system supports more than one physical variable, the query on NDVI is chosen as an example. Before making a query, a user needs to make following selections: a variable, NDVI; a specific time; an accuracy for the approximation, say 95%; a value range, 0.04 NDVI 0.26; an area size condition, which is the number of total points being greater or equal to 200; and less than 15% missing value ratio. After the user starts the query by clicking the "Begin" button as shown in Fig. 17.4, the system will return the results step by step as those shown in Fig. 17.6.

As the first step, the server parses the query, perform the computation on the representative histograms, and return the selected cells to the users. Data transferring from the server to the client is cluster by cluster. Figure 17.6 (a) displays all cells selected in three clusters together.

The step-wise procedure continues on higher resolution data. In this step, the server checks data values in selected cells only and returns the selected data points associated with data values in the given range. Figure 17.6 (b) is the result after all higher resolution cells are counted.

17.4.4 Results from the Prototype System

By now, the type I query is answered. If the query is a type II query, the algorithms for checking area sizes and missing value ratios are employed on the selected higher resolution data points. The output of this stage is convex hulls satisfying the additional constrains. The output procedure is still step by step, i.e. polygon by polygon in this case. Figure 17.6 (c) shows the final results. One may easily notice that there exist small isolated areas over which the NDVI values falling in the given range but not encompassed by any polygon due to the violation of the area size conditions.

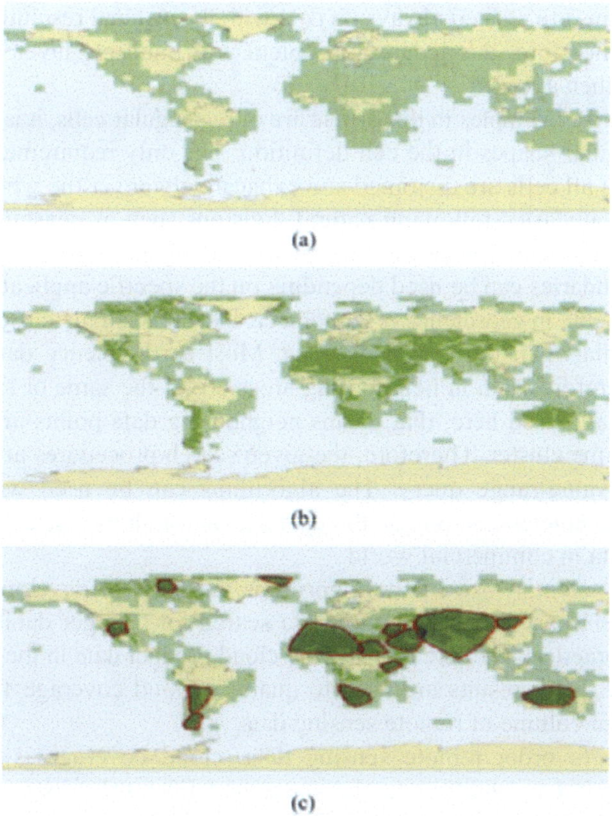

Figure 17.6 Searching results from the prototype system (a) candiate cells, (b) result pixels, (c) result regions

17.5 Conclusions and Future Work

The methods described in this article works well for the gridded Level-3 EOS data. The clustering method based on histograms significantly reduces the time needed to answer a value-range content-based query. Certainly, the trade-off is that the answers are approximate only. With the area size and missing value ratio conditions, such approximate answers may be more reasonable for real world applications. The prototype system provides a user-friendly interface for casual users to conduct content-based queries.

The time saving rates depend on the data compression ratio if the histograms are considered as a lossy data compression algorithm. When the original data are of high resolution such as data from the well known Moderate Resolution Imaging Spectroradiometer (MODIS) with 1 km by 1 km spatial resolution, the compression ratio is much larger than that with 1 degree (latitude) by 1 degree (longitude) data.

Since histograms are created always on data with the original resolution, no middle layers exist, therefore, the accuracy problem with multiple layer drilling-down algorithms is then avoided.

Although the examples in this article are of rectangular cells, it is not necessary to use any regular shapes in the cell definition. The only requirement for the cell division is that all cells are disjoined and exhaustively cover the whole study area. As a result, boundaries following natural divisions such as coastal lines and the climate division boundaries or even the political boundaries such as country, state, or county boundaries can be used depending on the specific applications.

One principle in geoscience is that objects that are spatially close to each other are more similar than those at a distance. Most earth science data follow this principle. Therefore, data in neighboring areas are of the same or similar values. Use the language used here, that means neighboring data points are more likely to be in the same cluster. Therefore, the given search procedures are efficient for answering a value-range query. The algorithms can be used with any other geophysical parameters as long as the data are not randomly distributed as those transaction data in commercial world.

Another potential use of the algorithms and system developed here is to treat the content-based search as metadata search for a larger database to refine queries. One practical example is to use the cloud product data in the data pyramid. The statistical query results are used to quantify cloud coverage for supporting queries of large volume of remote sensing data.

When users order remote sensing data, cloud coverage is an important selection criterion in addition to the space and time selections. For example, users working on land products or ocean surface products want to obtain remote sensing images without cloud cover or with limited percentage of cloud coverage. On the other hand, people who are studying cloud are more interested in retrieving data only with high cloud coverage. To support such "content-based queries", one can use cloud detection algorithms (Jedlovec and Laws, 2001) on each image before fetching back the image. This approach is necessary for a real time system but obviously not efficient for systems supporting historical research data. Another strategy is to give a cloud coverage measure for each granule as an attribute and make the cloud coverage values part of the query criterion. Nevertheless, when a sophisticated data server supports subsetting and mosaicing functions, a fixed value for each granule is not sufficient. In that case, the technology described above could provide a more useful algorithm for such a system. Of course, there are some challenges to be overcome before a practical system is developed based on the algorithms.

One challenge for a practical system is to have the right data to populate the data pyramid. The solution evidently depends on specific applications. Suppose a data system supports data search and access on data retrieved from measurements by the NASA Terra and Aqua satellites, the standard cloud data products such as cloud mask from MODIS, MOD35 (Ackerman et al., 2003) and values of cloud cover fraction in MOD06 (Menzel et al., 2002) can be used. For other purposes, even cloud data from models can also be helpful for creating the data pyramid.

One technical hurdle is an efficient algorithm for the histogram creation. When the data are of high resolution, one cannot count all data values in a selected cell. One efficient and reasonable technique is sampling. Sampling was used in the general databases for online aggregations (Hellerstein et al., 1997; Hass, 1999). It should be more efficient on earth science data because of the closeness feature depicted above.

Temporal variation of data may represent another obstacle. Because the time scale of cloud presentation is not very long, high temporal resolution, in hours, should be resolved. It is not expected that data pyramid is built in time dimension. In spite of that, natural variations of certain degree of accuracy such as seasonal variation and even diurnal variation could be utilized to achieve better cluster division.

The clustering algorithm itself may require improvement for larger data sets. New algorithms (Jain et al., 1999) are studied with an operational system under consideration. It should be pointed out that with initial clusters and representative histograms, the clustering procedure will become a classification procedure, which is computationally more efficient.

Other software components such as database management systems could be incorporated into such a system to reduce the development workload and to improve performance. Since the "content-based search" is a component in a large data server, communications between this component with other components must be efficient and easy to implement. That is why standard geoinformation presentation such as GML was counted in the design of the prototype.

Acknowledgements

We acknowledge partial prototype funding support from the NASA ESDIS Projects (NAG5-8607 and NAG5-10177), from the Earth Science Enterprise WP-ESIP CAN Program as well as from George Mason University.

References

Ackerman S, Strabala K, Menzel P, Frey R, Moeller C, Gumley L, Baum B, Seeman SW, Zhang H (2003) Discriminating Clear-Sky From Cloud With Modis Algorithm Theoretical Basis Document (Mod35)

Cracknell AP (1997) The Advanced Very High Resolution Radiometer. Taylor & Francis Inc

Everitt B (1993) Cluster Analysis. John Wiley & Sons

Ferraiolo J (2001) Scalable Vector Graphics (SVG) 1.0 Specification

Gary J, Bosworth A, Layman A, Pirahesh H (1996) Data Cube: A relational aggregation operator generalizing group-by, cross-tabs, and sub-totals. In: Proceedings of IEEE Conf on Data Engineering pp. 152 – 159

Haas PJ (1999) Techniques for online exploration of large object-relational databases. In: Ozsoyoglu ZM, Ozsoyoglu G, Hou W-C (eds) Proceedings of the 11th International Conference on Scientific and Statistical Database Management. IEEE Computer Society pp. 4 – 12

Han J, Kamber M (2001) Data Mining: Concepts and Techniques. Morgan Kaufmann Publishers

Hellerstein JM, Haas PJ, Wang HJ (1997) Online aggregation. In: Proce 1997 ACM SIGMOD Intl Conf Management of Data ACM Press pp. 171 – 182

Jain AK, Murty MN, Flynn PJ (1999) Data clustering: a review. ACM Computing Surveys 31(3): 264 – 323

Jedlovec GJ, Laws K (2001) Operational Cloud Detection in GOES Imagery. In: 11th Conference on Satellite Meteorology and Oceanography pp. 3 – 57 (American Meteorological Society, Madison, WI)

King MD, Greenstone R (eds) (1999) 1999 EOS Reference Handbook. NASA (Washington, DC)

Kyle H, McManus J, Ahmad S et al. (1998) Climatology Interdisciplinary Data Collection, Volumes 1 – 4, Monthly Means for Climate Studies. NASA Goddard DAAC Science Series, Earth Science Enterprise, National Aeronautics & Space Administration, NP-1998(06)-029-GSFC

Lake R, Cuthbert A (2000) Geography Markup Language (GML) V1.0. eds

Li Z, Wang XS, Kafatos M, Yang R (1998) A pyramid data model for supporting content-based browsing and knowledge discovery. In: Rafanelli M, Jarke M, (eds) Proceedings of the 10th International Conference on Scientific and Statistical Database Management. IEEE Computer Society pp. 170 – 179

Menzel WP, Baum BA, Strabala KI, Frey RA (2002) Cloud Top Properties and Cloud Phase Algorithm Theoretical Basis Document

NASA (2003) Earth Observing System Data Gateway. URL: http://redhook.gsfc.nasa.gov/imswww/pub/imswelcome/

NASA (GCMD) (2003) NASA's Global Change Master Directory (GCMD). URL: http://gcmd.nasa.gov/

O'Rourke J (1994) Computational Geometry in C. Cambridge University Press

Pearson F (1990) Map Projection: Theory and Applications. CRC Press, Inc

Poosala V, Ganti V (1999) Fast Approximation Answers to Aggregate Queries on a Data Cube. In: Ozsoyoglu ZM, Ozsoyoglu G, Hou W-C (eds) Proceedings of the 11th International Conference on Scientific and Statistical Database Management. IEEE Computer Society pp. 24 – 33

Poosala V, Ioannidis Y, Haas P, Shekita E (1996) Improved Histograms for Selectivity Estimation of Range Predicates. In: Proceeding 1996 ACM SIGMOD Intl.Conf.Management of Data. ACM Press pp. 294 – 305

Preparata FP, Shamos MI (1985) Computational Geometry. Springer-Verlag

Samet H (1990) Applications of Spatial Data Structures. Addison-Wesley Publishing Company

Tang J, Yang R, Yang K-S, Kafatos M (2003) Prototype of a value-range query technique on earth science data. In: Dimitrios Gunopulos SN (ed) Proceedings of the 15th International Conference on Scientific and Statistical Database Management. IEEE Computer Society pp. 229 – 232

Yang R, Kafatos M, Wang X (2002) Managing Scientific Metadata Using XML. IEEE Internet Computing 6(4): 52 – 59

Yang K-S, Yang R, Kafatos M (2001a) A feasible method to find areas with constraints using hierarchical depth-first clustering. In: Kerschberg L, Kafatos M (eds) Proceedings of the 13th International Conference on Scientific and Statistical Database Management. IEEE Computer Society pp. 257 – 262

Yang R, Yang K-S, Kafatos M, Wang X (2001b) Value-range queries on Earth science data via histogram clustering. In: Hornsby K, Roddick JF (eds) Interim Proceedings of International Workshop in Temporal, Spatial and Spatio-Temporal Data Mining. TSDM2000 Lyon, France, Lecture Notes in Artificial Intelligence 2001 Springer

18 The Impact of Precipitation and Temperature on Net Primary Productivity in Xinjiang, China from 1981 to 2000

Wei Gao, Zhiqiang Gao, Xiaoling Pan, James Slusser, Hamid Yimit and Yingjun Ma

18.1 Introduction

The response process and mechanism research of terrestrial ecosystem on global changes (including climate changes and human activities) are important components of research on global changes. The global warming and intensification of human activities over the past 100 years have aroused great attention of scientific communities and policymakers to the impact of climate changes on the internal process and feedback of terrestrial ecosystems, as well as research on ecological security under future climate scenarios (Cambridge University Press, 2000). The continuing rise in atmospheric CO_2 is considered as a main cause of the future changes in global climate (Melillo et al., 1993; McGuire and Melillo, 1995; Cao and Woodward, 1998a). Predicted climate changes include an increase in mean annual air temperature and alterations in precipitation pattern and cloud cover. Elevated atmospheric CO_2 and climate changes are expected to influence the ecosystems (Riedo et al., 2000). Terrestrial Net Primary Productivity (NPP) is regarded as the greatest annual carbon flux from the atmosphere to the biosphere and is considered as a main cause of seasonal fluctuations in atmospheric CO_2 concentrations (Law and Waring, 1994; Runyon and Waring, 1994; Hall and Ojima, 1995; McGuire and Melillo, 1995; Churkina and Running, 1998). The terrestrial biospheric productivity plays a very important role to the life of humankind and all heterotrophic organisms on Earth (Melillo et al., 1993; Gilmanov and Walter, 1995). It not only provides potential food resources, but a source of wood for construction, fabrication, and fuel. However, increased population and industry have potentially adverse effects on biospheric productivity. Consequently, estimating terrestrial primary productivity and monitoring its changes will be important in detecting the state of biospheric health (Churkina and Running, 1998). Spatial and temporal patterns of terrestrial vegetation have been closely linked with climate patterns (Schultz and Halpert, 1993). The vegetation's response to precipitation and temperature has been detected from seasonal and interannual variations in satellite measures of terrestrial vegetation (Goward and Prince, 1995; Braswell et al., 1997). It is needed to estimate quantitatively the relation between terrestrial vegetation and precipitation and temperature. Many models have been developed to study the responses in terms of primary productivity (Emanuel and Killough, 1984; Prince, 1991; Starfield and Chapin, 1996; Haxeltine and Prentice, 1996; Friend et al., 1997). While biogeochemical models address the problems related to primary productivity response in light of detailed eco-physiological mechanisms, they tend to simplify the structural aspect of responses (Gao et al., 2000a, 2000b). It has also become possible to investigate the global spatial and temporal dynamics of NPP by a combination of ecosystem process modeling and monitoring by remote sensing. Satellite observations provide a useful and powerful database for evaluating the dynamics of terrestrial biosphere because of its global coverage and short revisit interval (Cramer, 1999). NPP is a quantifiable characteristic of the biosphere and is a fundamental ecosystem variable modeled at the global scale. Practical considerations for estimating NPP exist in its utility to measure crop yield, forest production and other economically and socially significant products of vegetation growth. In this study, we used Global Production Efficiency Model (GLO-PEM) (Prince and Goward, 1995; Goetz et al., 2000; Cao et al., 2003a,b), comparing with Carbon Exchange in the Vegetation-Soil-Atmosphere system (CEVSA) (Cao and Woodward, 1998a, 1998b; Cao et al., 2003b), estimating interannual variations and spatial patterns of NPP caused by precipitation and temperature in Xinjiang between 1981 and 2000.

18.2 Material and Methods

18.2.1 Study Area

Xinjiang (70.104°E to 97.08°E, and 33.339°N to 49.925°N, see Fig. 18.1) is located in northwest China with an arid climate. The inhabitation of the human species along the Alluvial Fan River in the oasis areas reflects upon the typical oasis agriculture of the area. On the high elevations of mountains the distributions of ice and snow are accumulated in a vast area, which provide the required water resources for the oasis areas. As the endemic landscape and peculiar ecosystem types in arid land, the oasis is the basic habitation for human existence and development in an arid area. The mean annual precipitation varies from 8.2 mm in the desert to 581 mm in the Ili River Valley. The mean annual temperature ranges from – 25.2°C above the snow line to 15.8°C in desert. The elevation varies from – 159 m of Turpan Basin to 8,113 m of the Pamirs. The majority of this area consists of barren and sparsely vegetated land (62%). The croplands and urban areas are distributed in the oasis areas along the Tianshan Mountains (4%). There are forestlands (2%) and grasslands (30%) scattered in the area. The water and glacier cover about 2% – 3% of the area.

Figure 18.1 Geographical location of Xinjiang and the distribution of land cover types in Xinjiang

18.2.2 The NPP Estimation with GLO-PEM Model

NPP was simulated using the GLO-PEM, a semimechanistic plant photosynthesis and respiration model driven entirely with satellite Advanced Very High Resolution Radiometer (AVHRR) observations. These observations included both the Normalized Difference Vegetation Index (NDVI) and meteorological variables. GLO-PEM consists of linked components that describe processes of canopy radiation absorption, utilization, autotrophic respiration, and the regulation of these processes by environmental factors such as temperature, water vapor pressure deficit, and soil moisture:

$$\mathrm{NPP} = \sum_t \left[(S_t N_t) \varepsilon_{\mathrm{g}} - R \right]. \tag{18.1}$$

S_t is the incident PAR in time t. N_t is the fraction of incident Photosynthetically Active Radiation (PAR) absorbed by vegetation canopy calculated as a linear function of NDVI. ε_{g} is the light utilization efficiency of the absorbed PAR by vegetation in terms of gross primary production. R is autotrophic respiration calculated as a function of standing above-ground biomass, air temperature, and photosynthetic rate. Detailed descriptions of the algorithms for these variables are given in (Prince and Goward, 1995; Goetz et al., 2000; Cao et al., 2003a).

Plant photosynthesis depends on the capacity of the photosynthetic enzyme system to assimilate CO_2 and stomatal conductance to CO_2 from the atmosphere into the intercellular spaces. Both of which are affected by environmental factors, such as air temperature, water vapor pressure deficit, soil moisture and atmospheric CO_2 concentration. Therefore, ε_g is given as

$$\varepsilon_g = \varepsilon_g^* \delta \tag{18.2}$$

where ε_g^* is the maximum possible efficiency implied by photosynthetic enzyme kinetics that is affected by, among other factors, photosynthetic pathway, temperature and the CO_2/O_2 specificity ratio. δ is the reduction of ε_g^* caused by environmental factors that control stomatal conductance.

$$\sigma = f(T) f(\sigma_q) f(\sigma_\theta) \tag{18.3}$$

where $f(T)$, $f(\sigma_q)$ and $f(\sigma_\theta)$ represent the stresses of air temperature, water vapor pressure deficit and soil moisture to stomatal conductance, respectively. The temperature stress is estimated based on the assumption that plants are tuned through evolution to grow efficiently at the prevailing temperature. $f(T)$ reaches the maximum at the optimum temperature and tapers off for warmer or cooler temperatures.

$$f(T) = \frac{(T - T_{min})(T - T_{max})}{(T - T_{min})(T - T_{max}) - (T - T_{opt})^2} \tag{18.4}$$

T is air temperature (°C) and T_{min}, T_{opt} and T_{max} are the minimum, optimum and maximum temperatures for photosynthesis. We define the T_{min} and T_{max} as – 1°C and 50°C for C3 plants and 0°C and 50°C for C4 plants, respectively. The optimum temperature is estimated using the long-term, mean temperature for the growing season.

The water status of leaf and soil are critical in controlling stomatal conductance and hence photosynthetic rates. As the air dries, stomata progressively close for protecting the leaf from desiccation and so $f(\sigma_q)$ decreases.

$$\begin{aligned} f(\sigma_q) &= 1 - 0.05\sigma_q & 0 < \sigma_q \leq 15 \\ &= 0.25 & \sigma_q > 15 \end{aligned} \tag{18.5}$$

$$\sigma_q = Q_w w(T) - q \tag{18.6}$$

where σ_q is specific humidity deficit (g • kg^{-1}), $Q_w w(T)$ is the saturated specific humidity at the air temperature, q is the specific humidity of the air. Soil moisture affects stomatal conductance through both hydraulic and non-hydraulic linkage between roots and leaves even if the leaf water status remain unchanged, stomatal conductance decreases with decreases in soil moisture in response to increases in abscisic acid in roots.

$$f(\sigma_\theta) = 1 - \exp[0.081(\sigma_\theta - 83.03)] \tag{18.7}$$

σ_θ is the soil moisture deficit (mm) in the top 1.0 m of soil. Soil moisture deficit is calculated by establishing saturated water content from soil texture and subtracting actual water content (Woodward et al., 1995).

Satellite data sources and processing were described in Cao et al. (2003a). GLO-PEM was driven with the Pathfinder AVHRR Land (PAL) data at resolutions of 8 km and 10 days derived from channels 1, 2, 4 and 5 of AVHRR sensors aboard the NOAA-7, 9, 11, and 14 satellites (James and Kalluri, 1994). Radiances in channels 1 and 2 and brightness temperatures in channels 4 and 5 were used. While the effects of Rayleigh scattering and ozone were corrected in producing the Pathfinder data sets, cloud screening using the visible and thermal channels, and water vapor correction to the visible and near-infrared channels, using the two thermal channels, were carried out in GLO-PEM. Total column water vapor was estimated from the Television Infrared Observation Satellite (TIROS) Operational Vertical Sounder (TOVS) Pathfinder Path A data set. The equatorial crossing time of AVHRR changed substantially both within and between the satellite platforms, the orbital drift effects on the thermal bands were corrected here using the approach of Gleason et al. (2002). The incident PAR was obtained from the International Satellite Cloud Climatology Project (ISCCP). Monthly atmospheric CO_2 concentrations came from the measurements in the Mauna Loa Observation, Hawaii. The data of precipitation were derived from Global Precipitation Climatology Program (GPCP) data sets, inferred from infrared and microwave satellites.

18.2.3 The NPP Estimation with CEVSA Model

CEVSA was used to calculate NPP in this study to provide an independent estimate to evaluate the results of GLO-PEM. CEVSA consists of four modules: a biophysical module calculates the transfers of radiation, water, and heat; a plant growth module describes photosynthesis, autotrophic respiration, carbon and nitrogen allocation and accumulation among plant organs, Leaf Area Index (LAI) and litter production; and a soil module simulates decomposition of organic carbon and the transformations, inputs, and outputs of nitrogen in the soil. CEVSA calculates rates of plant photosynthesis, stomatal conductance, and autotrophic respiration for determining NPP. Detailed description about CEVSA was provided by Cao et al. (2003a; 2003b). We ran CEVSA with observation-based data of climate, atmospheric CO_2, and vegetation distribution at a spatial resolution of

8 km and a time-step of 10 days. Observed meteorological data were collected from local weather stations and interpolated spatially in ARC/INFO environment to produce needed spatial and temporal resolution.

18.3 Results and Discussion

18.3.1 A Comparison Between Estimated NPP with the GLOPEM and CEVSA

The NPP distribution over Xinjiang was simulated by both GLO-PEM and CEVSA models (Fig. 18.2) and interannual variations in NPP were compared (Fig. 18.3). The interannual variations of NPP estimated by GLO-PEM were close to the results of CEVSA. There were relatively large differences between two estimated NPP values in 1985, 1993, and 1996. The correlation between them was 0.47 (P<0.01). The spatial distributions of NPP simulated from both models also agreed generally well. The Ili River Valley and the forest areas of the Altay Mountain and the Tianshan Mountains showed the maximum NPP values. NPP decreased over oasis cropland areas along Tianshan Mountains. The smallest NPP values were showed in surrounding grasslands and desert areas of the Tarim Basin and the Junggar Basin. The mean annual NPP calculated from GLO-PEM model was 27 gC/m^2 with a deviation of 52 while the mean annual NPP calculated from the CEVSA model was 51 gC/m^2 with a deviation of 84 in Xinjiang. The estimated NPP from the CEVSA model was higher than estimated NPP from the GLO-PEM model partly because the CEVSA just accounted for the influences of changes of climatic factors while the GLO-PEM had the capacity to detect the effects of various non-climatic factors on NPP, including land use changes, fire, disease and insects.

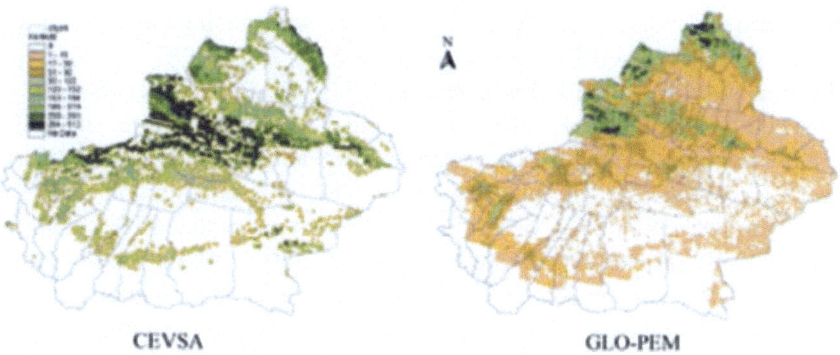

Figure 18.2 Mean NPP distribution in Xinjiang over 1981 – 2000 simulated by CEVSA and GLO-PEM

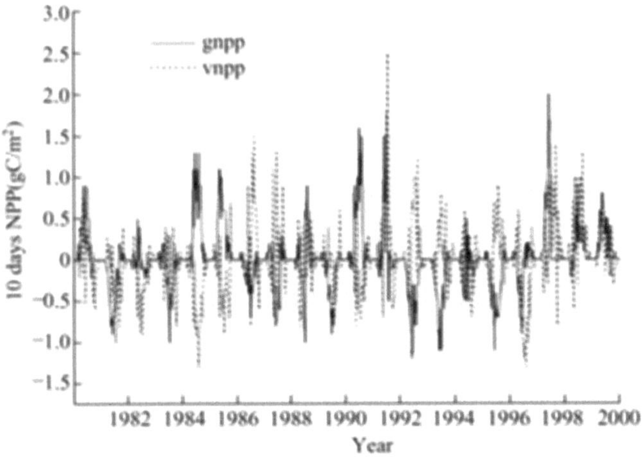

Figure 18.3 Interannual variations in Xinjiang mean NPP simulated by GLO-PEM (gnpp) and CEVSA (vnpp). NPP values were the difference between a 10-days simulated by models and the mean value for that 10 days over 20 years (1981 – 2000)

18.3.2 Precipitation and Temperature Impact on NPP

In this section we discuss the NPP trends and their relationships with climatic variables. There were large changes in the interannual variations of NPP, precipitation and temperature during 1981 and 2000 in Xinjiang (Fig. 18.4). The mean annual precipitation was 152 mm with an interannual variability 14.4%. Annual precipitation had an increasing trend by 1.13 mm yearly during the period 1981 – 2000. Precipitation increased 18.61 mm during 1981 – 1990 and only 2.83 mm during 1991 – 2000. The mean annual temperature was 5.77°C with an interannual variability 10.1%. Annual temperature had an increasing trend by 0.056°C yearly during the period 1981 – 2000. Xinjiang had become warmer obviously and the temperature increased 0.95°C in 20 years. The mean annual NPP over 20 years was 41.06 Tg • C (1 Tg = 10^{12} g) with an interannual variability 16%. The NPP decreased 1.6 Tg • C in the first 10 years (1981 – 1990) over the 1981 value while the NPP increased 2.38 Tg • C for the second 10 years (1991 – 2000) over the 1991 value. From Fig. 18.4 the trend line of NPP was almost horizontal while the trends of precipitation and temperature over 20 years. Table 18.1 provided 5-years mean values of NPP, precipitation and temperature in Xinjiang. The interannual variation in NPP was correlated with precipitation (R^2=0.36, P<0.01) and temperature (R^2=0.31, P<0.01). About 71% of Xinjiang area is arid land with sparse vegetation. In this area the interannual variation in NPP was not significantly affected by precipitation and temperature. These years when peak NPP occurred were not concurrent with those years when peak precipitation and temperature occurred.

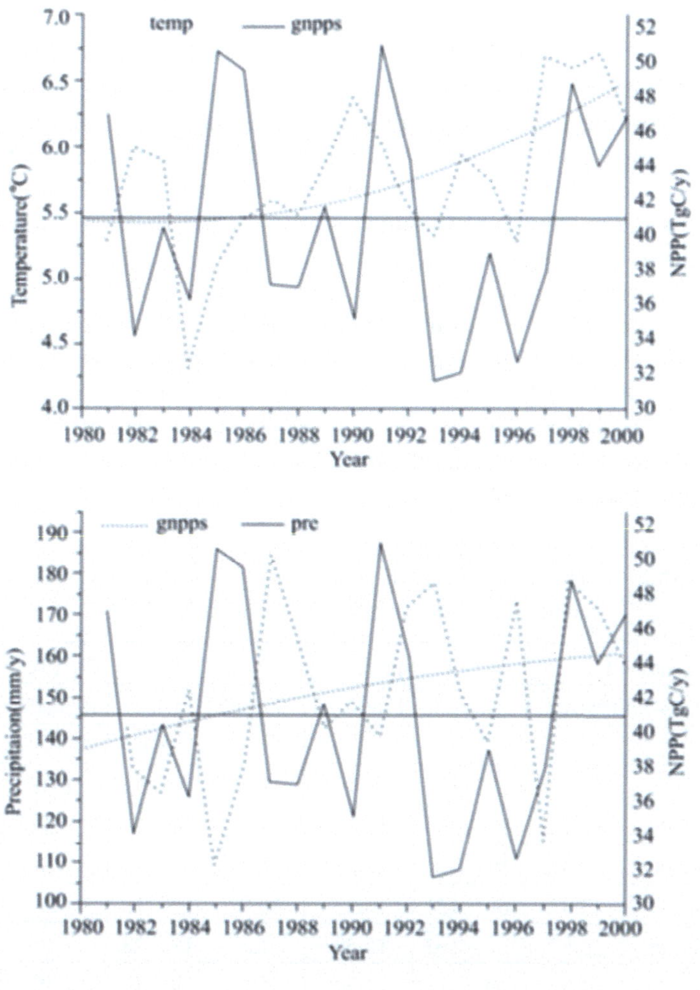

Figure 18.4 Changes in the interannual variations of NPP, precipitation and temperature

Table 18.1 5-years mean NPP, precipitation and temperature during 1981 – 2000 in Xinjiang

Year	Temperature (°C)	Precipitation (mm)	NPP (Tg • C)
1981 – 1985	5.33	136.81	41.88
1986 – 1990	5.75	155.42	40.28
1991 – 1995	5.72	156.37	39.86
1996 – 2000	6.28	159.20	42.24

The trends of increased precipitation and temperature over 20 years were presented in Table 18.2 and Fig. 18.5. Compared to the 1980s, the region with a mean annual temperature > 13°C was expanded 4% in 1990s, mainly distributed in areas of the Tarim Basin and Turpan Basin. The region with a mean annual temperature < 0°C was decreased 2% for areas of southern sides of the Tianshan and Altay Mountains, and the Tarim Basin. The regions with temperature between 0°C and 13°C in the 1980s also increased at different levels. The regions with a mean annual precipitation >200 mm increased about 5% from the 1980s to the 1990s and the increased areas covered both southern and northern foothills of the Tianshan Mountains, and the surrounding areas of the Ili River Valley and the Junggar Basins. The regions with a mean annual precipitation < 80 mm was decreased 3%, mainly distributed in surrounding areas of the Tarim Basin. There were negative trends of NPP in the 1980s while in the 1990s the trends of NPP were positive. The spatial patterns showed 89% of Xinjiang area with NPP less than 80 Tg • C had NPP increases in the 1990s at a different degree compared with those in the 1980s. The regions with NPP greater than 80 Tg • C shrank. These regions were located in the Ili River Valley and northern and western sides of the Junggar Basin.

Table 18.2 Comparisons of classifications of regions in NPP, precipitation and temperature between 1980s and 1990s in Xinjiang

Temperature (°C)			Precipitation (mm)			NPP (Tg • C)		
Class	1980s(%)	1990s(%)	Class	1980s(%)	1990s(%)	Class	1980s(%)	1990s(%)
						= 0	43.92	43.69
			36	11.96	11.26	0 – 3	6.48	6.67
0	17.09	15.18	37 – 56	12.06	11.71	4 – 6	6.13	6.14
1 – 2	12.00	12.47	57 – 79	13.58	11.79	7 – 11	6.56	6.90
2 – 5	3.33	2.87	80 – 121	10.32	10.92	12 – 18	6.51	6.89
6 – 9	16.25	15.87	122 – 163	12.50	11.37	19 – 28	6.09	6.13
10 – 11	21.70	21.80	164 – 199	11.50	11.19	29 – 46	5.92	6.21
12 – 13	27.02	25.23	200 – 246	9.74	12.71	47 – 78	6.18	6.40
14 – 15	2.37	6.11	247 – 296	11.01	9.84	79 – 142	6.44	5.84
>15	0.23	0.47	>296	7.33	9.19	>142	5.76	5.13

These spatial patterns were caused partly by the inconsistence between increased precipitation and increased temperature in one region. The precipitation was increased in the north of the Tianshan Mountains and the temperature was increased in the south of the Tianshan Mountains. The vegetation growth is depending on precipitation more than the temperature in the desert environment. The climate plays a role in changing NPP but the spatial pattern of NPP was not always consistent with patterns of precipitation and temperature in these areas.

We investigated 10-days trends in NPP and precipitation and temperature to determine whether climate plays a role in changing NPP between regions (Fig. 18.6).

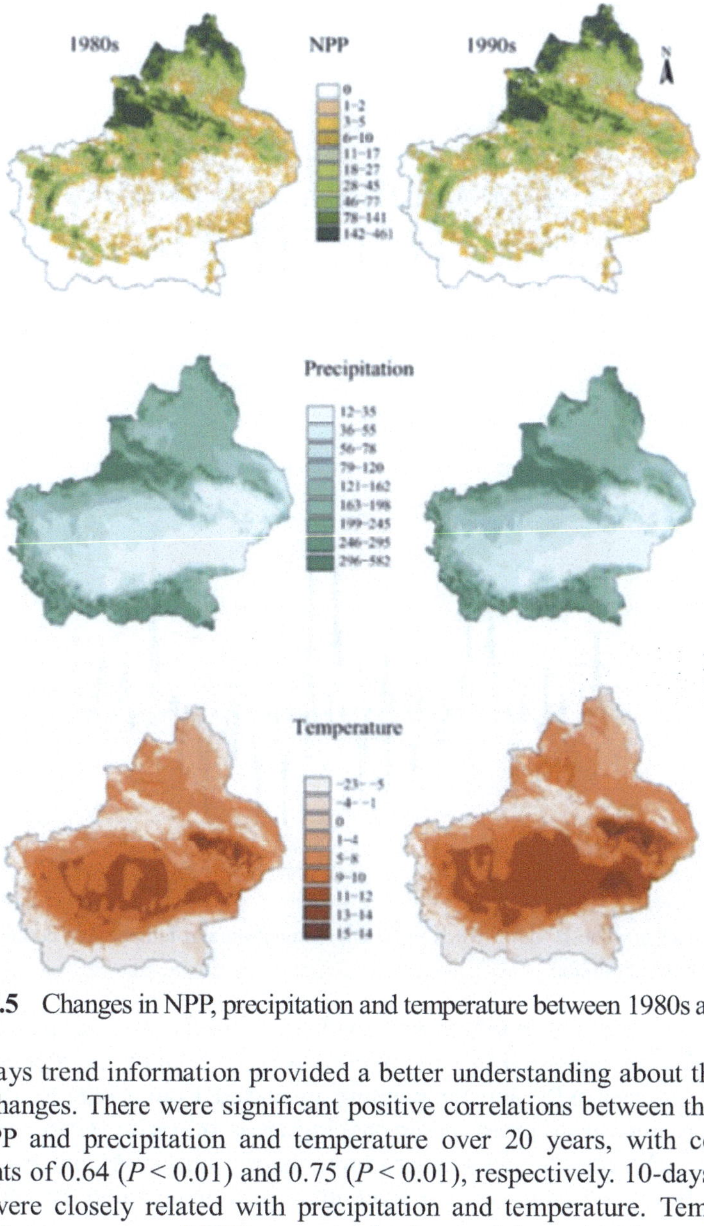

Figure 18.5 Changes in NPP, precipitation and temperature between 1980s and 1990s

The 10-days trend information provided a better understanding about the pattern of NPP changes. There were significant positive correlations between the 10-days mean NPP and precipitation and temperature over 20 years, with correlation coefficients of 0.64 ($P < 0.01$) and 0.75 ($P < 0.01$), respectively. 10-days changes in NPP were closely related with precipitation and temperature. Temperatures affected the responses of NPP to precipitation and increased in precipitation always caused an increase in NPP.

Figure 18.7 and Table 18.3 provide a spatial distribution of correlation coefficients between NPP and precipitation and temperature. The spatial changes of the correlation coefficient between NPP and precipitation from the 1980s to 1990s showed the regions with correlation coefficient greater than 0.28 increased. The increased regions distributed in surrounding areas of the Tarim Basin and

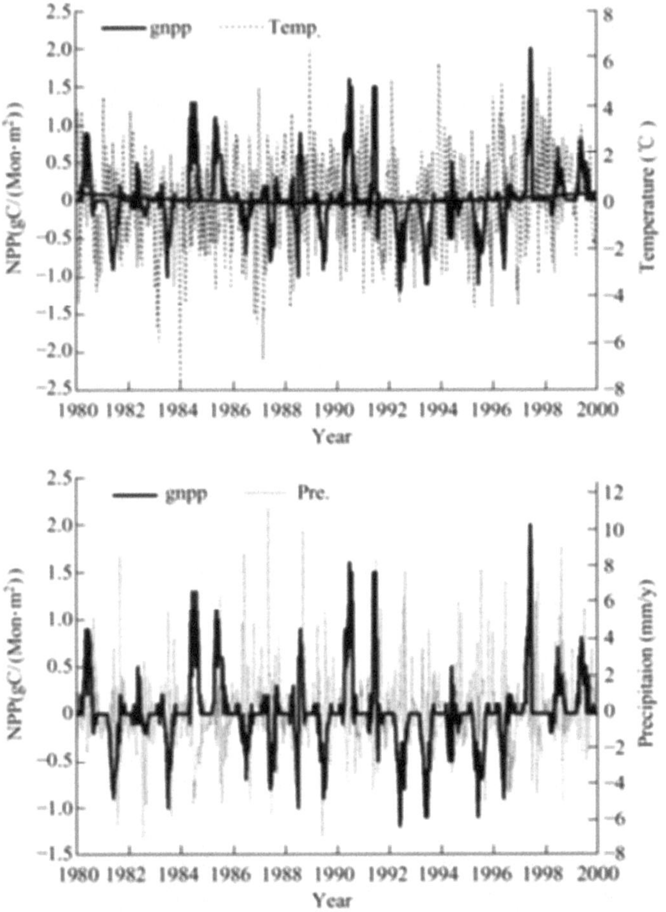

Figure 18.6 The 10-days trends in NPP, precipitation and temperature during 1981 and 2000 in Xinjiang

Junggar Basin and both northern and southern foothills of Tianshan Mountains. Figure 18.7 also suggested that the regions with correlation coefficients greater than 0.58 between NPP and temperature increased from 1980s to 1990s. The expanded regions overlapped in surrounding areas of Tarim Basin and Junggar Basin and both northern and southern foothills of Tianshan Mountains. With the available data it is hard to say that precipitation and temperature had driven NPP increases across Xinjiang, but it was obvious that both variables were factors that must be considered when explaining NPP increases in Xinjiang. The changes of NPP in Xinjiang were closely related with changes of 10-days precipitation and temperature. Concurrent increases in precipitation and temperature in growing period of the year may lead to an increase in vegetation growth. This may account for the NPP increase in above regions. Under the proper temperature condition desert vegetation could have a higher NPP with improved moisture availability.

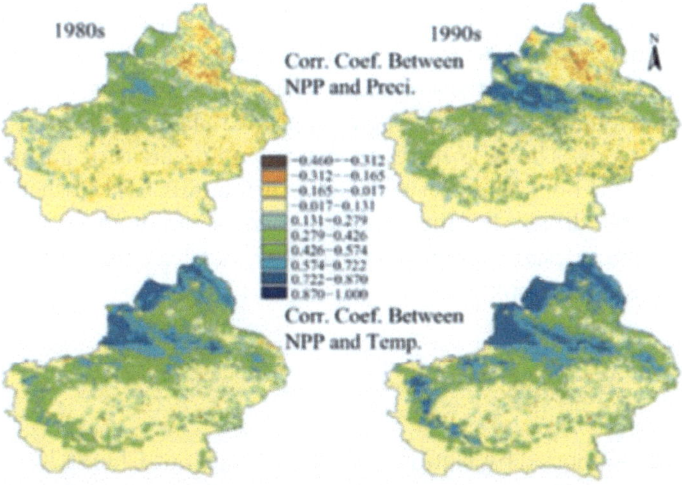

Figure 18.7 Spatial distribution of correlation coefficients between NPP and precipitation and temperature during 1980s and 1990s in Xinjiang

Table 18.3 Comparisons of classifications of regions in correlation coefficients between NPP(N) and precipitation(P) and temperature(T) during 1980s and 1990s in Xinjiang

Class	1980s(N_T)	1990s(N_T)	1980s(N_P)	1990s(N_P)
<(− 0.31)	0.06	0.06	0.05	0.08
(− 0.32) − (− 0.16)	0.26	0.18	1.07	1.12
(− 0.17) − (− 0.01)	3.81	3.64	12.44	11.32
(− 0.02) − 0.13	39.87	39.17	52.11	46.07
0.14 − 0.27	9.85	9.48	13.65	12.47
0.28 − 0.42	11.61	9.67	10.06	11.19
0.43 − 0.57	15.49	12.34	7.70	8.22
0.58 − 0.72	13.21	15.37	2.79	5.33
0.73 − 0.87	5.83	10.05	0.11	4.13
>0.87	0.01	0.04	0.02	0.05

18.4 Conclusions

The study was conducted for assessing the responses of NPP to precipitation and temperature in Xinjiang (semi-arid and arid desert ecological environment). GLO-PEM model was selected to calculate NPP during 1981 and 2000 at high spatial resolution. We investigated where and when NPP increased, identified regions showing similar NPP changes, and evaluated the contribution of changes in precipitation and temperature could account for estimated NPP changes.

During the 20-years time series mean annual NPP decreased in the first 10-year's period while mean annual NPP increased in the second 10-year's period. These results were reinforced by the simulations based on 10-days NPP trends. The changes in precipitation and temperature appear to play an important role in the 10-days NPP trends. The NPP increase in surrounding areas of Tarim Basin and Junggar Basin and both southern and northern foothills of Tianshan Mountains was likely to be driven by the seasonal increase in both precipitation and temperature. Other desert regions also had NPP increases that were obviously a result of increases in precipitation.

Acknowledgements

This work was supported by National 973 Key Project of China (2002CB412507 and G19990435), "Chuihui Plan" of Ministry of Education of China (Z2004-1-65002), Outstanding Overseas Chinese Scholars Fund of Chinese Academy of Sciences (2004-7-1), and the USDA/CSREES (Agreement 2001-34263-11248). Lee Wiedeman assisted in editing this manuscript.

References

Braswell BH, Schimel DS, Linder E, Moore B (1997) The response of global terrestrial ecosystems to interannual temperature variability. Science 278: 870 – 872

IPCC land-use change and forestry: a special report of the IPCC. Cambridge University Press, 2000

Cao MK, Prince SD, Li KA, Tao B, Small J, Shao XM (2003a) Response of terrestrial carbon uptake to climate interannual variability in China. Global Change Biology 9: 536 – 546

Cao MK, Tao B, Li DR, Shao XM, Prince SD (2003b) Interannual variation in terrestrial ecosystem carbon fluxes in China from 1981 to 1998. Acta Botanica Sinica 45(5): 552 – 560

Cao MK, Woodward FI (1998a) Dynamic responses of terrestrial ecosystem carbon cycling to global climate change. Nature 393: 249 – 252

Cao MK, Woodward FI (1998b) Net primary and ecosystem production and carbon stocks of terrestrial ecosystems and their response to climatic change. Global Change Biology 4: 185 – 198

Churkina G, Running SW (1998) Contrasting climatic controls on the estimated productivity of global terrestrial biomes. Ecosystem 1: 206 – 215

Cramer W (1999) Comparing global models of terrestrial Net Primary Productivity (NPP): Overview and key results. Global Change Biol. 5: 1 – 15

Emanuel WR, Killough GG (1984) Modelling terrestrial ecosystem in the global cycle with shifts in carbon storage capacity by land use change. Ecology 65(3): 970 – 983

Friend AD, Stevens AK, Knox RG, Cannell MGR (1997) A process-based, terrestrial biosphere model of ecosystem dynamics (Hybrid v. 3.0). Ecological Modelling 95(2 – 3): 249 – 287

Gao Q, Yu M, Yang XS (2000a) An analysis of sensitivity of terrestrial ecosystems in China to climatic change using spatial simulation. Climate Change 47: 373 – 400

Gao Q, Yu M, Yang XS (2000b) A simulation analysis of the relationship between regional primary production and vegetation structure under climatic change scenarios. Ecological Modelling 131(1): 33 – 45

Gilmanov TG, Walter C (1995) New estimate of organic matter reserves and net primary productivity of North American tundra ecosystems. Journal of Biogeography 22: 723 – 741

Gleason ACR, Prince SD, Goetz SJ (2002) Effects of orbital drift on land surface temperatures received from AVHRR sensors. Remote Sensing of Environment 79: 147 – 165

Goetz SJ, Prince SD, Small J (2000) Interannual variability of global terrestrial primary production: reduction of a model driven with satellite observations. J. of Geophysical Research 105: 20,007 – 20,091

Goward SN, Prince SD (1995) Transient effects of climate on vegetation dynamics: Satellite observations. J. Biogeogr. 22: 549 – 563

Hall DO, Ojima DS (1995) Response of temperate and tropical grasslands to CO_2 and climate change. Journal of Biogeography 22: 537 – 547

Haxeltine A, Prentice IC (1996) A general model for the light-use efficiency of primary production. Functional Ecology 10: 551 – 561

James M, Kalluri S (1994) The Pathfinder AVHRR land data set: an improved coarse resolution data set for terrestrial monitoring. International J. of Remote Sensing 15: 3,347 – 3,364

Law BE, Waring RH (1994) Combining remote sensing and climatic data to estimate net primary production across Oregon. Ecological Applications 4(4): 717 – 728

McGuire D, Melillo JM (1995) Equilibrium responses of soil carbon to climate change: empirical and process-based estimates. Journal of Biogeography 22: 785 – 796

Melillo JM, McGuire AD, Kicklighter DW, III BM, Vorosmarty CJ, Schloss AL (1993) Global climate change and terrestrial net primary production. Nature 363: 234 – 240

Prince S (1991) A model of regional primary production for use with coarse-resolution satellite data. International Journal of Remote Sensing 12: 1,313 – 1,330

Prince SD, Goward SN (1995) Global primary production: a remote sensing approach. J. Biogeography 22: 815 – 835

Riedo M, Gyalistras D, Fuhrer J (2000) Net primary production and carbon stocks in differently managed grasslands: simulation of site-specific sensitivity to an increase in atmospheric CO and to climate change. Ecological Modelling 134(2 – 3): 207 – 227

Runyon J, Waring RH (1994) Environmental limits on net primary production and light-use efficiency across the Oregon transect. Ecological application 4(2): 226 – 237

Schultz PA, Halpert MS (1993) Global correlation of temperature, NDVI and precipitation. Adv. Space Res. 13: 277 – 280

Starfield AM, Chapin FS (1996) Model of transient changes in arctic and boreal vegetation in response to climate and land use change. Ecological Applications 6(3): 842 – 864

Woodward FI, Smith TM, Emanuel WR (1995) A global land primary productivity and phytogeography model. Global Biogeochemical Cycles 9: 471 – 490

Appendix A Earth Science Remote Sensing Data and Services and Information Technology at the NASA/GES DISC DAAC

Steven J. Kempler

A.1 Introduction

A discussion on the philosophy of the organization would provide a better understanding of the GSFC Earth Sciences (GES) Data and Information Services Center (DISC) approaches described here.

Due to the long passage of time between this original presentation at GMU and the publication of this chapter, much material has been updated to reflect newer information, more relevant to today's evolving data management landscape.

A.1.1 What is the DISC—Mission Statement

The GES DISC provides data and information management services for global change research and education.

The GES DISC mission is to maximize the investment of NASA's Earth science by providing data and services that enable people to realize the scientific, educational, and application potential of global climate data.

As a source of data and information services for atmospheric and hydrologic science, the GES DISC aim is to facilitate the study the natural and human processes that influence Earth's climate.

A.1.2 What else is the DISC

The DISC is comprised of two parts:

(1) The Data Center (DC)—Traditionally known as the Distributed Active Archive Center (DAAC), the Data Center is responsible for the execution of data management functions, including: System execution (data ingest, archive, production, distribution, management, and user services); System Engineering and Infrastructure (data support, mission support, system administration, etc.), and; Initiation and maintenance of discipline unique services.

The Data Center goal is to efficiently and cost effectively ingest, produce, archive, and distribute Earth Science data and information, related to hydrology, and atmospheric dynamics and chemistry, and perform the data management functions and services that maximize the use and usefulness of this data and information.

Success is measured by the amount of use NASA Earth science data (in this case, located at the GES DISC) is getting in science research, applications, and education.

(2) The Products and Services Cooperative (PSC or the CO-OP)—Responsible for new products and services in cooperation with, and support of, the Earth science data community, including: Development and deployment of products and data management services for new projects; Development of new data information services; (Further) development of newer information science technologies that facilitate the use of Earth science data; New project initiatives; Services and technology reuse.

The GES CO-OPs goal is to, in complete collaboration with individuals being served, explore and develop information technologies and implement data management products and services in response to the changing data management requirements of the Earth science community, and the development of new applications

Success is measured by the amount of use and usability of new products generated from current and future missions, and...

Success is measured by the amount of information management services and technology nurtured at the GES DISC, that are reused by other data centers.

The GES DISC approach to new and existing activities is:

(1) Engage Users

(2) Communicate frequently

(3) Provide dedicated points of contact for systems and interfaces

(4) Build economically

(5) Look for reuse, ways to save funds

(6) Be willing to take calculated risks; Otherwise be low risk

(7) But also, build to integrate new technologies

(8) Engage employees—ensure that employees realize the value of their contributions

The bottom line:

We will not: Build it and they will come.

We will: Build it because they came (collaborating on a mutual interest)

The organizational breakout above has allowed the DISC to be responsive to science and application user needs, as well as the evolution of information systems.

A.1.3 Disciplines, Measurements, Missions, and Applications

The GES DISCs primary customers are the Earth science pure research, applied research, and applications communities, although the GES DISC has never been known to turn anybody away. Currently, several methods are available for accessing data located at the DISC through the many services offered. Data can be searched for and found through on-line/FTP services, EOS Data Gateway (EDG), or the DISCs newest and fastest interface, Mirador. Users of GES DISC data may access data and services starting from the GES DISC home page (http://disc.gsfc.nasa.gov/) via one of four "views" into the data.

A.1.3.1 Disciplines

The GES DISC is responsible for the stewardship and active accessibility of NASA Earth science data related to the following disciplines:

(1) Atmospheric Chemistry
(2) Atmospheric Dynamics
(3) Precipitation
(4) Oceans
(5) Radiance data

This view into the data provides users with scientific measurements and missions pertinent to the discipline.

A.1.3.2 Measurements

Once selecting the desired measurement from the Parameter Information Locator Tool (PILoT) or Parameter Information Pages (which provide complete parameter documentation), the user is given a choice of sensors that provide the requested measurement. Descriptions of each sensor's version of the measurement are also given. Go directly to: http://disc.gsfc.nasa.gov/get_data.shtml.

A.1.3.3 Satellites and Instruments

A user can also access data by indicating a specific mission/sensor. Table A.1 lists missions/sensors from which data is served by the GES DISC:

Table A.1 Missions/sensors from which data is searved by the GES DISC

Data Dates	Mission Sensor
1978 – 1994&	
1996 – 2000	TOMS—Total Ozone Mapping Spectrometer
1991 – 2001	UARS—Upper Atmosphere Research Satellite
1980 – 1993&	
2006 –	GMAO—Global Modeling and Assimilation Office
1978 – 1991	TOVS—Total Ozone Mapping Spectrometer
1997 –	TRMM—Tropical Rainfall Measuring Mission
2002 –	Aqua AIRS—Atmospheric Infrared Sounder

Continued

Data Dates	Mission Sensor
2003 –	SORCE—Solar Radiationand Climate Experiment
2004 –	Aura MLS—Microwave Limb Sounder
2004 –	Aura HIRDLS—High Resolution Dynamics Limb Sounder
2004 –	Aura OMI—Ozone Monitoring Instrument
Future	Glory—Irradiance measurements
Future	GPM—Global Precipitation Measurement

A.1.3.4 Applications

As the GES DISC reaches out to specific applications research areas, this view will provide relevant data directly. Existing and future targeted applications and research include agricultural decision support, aerosols research, and the atmospheric data relevant to specific multi-mission science (e.g. data relevant to the A-Train; The A-Train Data Depot).

A.2 An Integrated Organization

Traditionally, the GES DISC has been comprised of four highly interoperable groups of skills: Each playing a vital role in the GES DISC. Each skill group is essential to GES DISC success, as well as the success of each other. The four GES DISC skill groups include:

(1) Engineering (a.k.a., System development; infrastructure)
(2) System Execution
(3) Data Support
(4) Mission Support (a.k.a. Science Integration)

A.2.1 Engineering

The GES DISC engineering staff, is made up of highly skilled systems engineers, software engineers, system administrators, and database administrators. The engineer's role is to maintain the integrity, currency, and usability of GES DISC operational systems. Specifically, refreshing GES Systems with new technologies, improved procedures, and new capabilities all fall on the shoulders of the engineers. In addition, the engineers are very involved with examining advanced technologies and their potential integration into Earth science information systems. Forward looking research and affordable development is focused on facilitating the goals of Earth science data management systems.

A.2.2 Systems Execution

The GES DISC data systems operations personnel are responsible for the 24X7 operations of the GES DISC active archives, processing, and distribution. Personnel include ground controllers, operators, and distribution, ingest and production technicians. This group ensures that the system archives and distributes data properly, monitors system health, and distributes information on hard media.

A.2.3 Data Support

Science-trained and engineer-trained staff, discipline specific data support personnel focus on providing support to data users interested in data in their area of expertise. These personnel are equipped to answer any questions regarding the data ranging from data access questions to characteristics of the data. In addition, they are at the front line to acquire science requirements so that tools and services can be best developed to serve the broader community. Also, data support staff actively develop specific value-added data products to satisfy the 'heard' needs of GES DISC customers.

Last, but not least, the WWW specialists, the helpdesk and data usage/ statistics coordinator, and outreach specialists, also play essential roles at the GES DISC.

A.2.4 Mission Support

The Mission Support personnel address new and existing mission activities associated with ensuring successful data throughput. Specifically, this group interacts with new projects and their associated science teams to perform the groundwork for GES DISC data system support. This early involvement is extremely beneficial to both the project and the DISC. In addition, the Mission Support folks interacts with current science teams to ensure that throughput infrastructure is performing optimally.

A.2.5 An Integrated GES DAAC

As mentioned, the four skill groups are highly dependent on each other. Historically, each skill group had a specific set of jobs they had to do that benefited the data center. Just as the tasks were clear, the interactions between the groups were clear, leading to great data center success and advances in data management.

As you will see in Section A.4, the integration of GES DISC skills have drastically changed from macro integration to micro integration.

A.3 Utilizing Information Technology: Data, Information, Services

A.3.1 Data Access, Visualization and Analysis Tools

The GES DISC has a long history of providing a very high level of user support and data expertise, thus continuously improving the usability and accessibility of its data holdings. Basic data support services include:

(1) Providing ancillary data for science research, applied research, and applications users (e.g., education, policy makers, etc.)

(2) Distributing data products electronically

(3) Providing online data search/browse/order

(4) Maintaining a full suite of documentation (detailed guide, summary guide, readme)

(5) Providing user support via dedicated data support personnel (data access, HDF-EOS, documentation, data usage, etc)

(6) Providing subsetting, on-demand subsetting, subsampling, and data mining capabilities for public use

(7) Making data available in GIS and binary formats

(8) Providing data publication tools

(9) Providing a full suite of user services and outreach

In recent years, as Earth science data users have become more sophisticated in their access and use of data and information, the GES DISC has reacted by studying evolving user patterns and needs, resulting in the implementation of innovative approaches to more advanced value added services. These include:

(1) Services Oriented Architecture

(2) Alternative distribution mechanisms (e.g., OpenDAP)

(3) A Near Archive Data Mining (NADM), renamed S4PM-DM, environment ready for community supplied algorithms that would reduce the data distribution volume to include just the data desired by the algorithm developer

(4) A DISC On-line Visualization and Analysis System, Giovanni

(5) A Google like search tool into Earth science data, called Mirador

It is worth providing expanded information about NADM, Giovanni, and Mirador.

A.3.1.1 NADM (S4PM-DM)

Figure A.1 shows Near Archive Data Mining (NADM) system flow chart.

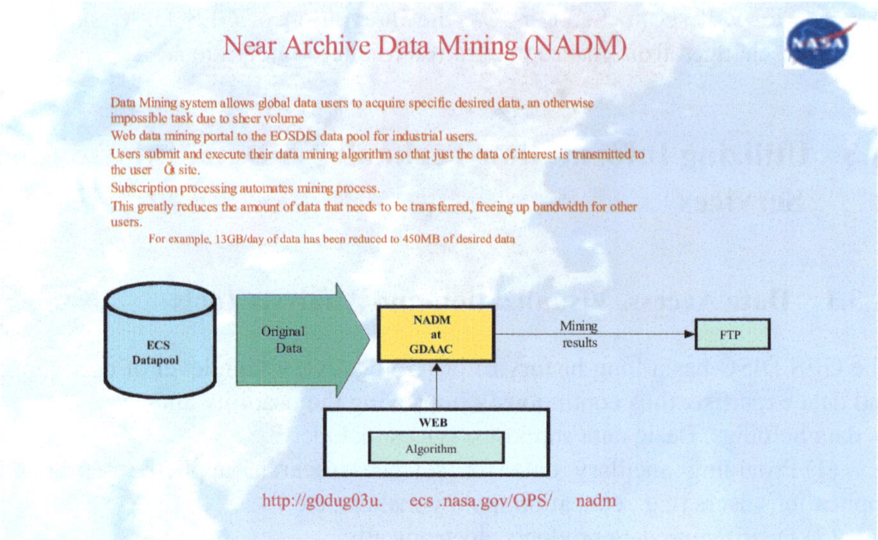

Figure A.1 Near Archive Data Mining system flow chart

A.3.1.2 Giovanni

Giovanni (Fig. A.2) is an on-line data exploration, visualization, and analysis tool. It has been reused for nine measurement specific implementations. The following illustrates its atmospheric composition instantiation.

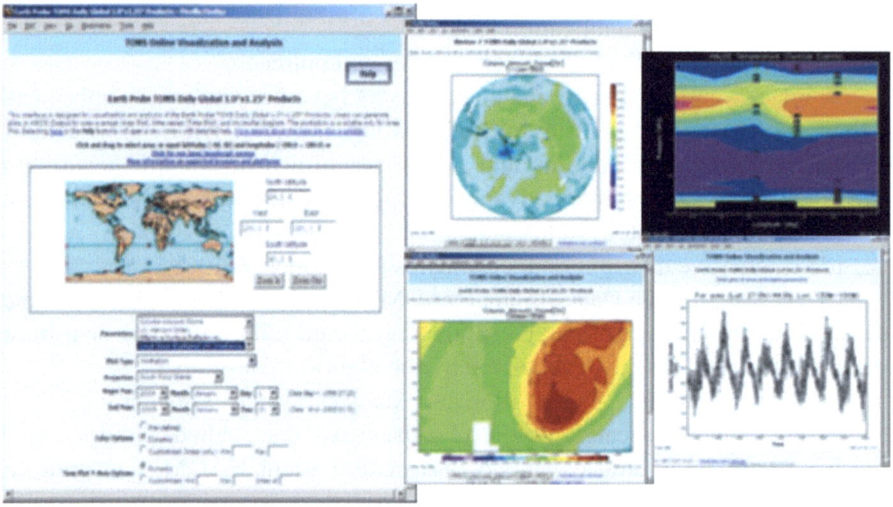

Figure A.2 Example products of Giovanni

A.3.1.3 Mirador

Mirador (Fig. A.3) is one of the newest tools developed at the GES DISC, created in response to making information as easily accessible as done by google ™. The interface is quite simple:

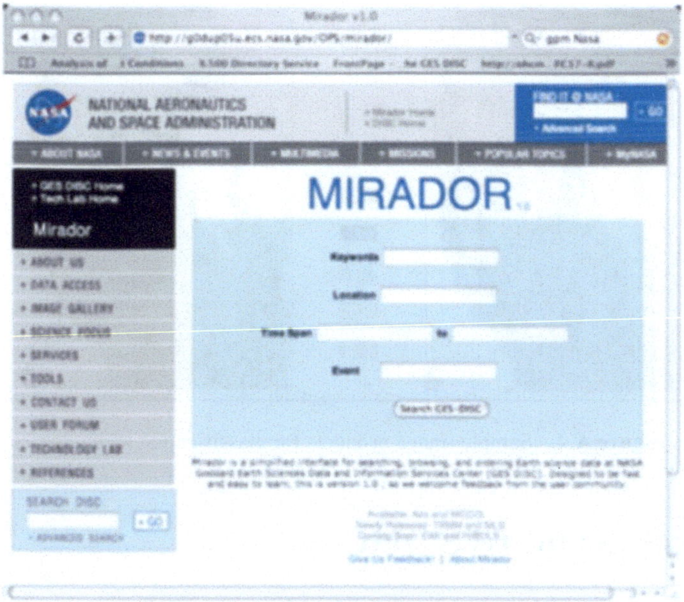

Figure A.3 Example of Mirador interface

A.3.2 Examining Advanced Technologies

The GES DISC continues to experiment with new technologies that show promise in furthering data accessibility and information extraction. Currently, the GES DISC has expended varying degrees of effort in the following areas:

(1) All on-line archives data management (S4PA)

(2) Content based data management utilizing Bayesian classification

(3) On demand data processing architecture

(4) Intelligent data understanding

The Bayesian content classification work has found much success as depicted in the likeness of Fig. A.4.

Data and information services are becoming increasingly more significant in managing large volumes of data. The GES DISC strives to implement services such that can be reused for other missions and/or different measurements, likely through collaborations with potential partners seeking to implement/utilize such services.

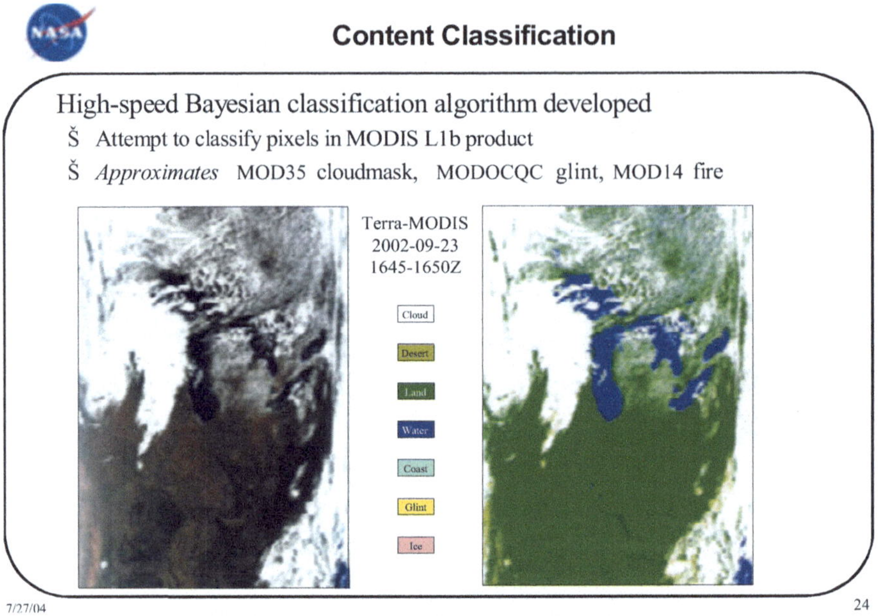

Figure A.4 Sample image products for classification technologies

A.4 Evolving the GES DISC

The GES DISC, as a whole, recognizes that the general ability to process, archive and distribute increasing amounts of information, the advances in science research, the exploding accessibility to information, and the quantum leaps in computer technologies, all have lead to (i.e., required) the need to evolve information management capabilities and techniques to NOT inhibit the progress of the user population. Individually, the employees of the GES DISC view this as the most interesting of challenges.

A.4.1 Why Evolve Earth Science Data Systems

Several drivers exist that evolve GES DISC, naturally or by decision. Natural drivers are those that result from new discoveries, lead to new needs to further explore these discoveries, and thus new accommodations for the new needs. Natural evolution drivers includes:

(1) Science discoveries that need new ways to analyze data and information

(2) Technology paradigm shifts that enable increases in computing power and/or decreases in computing prices

(3) New and more clever ways to represent data and information enabled by

new software products or the integration of software products or software/hardware products.

(4) The ability to serve the increasing diversity of user needs that accommodate users technical sophistication, areas of interest, knowledge of data, and willingness to better understand data management techniques not is their scope of primary interest

In addition, NASA Earth science data management has emphasized a:

(1) Movement towards measurement based data management systems

(2) Movement towards distributed data management, processing, archive, and distribution

(3) The importance of information technology as a facilitating tool

A.4.2 GES DISC Evolution

The GES DISC continues to evolve in ways that best addresses the aforementioned drivers. Specifically, the GES DISC plans to move in the following directions:

Essential to data stewardship:

(1) Continue to move (more so) data support focus from missions to disciplines/measurements

(2) Complete data set documentation; Ensure data safe keeping via DISC archives

(3) Continue to explore and implement promising technologies that support users in extracting knowledge from our data and information holdings

(4) Develop knowledge based support for general users, while maturing support for measurement based users

Providing resident expertise to further facilitate the use and access of Earth science data:

(1) Seek opportunities to partner with: scientists, applications

(2) Become experts in tools that support interdisciplinary studies (e.g., GIS, interoperability)

(3) Become experts in specific measurements (e.g., atmospheres: aerosol, ozone, hydrologic applications such as agriculture decision support systems, etc.)

(4) Implement smaller, measurement specific, on-line archives for heritage and current data products (and supporting subsystems)

(5) Further expand access to specific measurements, applications, atmospheric correction, etc. (e.g., reuse GES DISC on-line visualization and analysis tool {GIOVANNI})

(6) Automate routine procedures

(7) Collaborate with science teams on new smaller missions that lend themselves to measurements consistent with resident measurements of interest

(8) Collaborate on utilizing NASA investments in advanced data management technologies

A.4.3 The Evolved GES DISC

NASA's investment in its institutional data management facilities has resulted in a collection of high level data management expertise that are not easily matched, and essential to the increasingly sophisticated needs of the Earth science community that NASA serves. Coupled with the opportunities to meet data user needs through the innovative use of advancing information technologies, the GES DISCs evolved state is situated at this intersection: Understanding and applying information technology, and the science scenarios that require the technology to succeed. The GES DISC understands:

(1) Data services are becoming more and more relevant for supporting science

(2) Incubating innovative data management and analysis services can lead to facilitating the use of these advanced services at other data centers

(3) Collaborations with NASA funded information technologists can further infuse promising technologies into a system with real data and real end users, providing an essential avenue for data,, through advanced information science technologies, to reach its targeted audiences

In order to take action on this understanding, the GES DISC has re-established its infrastructure for continuous evolution:

(1) GES DISC data will be accessible through a series of discipline specific, disk based portals that will contain familiar and new services relevant to the needs of the particular discipline community being served by that portal ("data center"). Besides being discipline specific, portals can be application specific, research specific, etc. Previously, the DISC always covered many disciplines but now each discipline's data will be more easily accessible. Current portals (and planned portals) include: Atmospheric Composition (ACDISC), Precipitation (PDISC), Atmospheric Dynamics (ADDISC), A-Train Data Depot (ATDD), Oceans (CAN), Agriculture (AgDISC), NEESPI

(2) GES DISC will continue to exemplify automating operational activities for cost and performance efficiency

(3) GES DISC will work with new and existing collaborators to continue to innovate data management services and technologies. These services and technologies are as likely to be initiated (and developed) external to the GES DISC as they are to be initiated and developed internal to the DISC. The gems will shine

GES DISC personnel skills integration will occur at a micro level. With evolution, smaller teams of various skills will work closely together on "projects". Projects can be a discipline specific data center; Cross cutting technology development (e.g., OGC, Data Fusion, Giovanni, etc.), or; Internal tasks (e.g., science support, system execution, system administration, etc.)

A.5 Summary

The GES DISC:

(1) Focuses on intelligent use of NASA's information through sound data management

(2) Ensures unencumbered distribution of NASA Earth science data to all users

(3) Provides complete user services and data expertise services

(4) Exploits advanced technologies to further facilitate the Earth science missions

Thanks to the active participation and dedication of GES DISC employees, this organization has been able to efficiently focus on the "mission" it was given. Not only has the GES DISC organized and reorganized itself to faithfully steward its data holdings by ensuring that the data is processed, archived, and made available, with full documentation and user support...but what makes the GES DISC a standout organization is the experience its gained in technically and scientifically managing Earth science data, and its ability to apply its expertise in recognizing information science growth areas that further facilitate improving the management, accessibility, and usability of Earth science data for science research and applications.

Appendix B

Wenli Yang

B.1 A C Code Example for the HDF SD API

The following C code shows how to use the SD API to (1) initialize the SD interface and create an HDF file named myHDFfile; (2) create a SDS named mySDS in myHDFfile; (3) write into mySDS a 8-bit integer type, 5-row by 4-column array named myDataArray; (6) assign dimension names to the SDS; (7) set global, SDS, and dimension attributes; and (8) terminate access to the SDS and the SD interface.

```
#include"hdf.h"
#include"mfhdf.h"
#define RANK 2  /* define the number of dimensions of the SDS. */
#define ROW 5   /* define the row number of the SDS array */
#define COL 4   /* define the column number of the SDS array */
#define MAXATTR 1024 /* specify the maximum attribute length */
main()
    {
    int32 sdID;                 /* SD interface identifier */
    int32 sdsID;                /* SDS identifier*/
    int32 dimID;                /* dimension identifier */
    int32 dimSize[RANK];        /* the SDS's dimension sizes */
    char  attrValue[MAXATTR]; /* for storing attribute value */
    /* the data to be written to the SDS. */
    int8  myDataArray[ROW][COL]={1,2,3,4,5,6,7,8,9,10,11,12,13,
                                 14,15,16,17,18,19,20};
    int8 fillValue;              /* fill value for the SDS */
    /* the following three arrays specify the locations, intervals, and
    number of data points in the SDS to which data is to be written. The
    sizes of these arrays must be equal to the number of dimensions of the
    SDS so that each element of these array corresponds to one dimension
    in the SDS */
    int32 start[RANK],stride[RANK],edge[RANK];
    int32 returnStatus; /* return value of a SDS function call */

    /* Start SD interface and create an HDF file named myHDFfile. */
    sdID=SDstart("myHDFfile",DFACC_CREATE);

    /* set file level metadata (global attribute), to the HDF file. */
```

```
strcpy(attrValue,"This is a test HDF file.");
returnStatus=SDsetattr(sdID,"File metadata",
           DFNT_CHAR,strlen(attrValue),attrValue);

/* set dimension size of the SDS to be created. */
dimSize[0]=ROW;
dimSize[1]=COL;

/* create a SDS, mySDS, in the myHDFfile file (now referred to by the
SD interface identifier, sdID) and specify its data type, number of
dimensions and dimesion sizes. */
sdsID=SDcreate(sdID,"mySDS",DFNT_INT8,RANK,dimSize);

/* set the SDS fill value to 0 */
fillValue=0;
returnStatus=SDsetfillvalue(sdsID,&fillValue);

/* specify the star location in SDS from where the array data is to be
written.        (0,0) indicates that the SDS is to be filled from the
beginning. */
start[0]=0;
start[1]=0;

/* specify if the array data is to be written into the SDS's continuous
region or by skipping at certain intervals.  (1,1) indicates data is
to be written to the SDS's continuously while (m,n) says that data is
to be written to the SDS's every m elements at the first dimension and
every n elements at the second dimension.  Note that skipping always
refers to the SDS data array object in HDF file, not the data array in
memory, i.e., myDataArray in this example. */
stride[0]=1;
stride[1]=1;

/* specify how many data points need to be written out at each dimension.
Note that these values are not the end positions in the SDS. The end
positions are determined by the start position specified in the start
array and the stride values specified in stride array and the number
of points specified in the following edge array. */
edge[0]=dimSize[0];
edge[1]=dimSize[1];

/* write the array, my DataArray, into the SDS.  Note that the stride
array can, and should in order to increase efficiency, be replaced by
NULL if all the stride values are set to 1 (i.e., no skipping for all
SDS dimensions. */
returnStatus=SDwritedata(sdsID,start,stride,edge,
                     (VOIDP)myDataArray);

/* set metadata for the SDS */
```

```
    strcpy(attrValue,"This is a test SDS.");
    returnStatus=SDsetattr(sdsID,"SDS metadata",
                DFNT_CHAR,strlen(attrValue),attrValue);

    /* assign names and attributes to the two dimensions of the SDS by first
    get dimension identifier and then set name and attribute using the
    identifier. */
    dimID=SDgetdimid(sdsID,0);
    SDsetdimname(dimID,"Number of rows");
    strcpy(attrValue,"Row is along track direction.");
    returnStatus=SDsetattr(dimID,"Dimension Row",
                DFNT_CHAR,strlen(attrValue),attrValue);
    dimID=SDgetdimid(sdsID,1);
    SDsetdimname(dimID,"Number of columns");
    strcpy(attrValue,"Column is along scan direction.");
    returnStatus=SDsetattr(dimID,"Dimension Column",
                DFNT_CHAR,strlen(attrValue),attrValue);

    /*terminate access to the SDS, SD interface and close the file.*/
    returnStatus=SDendaccess(sdsID);
    returnStatus=SDend(sdID);
}
```

In the above example, the return values of SD interface functions (e.g. sdID, returnStatus) are recorded and some are used in subsequent calls (i.e. sdID, dimID, and sdsID). In most cases, when a SD interface function performs successfully, it returns either an identifier, an index, a reference value or a predefined SUCCEED (0) value. A return value of a predefined FAIL (−1) indicate that the corresponding function call does not perform correctly. Some SD interface functions may return different values, such as the predefined TRUE (1) or FALSE (−1) for the SDiscoordvar call and SD_FILL and SD_NOFILL for the SDsetfillmode call. It is a good practice to check the returned value after a SD function call so that bugs can be found at an early stage. Users can refer to the HDF reference manual corresponding to the library used.

B.2 A C Code Example for the HDF-EOS SW API

The following C code shows how to use the SW API to (1) initialize the SW interface and create an HDF-EOS file named myHDFfile; (2) create a swath named mySwath in myHDFfile; (3) define dimensions for the swath; (4) create a data field named myDataField; (5) create two geolocation fields named Latitude and Longitude in mySwath; (6) set dimension mapping relationship using offset values of 0 and 0 and increment values of 2 and 2 in the row and column directions; (7) write into myDataField a 8-bit integer type, 5-row by 5-column data array named myDataArray; (8) write two 32-bit floating point type, 3-row by 3-column arrays containing latitude and longitude values into the Latitude and Longitude geolocation fields, respectively; and (9) terminate access to the swath and close the HDF-EOS file.

```
#include"hdf.h"
#include"mfhdf.h"
#include"HdfEosDef.h"
#define RANK   2      /* define dimension number */
#define ROW    5      /* define data field row number */
#define COL    5      /* define data field column number */
#define GEOROW 3      /* define geofield row number */
#define GEOCOL 3      /* define geofield row number */
#define MAXATTR 1024 /* specify the maximum attribute length */
main()
{
    int32 fileID;                  /* HDF-EOS file identifier */
    int32 swID;                    /* swath identifier */
    int32 dimID;                   /* dimension identifier */
    int32 dimSize[RANK];            /* dimension sizes */
    /* set data field values */
    int8  myDataArray[ROW][COL]={1,2,3,4,5,6,7,8,9,10,11,12,
                    13,14,15,16,17,18,19,20,21,22,23,24,25};
    /* set latitude and latitude values */
    float32 latitude[GEOROW][GEOCOL]={30.03,30.03,30.03,
                    30.02,30.02,30.02,30.01,30.01,30.01};
    float32 longitude[GEOROW][GEOCOL]={-90.03,-90.02,-90.01,
                    -90.03,-90.02,-90.01,-90.03,-90.02,-90.01};
    int32 offset[2],increment[2]; /*dimension mapping values*/
    int32 start[RANK], stride[RANK], edge[RANK];

    /* create a new HDF-EOS file and a new swath within this file. */
    fileID=SWopen("myHDFfile",DFACC_CREATE);
    swID=SWcreate(fileID,"mySwath");

    /* define dimension names and size for the swath. All dimension names
    to be used in this swath, including those to be defined for both data-
    and geo-fields, must be defined in the swath. */
    SWdefdim(swID,"dataRow",(int32)ROW);
    SWdefdim(swID,"dataColumn",(int32)COL);
    SWdefdim(swID,"geoRow",(int32)GEOROW);
    SWdefdim(swID,"geoColumn",(int32)GEOCOL);

    /* define a data field and two geolocation fields in the swath */
    SWdefdatafield(swID,"myDataField",
          "dataRow,dataColumn",DFNT_INT8,HDFE_NOMERGE);
    SWdefgeofield(swID,"Latitude","geoRow,geoColumn",
              DFNT_FLOAT32,HDFE_NOMERGE);
    SWdefgeofield(swID,"Longitude","geoRow,geoColumn",
              DFNT_FLOAT32,HDFE_NOMERGE);
```

```
    /* define dimension mapping between data and geolocation fields */
    offset[0]=offset[1]=0;
    increment[0]=increment[1]=2;
    SWdefdimmap(swID,"dataRow","geoRow",
              offset[0],increment[0]);
    SWdefdimmap(swID,"dataColumn","geoColumn",
              offset[1],increment[1]);

    /* detach and then re-attach the swath before writing to assure that
    fields are properly established */
    returnStatus=SWdetach(swID);
    swID=SWattach(fileID,"mySwath");

    /* write data array to the data field */
    start[0]=start[1]=0;
    stride[0]=stride[1]=1;
    edge[0]=ROW;
    edge[1]=COL;
    SWwritefield(swID,"myDataField",start,stride,edge,
               (VOIDP)myDataArray);

    /* write geolocation array to the data field */
    edge[0]=GEOROW;
    edge[1]=GEOCOL;
    SWwritefield(swID,"Latitude",start,stride,edge,
               (VOIDP)latitude);
    SWwritefield(swID,"Longitude",start,stride,edge,
               (VOIDP)longitude);

    /* terminate access to the swath and close the file */
    SWdetach(swID);
    SWclose(fileID);
}
```

B.3 A C Code Segment for the HDF-EOS GD API

The following C code segment shows how to use the GD API to (1) open and access an existing grid, myGrid, in an HDF-EOS file, myHDFfile; (2) Perform various inquiries; (3) set a spatial subset region; (4) read field data from the subset region; and (5) terminate access to the grid and close the file.

```
main()
{
    /* declarations omitted here */

    /* Initializing the GD interface, open the file with read mode, and
```

```
        establish access to the grid named myGrid. */
        fileID=GDopen("myHDFfile",DFACC_READ);
        gdID=GDattach(fileID,"myGrid");

        /* obtain the grid's spatial coordinate information, get its X and Y
        dimension sizes, find out which corner is its origin, and retrieve its
        upper left and lower right corner's coordinate values. */
        GDprojinfo(gdID,&projectionCode,&zoneCode,
                &sphereCode,projParameters);
        GDorigininfo(gdID,&originCode);
        GDgridinfo(gdID,&xDimSize,&yDimSize,upLeft,lowRight);

        /* find out how many data fields are included in the grid and their names,
        number of dimensions, and data types.  And then get information about
        a particular data field, myDataFied, including data type, number of
        dimensions and dimension names and sizes */
        GDinqfields(gdID,fieldNames,fieldDimNumbers,
                fieldDataTypes);
        GDfieldinfo(gdID,"myDataField",&nDims,dimSizes,
                &dataType,dimNames);
        /* define a subset region on the grid using upper left (0,90) and lower
        right (90,0) longitude/latitude values and then read the region's data
        from the myDataField field into a dataBuffer. */
        longitude[0]=0.;
        longitude[1]=90.;
        latitude[0]=90.;
        latitude[1]=0.;
        regionID=GDdefboxregion(gdID,longitude,latitude);
        GDextractregion(gdID, regionID,"myDataField",
                (VOIDP)dataBuffer);
        /* terminate access to the grid and close the file */
        GDdetach(gdID);
        GDclose(fileID);
}
```

B.4 A C Code Example for the HDF5 API

The C program shown bellow is an example of writing a 2D array in memory to a dataset in an HDF5 file, attaching an attribute to the dataset as metadata, and reading data back to the memory from the dataset in the HDF5 file. The H5D interface provides very flexible subset options for data R/W. This example shows how to write a complete array to a dataset and to read a subset of dataset to a position offsetting from the array's starting.

```
#include "hdf5.h"
#define RANK      2
#define ROW       5
```

```
#define COL       4
#define MAXSTR    64
#define SUCCEED   0
int main(void)
{
    hid_t     fileID,dspaceID,dtypeID,dsetID,attrID;
    hid_t     attrSpaceID,attrTypeID,memspaceID;
    hsize_t   dimSize[RANK],offset[RANK],count[RANK];
    short     myDataArray[ROW][COL]={1,2,3,4,5,6,7,8,9,10,11,
              12,13,14,15,16,17,18,19,20};/* data to be written */
    char      fileName[MAXSTR]={"myHDF5file"};
    char      datasetName[MAXSTR]={"myTestDataset"};
    char      attrName[MAXSTR]={"myAttribute"};
    char      attrValue[MAXSTR];
    int       i,j;
    herr_t    returnStatus;  /* return status code */

    /* create a new HDF5 file. erase all data if it already exists. */
    fileID=H5Fcreate(fileName,H5F_ACC_TRUNC,
                   H5P_DEFAULT,H5P_DEFAULT);

    /* set dimensionality and data type of the dataset to be created*/
    dimSize[0]=ROW;
    dimSize[1]=COL;
    dspaceID=H5Screate_simple(RANK,dimSize,NULL);
    dtypeID=H5Tcopy(H5T_NATIVE_SHORT);

    /* create the dataset */
    dsetID=H5Dcreate(fileID,datasetName,
                  H5T_NATIVE_SHORT,dspaceID,H5P_DEFAULT);

    /* write all data in memory, myArrayData, to the dataset */
    returnStatus=H5Dwrite(dsetID,dtypeID,
              H5S_ALL,H5S_ALL,H5P_DEFAULT,myDataArray);

    /* the followings are to create an attribute for the dataset */
    /* create attribute dataspace */
    attrSpaceID=H5Screate(H5S_SCALAR);
    /*set attribute datatype as string whose size is in unit of byte*/
    attrTypeID=H5Tcopy(H5T_C_S1);
    /* set attribute value total size for the attribute */
    strcpy(attrValue,"this is my test dataset attribute");
    returnStatus=H5Tset_size(attrTypeID,
                        strlen(attrValue));
    /* attach the attribute to the dataset and then write the attribute value
    */
    attrID=H5Acreate(dsetID,attrName,attrTypeID,
                   attrSpaceID,H5P_DEFAULT);
    returnStatus=H5Awrite(attrID,attrTypeID,attrValue);
```

```
/* close attribute IDs */
returnStatus=H5Sclose(attrSpaceID);
returnStatus=H5Tclose(attrTypeID);
returnStatus=H5Aclose(attrID);

/* close dataset IDs */
returnStatus=H5Sclose(dspaceID);
returnStatus=H5Tclose(dtypeID);
returnStatus=H5Dclose(dsetID);

/* In the following, a subset of the dataset back into the memory. The
subset starts from the second row and second column of the dataset and
is of 3 rows and 2 columns from the starting point.  These data will
be put back to the myDataArray array, starting from the third row and
third column. Note that dimensionality and datatype are queried here
because they are already known.  The information must be obtained if
it is not previously know in order to correctly allocate define memory
dataspace. */

/* open dataset and get its dataspace and datatype IDs */
dsetID=H5Dopen(fileID,datasetName);
dspaceID = H5Dget_space(dsetID);

/* define subset region.  The offset/count are equivalent to start/edge
in HDF4. Stride is default to NULL. The last parameter, NULL, specifies
no blocking, i.e., each single data points being as one block. */
offset[0]=1;
offset[1]=1;
count[0]=3;
count[1]=2;
returnStatus=H5Sselect_hyperslab(dspaceID,
              H5S_SELECT_SET,offset,NULL,count,NULL);

/* define the dimensionality in memeory *
memspaceID=H5Screate_simple(RANK,dimSize,NULL);

/* specify that the data is to be stored in the array starting from the
third row and their column. */
offset[0] = 2;
offset[1] = 2;
count[0]  = 3;
count[1]  = 2;
returnStatus=H5Sselect_hyperslab(memspaceID,
              H5S_SELECT_SET,offset,NULL,count,NULL);

/* Initialize the memory with zero and read data from dataset to the
memory using their corresponding dataspace IDs */
for(i=0;i<ROW;++i)
  for(j=0;j<COL;++j)
```

```
        myDataArray[i][j]=0;
    returnStatus=H5Dread(dsetID,H5T_NATIVE_SHORT,
            memspaceID,dspaceID,H5P_DEFAULT,myDataArray);

    /* close the datset space, dataset, and memory space */
    returnStatus=H5Sclose(dspaceID);
    returnStatus=H5Dclose(dsetID);
    returnStatus=H5Sclose(memspaceID);

    /* close file */
    returnStatus=H5Fclose(fileID);

    return SUCCEED;}
```

B.5 A C Code Example for the HDF-EOS5 SW API

The following C code example demonstrates some Swath operations in an HDF-EOS5 file: (1) open a file, myHDFfile, and clear any data if the file already exists; (2) create a swath structure, mySwath; (3) define dimensions; (4) define geolocation and data fields; (5) define both regular and index mappings. Note that these two are the same in this example and in reality the index mapping is not needed when regular mapping is available; (6) write data to the geolocation and data field; (7) write file, swath, data field group, and dataset level attributes with either string or double data types; and (8) terminate access to the swath and close the file. The resultant swath structure is shown in Fig. 10.6 starting from the mySwath object, except that the two index mapping objects, named _INDEXMAP: geoColumn, dataColumn and _INDEXMAP: geoRow, dataRow, are not drawn.

```
#include  <HE5_HdfEosDef.h>
#define ROW     5
#define COL     5
#define GEOROW  3
#define GEOCOL  3
#define MAXSTR  128
#define MAXNUM  5
#define SUCCEED 0

int main(void)
{
    hid_t   fileID,swID; /* file and group identifies */
    /* arrays for regular and index dimension mappings */
    hsize_t increment[2],offset[2];
    long indexRow[GEOROW]={0,2,4}, indexCol[GEOCOL]={0,2,4};
    /* arrays image dataset and geolocation datasets */
    double  myDataArray[ROW][COL];
    float   latData[GEOROW][GEOCOL]={30.03,30.03,30.03,30.02,
                                     30.02,30.02,30.01,30.01,30.01};
```

```
float   lonData[GEOROW][GEOCOL]={-90.03,-90.02,-90.01,
                -90.03,-90.02,-90.01,-90.03,-90.02,-90.01};
/* a string type attribute and a double type attribute */
char    attrValueChar[MAXSTR];
double  attrValueDouble[MAXNUM]={1.,2.,3.,4.,5.};
/* other variables */
hsize_t attrCount[1];
hsize_t start[3],edge[3];
int     i, j;

for(i=0;i<ROW;++i) /* set data values to myDataArray */
  for(j=0;j<COL;++j)
    myDataArray[i][j]=(float)i*(float)COL+j;

/* Open a "myHDFfile" file */
fileID=HE5_SWopen("myHDFfile",H5F_ACC_TRUNC);

/* Create a swath structure */
swID=HE5_SWcreate(fileID,"mySwath");

/* define dimension names and sizes for the swath */
HE5_SWdefdim(swID,"dataRow",ROW);
HE5_SWdefdim(swID,"dataColumn",COL);
HE5_SWdefdim(swID,"geoRow",GEOROW);
HE5_SWdefdim(swID,"geoColumn",GEOCOL);

/* define a data field and two geolocation fields in the swath */
HE5_SWdefgeofield(swID,"Latitude","geoRow,geoColumn",
            NULL,H5T_NATIVE_FLOAT,HE5_HDFE_NOMERGE);
HE5_SWdefgeofield(swID,"Longitude","geoRow,geoColumn",
            NULL,H5T_NATIVE_FLOAT,HE5_HDFE_NOMERGE);
HE5_SWdefdatafield(swID,"myDataField","dataRow,
dataColumn",NULL,H5T_NATIVE_DOUBLE,HE5_HDFE_NOMERGE);

/* define dimension mapping between data and geolocation fields. */
offset[0]=offset[1]=0;
increment[0]=increment[1]=2;
HE5_SWdefdimmap(swID,"geoRow","dataRow",offset[0],
            increment[0]);
HE5_SWdefdimmap(swID,"geoColumn","dataColumn",
            offset[1],increment[1]);
HE5_SWdefidxmap(swID,"geoRow","dataRow",indexRow);
HE5_SWdefidxmap(swID,"geoColumn","dataColumn",
            indexCol);

/*detach/reattach swath to assure fields are properly established*/
HE5_SWdetach(swID);
swID=HE5_SWattach(fileID,"mySwath");
```

```
    /* Write geolocation fields and data field */
    start[0]=start[1]=0;
    edge[0]=GEOROW;
    edge[1]=GEOCOL;
    HE5_SWwritefield(swID,"Longitude",start,NULL,edge,
                  lonData);
    HE5_SWwritefield(swID,"Latitude",start,NULL,edge,
                  latData);
    start[0]=start[1]=start[2]=0;
    edge[0]=ROW;
    edge[1]=COL;
    HE5_SWwritefield(swID,"myDataField",start,NULL,edge,
                  myDataArray);
    /* Write a file level attribute */
    strcpy(attrValueChar,"HDF-EOS5 file attribute.");
    attrCount[0]=strlen(attrValueChar);
    HE5_EHwriteglbattr(fileID,"fileCharAttr",
              H5T_NATIVE_CHAR,attrCount,attrValueChar);
    /* Write a double type attribute to the swath group */
    attrCount[0]=MAXNUM;
    HE5_SWwriteattr(swID,"swathGroupAttr",
              H5T_NATIVE_DOUBLE,attrCount,attrValueDouble);
    /* Write a string attribute to the "Data Fields" Group */
    strcpy(attrValueChar,"Data Field group attribute");
    attrCount[0]=strlen(attrValueChar);
    HE5_SWwritegrpattr(swID,"dataFieldGroupAttr",
              H5T_NATIVE_CHAR,attrCount,attrValueChar);
    /* Write a double type attributes to the myDataField */
    attrCount[0]=MAXNUM;
    HE5_SWwritelocattr(swID,"myDataField","localAttr",
              H5T_NATIVE_DOUBLE,attrCount,attrValueDouble);

    /* terminate access to the swath and close the file */
    HE5_SWdetach(swID);
    HE5_SWclose(fileID);
    return SUCCEED;
}
```

Appendix C Internet Links for Data Access (Search and Order)

John L. Dwyer

ASTER On-Demand Gateway	http://edcdaac.usgs.gov/asterondemand/index.html
Contact SCAR	http://mapping.usgs.gov/html/scarlibr.html
Earth Explorer	http://EarthExplorer.usgs.gov
EDC MODIS Direct Broadcast	http://modisdb.usgs.gov/obtaining.php
EOS Data Gateway	http://edcimswww.cr.usgs.gov/pub/imswelcome/
Global Land Cover Characteristics	http://edcdaac.usgs.gov/glcc/glcc.html
GloVis	http://glovis.usgs.gov/
GTOPO30	http://lpdaac.usgs.gov/gtopo30/gtopo30.html
HYDRO1k	http://lpdaac.usgs.gov/gtopo30/hydro/index.html
Hyperion and ALI Data Acquisition Request (DAR)	http://eo1.usgs.gov/dataproducts/dar.asp
LP DAAC Data Pool	http://lpdaac2.usgs.gov/datapool/datapool.asp
LP DAAC SIR-C	http://edcdaac.usgs.gov/sir-c/sir-c.html
LP DAAC User Services	http://LPDAAC.usgs.gov
MRLC 2001	http://www.mrlc.gov/index.asp
NASA's Earth Science and Image Analysis (Shuttle and ISS)	http://eol.jsc.nasa.gov/
National Hydrography Dataset	http://nhd.usgs.gov/data.html
National Land Cover Dataset 1992 (NLCD92)	http://landcover.usgs.gov/natllandcover.asp
NED Index Map	http://edcsns17.cr.usgs.gov/nedcd/
NOAA Archive	http://www.saa.noaa.gov/nsaa/products/welcome
PhotoFinder	http://edcsns17.cr.usgs.gov/finder/finder_main.pl?dataset_name = NAPP
The National Map	http://nationalmap.usgs.gov/
USGS Geographic Data	http://edc.usgs.gov/geodata/
USGS Seamless Distribution System	http://seamless.usgs.gov/
USGS Store	http://store.usgs.gov/

Index

Index

The manufacturer's authorised representative in the EU is Springer Nature Customer Service Centre GmbH, Europaplatz 3, 69115 Heidelberg, Germany. If you have any concerns regarding our products, please contact ProductSafety@springernature.com

Printed and bound by CPI Group (UK) Ltd, Croydon, CR0 4YY
30/09/2026
02225602-0012